Adaptations of Desert Organisms

Edited by J. L. Cloudsley-Thompson

Springer

Berlin
Heidelberg
New York
Barcelona
Budapest
Hong Kong
London
Milan
Paris
Tokyo

Volumes already published

Ecophysiology of the Camelidae and Desert Ruminants
By R. T. Wilson (1989)

Ecophysiology of Desert Arthropods and Reptiles
By J. L. Cloudsley-Thompson (1991)

Plant Nutrients in Desert Environments
By A. Day and K. Ludeke (1993)

Seed Germination in Desert Plants
By Y. Gutterman (1993)

Behavioural Adaptations of Desert Animals
By G. Costa (1995)

Invertebrates in Hot and Cold Arid Environments
By L. Sømme (1995)

In preparation

Energetics of Desert Invertebrates
By H. Heatwole (1995)

Ecophysiology of Desert Birds
By G.L. Maclean (1996)

Lauritz Sømme

Invertebrates in Hot and Cold Arid Environments

With 77 Figures

Springer

Professor Dr. LAURITZ SØMME
University of Oslo
Department of Biology
P.O. Box 1050 Blindern
N-0316 Oslo
Norway

Cover illustration: Photograph by J.L. Cloudsley-Thompson

ISBN 3-540-58985-6 Springer-Verlag Berlin Heidelberg New York

Die Deutsche Bibliothek - CIP-Einheitsaufnahme
Sømme, Lauritz:
Invertebrates in hot and cold arid environments / Lauritz
Sømme. - Berlin ; Heidelberg ; New York : Springer, 1995
(Adaptations of desert organisms)
ISBN 3-540-58985-6
CIP data applied for

Typesetting: K. Darms, Göttingen
SPIN 10125844 31/3137-5 4 3 2 1 0 - Printed on acid-free paper

Preface

The idea to write a book on the adaptation of invertebrates from both hot and cold arid environments in the series Adaptations of Desert Organisms originated from Professor J.L. Cloudsley-Thompson. My thanks are due to him for his encouragement during the preparation of the present volume, and for his many suggestions to improve the content. I am also most grateful to Professor K.E. Zachariassen, University of Trondheim, Dr. H. Ramløv, University of Copenhagen and Dr. M. Holmstrup, University of Aarhus for valuable comments to parts of the manuscript. The scanning electron microscope photographs of tardigrades have been reproduced by the courtesy of Drs. R.M. Kristensen and H. Ramløv, University of Copenhagen, and the photographs of freezing in nematodes by the courtesy of Dr. David Wharton, University of Otago.

Traditionally, the biology of plants and animals from the hot temperate and subtropical deserts is treated separately from those of cold polar deserts. Within the series *Adaptations of Desert Organisms*, edited by Professor J.L. Cloudsley-Thompson, most books are concerned with hot deserts, while particular attention is given to polar deserts in the present volume. To point out differences and similarities, an attempt is made to compare invertebrates from such different regions, in particular with regard to adaptations to their hostile environments. Questions are raised if species from hot and cold deserts exhibit similarities in their tolerance of drought, and how species from both regions are adapted to low temperatures. To provide a background to these problems, the climate and fauna of hot and cold deserts are described in the first two chapters.

In the Arctic and the Antarctic, precipitation is low and comparable to that of many deserts in warmer regions. In winter, snow and ice predominate, and liquid water is unavailable to terrestrial organisms. Clearly, invertebrates inhabiting these regions must be able to survive both aridity and low temperatures. Similarly, species from hot temperate deserts are not only adapted to dry conditions but to cold as well. Winter severity increases with latitude, altitude and continentality.

The fauna of insects, spiders and myriapods of hot deserts is fundamentally different from that of polar deserts, suggesting differences in historical development and specificity of adaptation. On the other hand, similarities are seen in the fauna of mites and springtails, comprising important elements in soil ecosystems. Nematodes are abundant in both hot deserts and polar regions. Like tardigrades, they are able to survive almost complete loss of water.

The following chapters discuss an array of adaptational strategies required for life in hot and cold deserts. In arthropods, several physiological mechanisms of water conservation and cold hardiness have been developed. Of equal importance to insects and other terrestrial arthropods is the ability to avoid the most stressful situations through their behaviour. Microclimatic niches, ameliorating extreme moisture and temperature fluctuations, are of paramount importance.

The unpredictable moisture conditions of hot deserts favour opportunistic activity and development when water is available. Dormancy may be an important strategy to survive adverse periods. In polar areas, seasonal changes are regular, but the summer is short and cool; the winter is long and cold, retarding the phenology of most species.

Recently, several authors have suggested that the tolerance of terrestrial arthropods to cold and drought may, in fact, be overlapping adaptations. Through their behaviour and physiology, many species are protected against both cold and drought. In tardigrades, nematodes and earthworms, tolerance to dehydration enhances resistance to cold.

Our present knowledge on the adaptations of invertebrates to hot and cold arid environments is far from complete. To understand the role of the organisms in their environment, more physiological and ecological studies are required. In many taxa, behavioural and phenological adaptations are poorly understood. Surprisingly, even information on the biodiversity of many hot and cold desert regions is still incomplete. It is hoped that the presentation of the different problems, as described in this book, may stimulate to further research on the variety of desert invertebrates and their biology.

The present volume covers a broad spectrum of topics, and relies heavily on the use of illustrations from other publications. I am most grateful for permission received from the different authors personally, and for the courtesy of the publishers in permitting reproduction of figures from their publications. Figures are reprinted with the kind permission of the following publishers and organizations: Academic Press Ltd., London (Figs. 2.1, 2.5, 5.5, 6.1); Academic Press Inc., Orlando (Figs. 1.5, 9.10, 9.12); American Physiological Society, Bethesda (Fig. 4.4); Annual Reviews Inc. (Annual Review of Physiology), Palo Alto (Fig. 5.6); Australian Museum, Sidney (Fig. 2.3); Blackwell Science Ltd., Oxford (Figs. 7.2, 11.4, Table 9.3); Canadian Museum of Nature, Ottawa (Fig. 2.10); Cambridge University Press, Cambridge (Figs. 1.1, 2.7, 7.3); Company of Biologists Ltd., Cambridge (Figs. 4.3, 6.3, 6.4); Cryo-Letters, Cambridge (Fig. 9.6); Czech Academy of Science, Ceské Budejovice (Fig. 3.7); Elsevier Science Ltd., Kidlington (Figs. 4.2, 4.6, 8.1, 8.2, 8.3, 9.11); Entomological Society of Canada, Ottawa (Figs. 1.2, 1.3); Gauthier-Villars Publishers, Montrouge (Figs. 3.1, 4.5); Japan Plant Protection Association, Tokyo (Fig. 8.4); John Wiley & Sons Inc., New York

(Figs. 5.2, 5.4, 7.7); National Research Council of Canada, Ottawa (Figs. 7.1, 11.2); Norwegian Institue of Nature Research, Trondheim (Fig. 4.1); Munksgaard International Booksellers and Publishers, Copenhagen (Fig. 2.4, 9.1, 9.5); Redacción de Vieraea, La Laguna (Fig. 3.3); Regents of the University of Colorado, Boulder (Figs. 1.9, 6.6, 6.7); Society of Nematologists, Washington D.C. (Fig. 5.1); Springer-Verlag, Heidelberg (Figs. 1.8, 2.6, 3.2, 6.5, 7.5, 7.6, 9.2, 9.13, 11.1, 11.3); University of Arizona Press, Tucson (Fig. 2.2).

Oslo, Spring 1995 Lauritz Sømme

Contents

1 The Climate of Arid Environments

1.1 Subtropical and Temperate Deserts

1.1.1 Definitions

Deserts are not always characterized by great heat and extensive sand dunes. A variety of biotopes and temperature regimes are found in temperate and subtropical deserts and extremely low temperatures are typical for polar deserts. As pointed out by Logan (1968), an important common characteristic of all deserts is their aridity.

In spite of this simple similarity, deserts are difficult to define and there is considerable disagreement in the interpretation of different authors. The classification system of W. Köppen is widely used (Trewartha and Horn 1980), and according to this, 26% of the continental areas is characterized by dry climates. Of the total area, 14% is steppe or semi-desert and 12% desert. The boundary between humid and semi-arid areas is drawn where evaporation of moisture equals the precipitation. Data on evaporation is not easily available, but formulas have been developed to calculate the potential evaporation from other climatic factors (Trewartha and Horn 1980).

As pointed out by Allan and Warren (1993), the most widely used definition of deserts and arid areas was developed by Meigs (1953) for UNESCO. This system classifies deserts as semi-arid, arid and hyper-arid. In total, the three categories cover one-third of the earth's land surface. Mean annual precipitation in hyper-arid areas is less than 25 mm, in arid areas less than 200 mm and less than 600 mm in semi-arid areas (Meigs 1953).

The use of boundaries based on evaporation or precipitation cannot give an exact picture of the degree of aridity. Evaporation increases with temperature, and the effectiveness of precipitation is affected by several factors (Logan 1968; McGinnies 1979). Rainfall in deserts often takes place as violent convectional showers. Desert wadis become torrents of water and because of great runoffs, only a proportion of the precipitation becomes available to plants and animals. The high temperature of the desert soils and dry surrounding air causes rapid evaporation, particularly during the summer. For living organisms, it is a problem that the time or season of rainfall in hot deserts often is unpredictable. An adaptation to this is seen in the opportunistic response of plants and animals when water is available for short periods. In polar deserts, water frozen as ice or precipitation in the form of snow cannot be directly utilized by plants and invertebrate animals as a source of moisture. In spite of these objections, the annual precipitation of an area gives a relative approximation of the dryness, and is useful for a general comparison of different geographical areas.

Aridity rises mainly from three general causes, acting individually or in combination (McGinnies 1979; West 1983a). Part of the desert areas of North America and the leeward side of the Andes Mountain Range results from the drying effect on air masses moving over a mountain barrier. Great distance from the ocean is a cause of aridity in inland deserts, such as the Turkestan, Takla Makan and Gobi deserts. Aridity also arises from the lack of storm systems by the formation of dry stable air masses that resist convective currents. They are mostly formed at latitudes of 25° to 35°N and S, where the majority of the world's hot deserts are situated. The Arabian and Australian deserts and the Sahara Desert are examples of areas located between major wind belts and storm systems.

1.1.2 The Distribution of Deserts

Desert may be classified in several ways on the basis of geographical location, seasonality, temperature and precipitation (e.g. Meigs 1953; Logan 1968; McGinnies 1979; Trewartha and Horn 1980; Evenari 1985; Cloudsley-Thompson 1991; Allan and Warren 1993). Several reviews on desert ecology are available, e.g. Goodall et al. (1979), Heathcote (1983), West (1983b), Evenari et al. (1985, 1986). Desert boundaries and subdivisions can be defined according to detailed floristic and ecosystem analysis in combination with macroclimatic patterns.

For the presentation here, the purpose is to give a broad description of the great diversity of biotopes inhabited by desert invertebrates. For more detailed classifications and descriptions of temperate and subtropical desert, the reader is referred to the references given in the preceding paragraph. For simplification, the term hot deserts is used in the following sections for all deserts with permanent or seasonal hot and dry conditions, i.e. for both temperate and subtropical deserts. In contrast, the term cold deserts is restricted to polar deserts of the Arctic and the Antarctic where winter temperatures are extremely low and summer temperatures are cool.

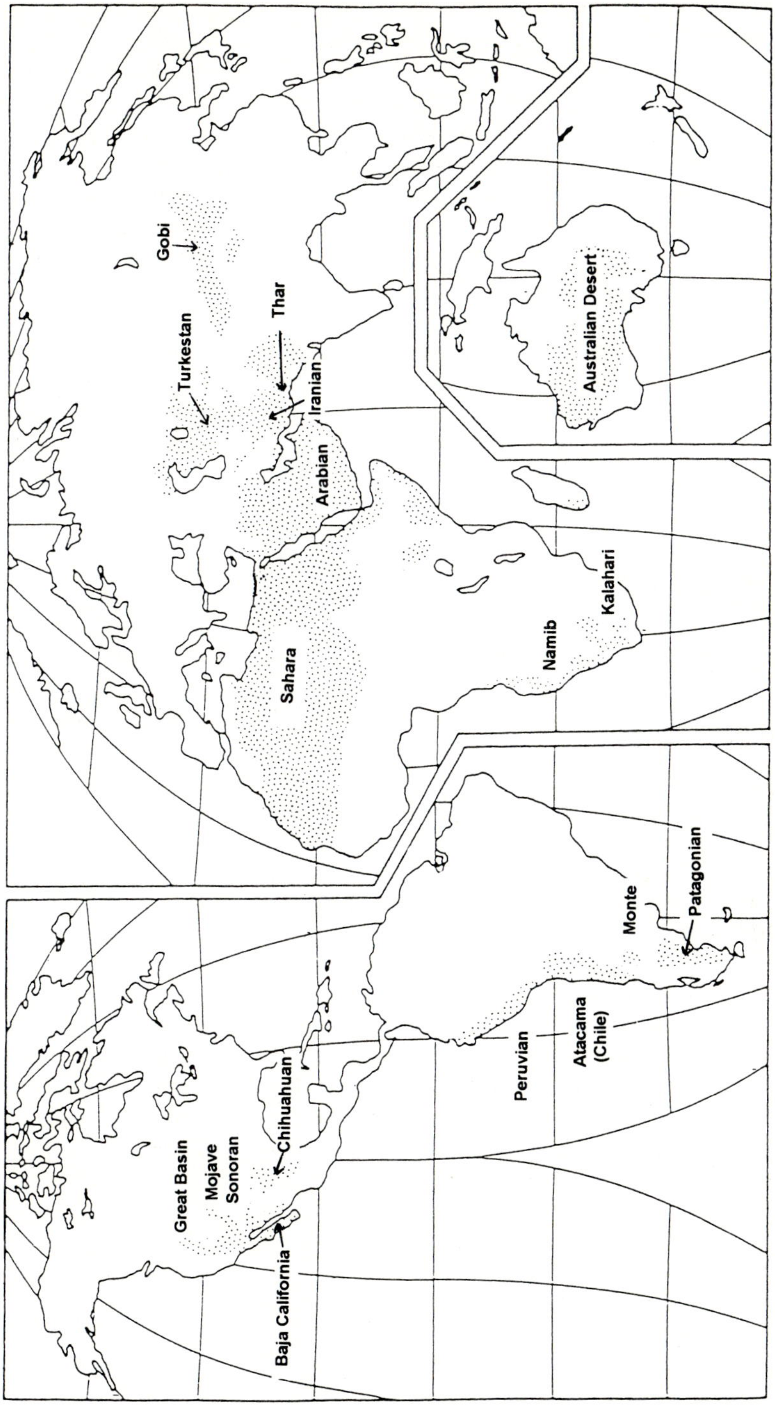

Fig. 1.1. World distribution of temperate and subtropical deserts. (After Fitzpatrick 1979)

1.1.2.1 Subtropical Deserts

Subtropical deserts are found in two broad belts around the world (Logan 1968; McGinnies 1979). These deserts are also termed hot deserts (Evenari et al. 1986), and distinguished from temperate deserts by the virtual absence of snowfall. The tropical deserts are characterized by considerable temperature fluctuation. As pointed out by Evenari (1985), one outstanding feature of subtropical hot deserts is the extreme quantitative, temporal and spatial irregularity, unreliability and variability of rainfall, which is therefore completely unpredictable. Occasional frost may occur, in particular at higher elevations. The large extension of the desert areas results from the formation of high pressure zones north and south of the equator (Logan 1968; Allan and Warren 1993).

In the northern hemisphere, subtropical deserts comprise the Sahara--Arabian desert region, the southern Iranian and the Thar deserts (Fig. 1.1), as well as the Sonoran, Mojave and Chihuahuan deserts of North America. The Sahara Desert covers some of the hottest and driest areas of the world. Average annual preciptiation ranges from almost zero in the most arid parts to 400 mm on the semi-arid fringes of the desert. Extreme air temperatures above 50 °C have been recorded, while annual mean varies from 20 to 30 °C. Most of the Arabian Peninsula is also arid or extremely arid. In large parts of Saudi Arabia mean annual precipitation is below 100 mm. The Thar Desert is relatively cold and windy and receives about 300 mm of rain on the average.

In the southern hemisphere, subtropical deserts comprise the Kalahari Desert of southern Africa, the deserts of Australia and the Monte Desert of Argentina (Fig. 1.1). The two belts are separated by areas with high precipitation and the formation of rain forests.

1.1.2.2 Cool Coastal Deserts

Cool coastal deserts are in principle subtropical, but are unusual due to their moist air and relatively low temperatures (Logan 1968; McGinnies 1979). The three main coastal deserts of the world include the Namib Desert on the coasts of southern Africa, the Peruvian-Chilean deserts and the desert on the Pacific coast of Baja California in Mexico (Fig. 1.1).

Seawater of the oceans outside the coastal deserts originates as currents from polar regions. Moist and cold air formed by contact with the cold currents moves inland as a thin layer below the dry and hot tropical air. In spite of the moist air, rainfall rarely occurs in coastal deserts, but condensation takes place every night. In the Namib Desert, the fog contributes 150 mm of moisture per year, and an estimated 30 mm in southern Israel (Heathcote 1983). Although the amounts are not large, condensation is of great biological importance, making possible the growth of scattered vegetation and providing a source of moisture for arthropods such as the 'fog-basking' beetles of the Namib Desert.

1.1.2.3 Temperate Deserts

Temperate deserts include arid areas, several of which have very low winter temperatures. Mountainous regions of North America and Asia also belong to this classification (West 1983b). Some of the temperate deserts, such as the Gobi Desert and other arid areas of central Asia, are characterized as continental deserts (Logan 1968). Due to their great distance from the coast and high elevation, some of these areas experience extreme dry and cold conditions. Temperature fluctuations are high. As an example, according to Walter and Box (1983), the Pamir Desert has maximum air temperatures in the range of 25 to 30 °C and minimum temperatures in the range -40 to -45 °C. Annual precipitation is about 100 mm.

Severe winter cold and great annual temperature fluctuations are also characteristic of the Great Basin Desert of North America. Average annual precipitation ranges from 160 to 420 mm (West 1983b). Minimum temperatures fall below -30 °C and snow cover may persist for several weeks during the winter.

In South America, the vast area of Patagonia is considered as a cool semi-desert (Soriano 1983). Precipitation decreases very rapidly east of the Andes Mountain Range, and the dryness of Patagonia is enhanced by strong winds causing high evaporative rates. Due to cold coastal streams, both summer and winter temperatures are greatly meliorated.

1.1.3 Microclimate

The microclimate is of particular importance in hot and arid environments. As pointed out by Heathcote (1983), microclimatic niches make it possible for living organisms to retreat for survival or spend their whole life cycle in more protected conditions. Microclimate in deserts is formed by small and large irregularities in the landscape. A small pebble or a large stone provides shadows and shelters that alter the temperature. Plants offer shadows and conserve humidity, forming microhabitats where invertebrates may survive. In the Sahara Desert, Pierre (1958) found reduction in temperature up to 17 °C below and at the side of bushes of *Tamarix aphylla*. Shelter from wind reduces the loss of moisture by evaporation from plants and soil surface. Higher and more rugged terrain results in even larger microclimatic variety (Heathcote 1983). Mountains offer great modifications of the climate due to the reduction of air temperatures with altitude, their shade and effect on wind and moisture.

In terrestrial arthropods, behavioural adaptation is of principle importance for survival in arid environments. Their body temperature fluctuates with the ambient temperature, and levels exceeding their potential of survival are avoided by seeking cooler microhabitats. The behaviour of desert insects and other arthropods has been thoroughly discussed by Crawford (1981) and Cloudsley-Thompson (1991) (Chap. 6.1). While avoiding the heat of summer days by sheltering in cooler microenvironments, desert arthropods emerge at dusk and dawn, or during the night. In many species, the emergence is timed by circadian rythms. By digging into sand and soil and hiding underneath rocks, a number of species lead a hidden life under microclimatic conditions which they are able to survive. In spite of large diel temperature fluctuations at the surface, the burrows of scorpions and spiders offer a more stable environment. In burrows of the desert woodlice *Hemilepistus reaumuri*, temperatures below a depth of 30 cm remain relatively constant at 35 to 40 °C, and air humidity is saturated due to soil moisture (Coenen-Stass 1989). In more favourable seasons, or in short periods following rainfall, many desert arthropods become active on the surface.

1.2 Polar Deserts

Abiotic factors are of major importance in determining the distribution and survival of organisms in cold and arid environments of the Arctic and the Antarctic. In polar regions most attention has been paid to the temperature regime and its effect on plants and animals. Although less investigated, moisture conditions may be of equal importance. Large parts of polar areas receive very little precipitation, and are considered as deserts or semi-deserts.

In temperate and subtropical areas, deserts may be defined according to the yearly amount of precipition, which falls almost exclusively as rain (Chap. 1.1). The precipitation of large arctic and antarctic areas is often as low as in hot arid and semi-arid deserts. In polar deserts, however, a great proportion of the precipitation falls as snow, and liquid water is not available to plants and animals for several months. This adds to the aridity. On the other hand, the snow cover protects overwintering invertebrates against low air temperatures, and the enclosed air is saturated with humidity in balance with the vapour pressure of ice and snow. Organisms that are able to resist dehydration, or able to adjust the vapour pressure of their body fluids in balance with their surroundings, are protected against desiccation, although water in the liquid phase is missing.

The summer arrives very abruptly in polar areas. Due to the permafrost, very little water penetrates to the deeper layers of the soil, but is flushed away in numerous streams and rivers. The spring flush of water from melting snow and ice provides favourable moisture conditions at the beginning of the growing season. Locally, the upper soil layer is saturated with water. Later, in the summer, when the ground becomes drier, terrestrial plants and animals depend on precipitation in the form of rain. To living organisms, the summer rain is more important than the winter snow in determining moisture conditions during the growing season. For this reason, polar deserts are not easily characterized by yearly amount of precipitation.

The general climate, or macroclimate, of an area gives only a superficial impression of the conditions experienced by living organisms. Temperature and humidity in the immediate environment of plants and invertebrates are influenced by several factors. At the ground level, favourable microclimatic niches are extremely important for the development and survival of the species.

1.2.1 The Arctic

The climate of the Arctic has been reviewed and described by several authors (e.g. McKay et al. 1970; Barry and Hare 1974; Danks 1981; Stonehouse 1989; Maxwell 1992). Atmospheric circulation in the Arctic is permanently dominated by a giant low pressure centre. Due to the circumpolar vortex, westerly winds are circulated anticlockwise around the globe at middle latitudes. In this zone, cold airstreams from arctic regions meet milder southern airstreams at the arctic front. The circumpolar vortex is most intense in the winter, strengthening the westerly winds. Distinct troughs are formed over Canada, Kamchatka and Novaya Zemlya. In the summer, the lowest pressure occurs over the pole. The eastward movements of low pressures are diverted by anticyclone pressure systems. The disturbances allow warmer weather to penetrate the Arctic, and further to the north in summer than in winter. Horizontal heat transport is of great importance for the arctic energy balance, and consequently for the temperature conditions of plants and animals.

Topography and coastal effects have great influence on the general climate (Danks 1981). The formation of clouds by mountain ranges and smaller hills affects the local climate by interrupting solar insolation. Favourable conditions are produced from katabatic warming of air flowing down mountain slopes, and examples of such föhn winds are known from several locations in the Arctic. Valleys may act as channels for local winds, sometimes transporting cold air into new areas. Low temperatures are also caused by the accumulation of cold stagnant air at the bottom of valleys. Coastal climates are generally characterized by more winds, higher precipitation and the formation of clouds and fog, resulting in lower summer temperatures compared with inland sites.

1.2.1.1 Energy Budget

According to Barry and Hare (1974), there is no lack of solar energy in the Arctic. Annual solar radiation along the northern coast of the Queen Elizabeth Islands and Siberia amounts to one-third of the energy received in subtropical deserts. In spite of strong radiative cooling during the winter, the energy budget is positive due to the long summer days and heat transferred horizontally by winds and ocean currents. On the other hand, energy is lost due to the reflection from snow and ice, even during the summer. Spring arrives considerably later in the Arctic than in temperate areas. As pointed out by Maxwell (1992), during the summer and autumn the climate is influenced by stretches of cold open water, which result in local winds, low clouds and frequent occurrence of fog.

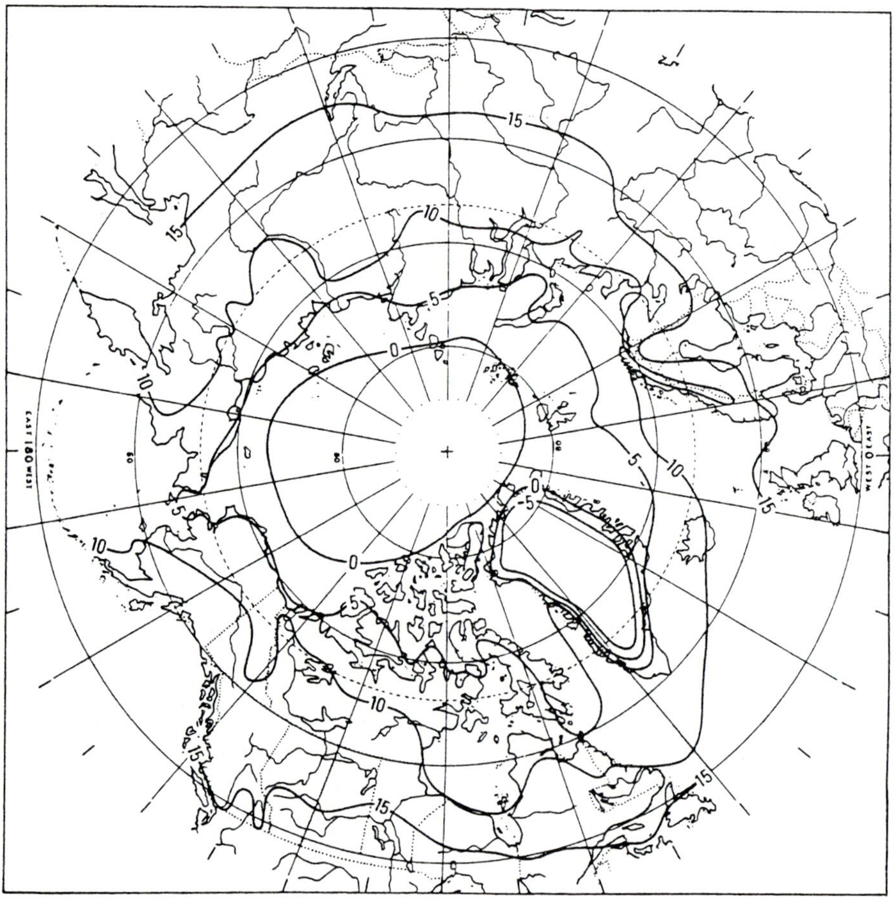

Fig. 1.2. Arctic summer isotherms for the month of August. (Danks 1981, after McKay et al. 1970)

1.2.1.2 Temperature

Summer isotherms in the Arctic (Fig. 1.2) reflect the distribution of land and sea (McKay et al. 1970; Danks 1981). Daily air temperatures are generally below 10 °C in the summer months, but may rise to 20 °C during brief sunny periods. Lowest summer temperatures are recorded from interior parts of Greenland, Ellesmere Island and other mountainous areas. Cold air over the Arctic Ocean remains at 0 °C. In addition to topography, temperature is greatly modified by local conditions such as snow depth, cloudiness, windshelter and insolation.

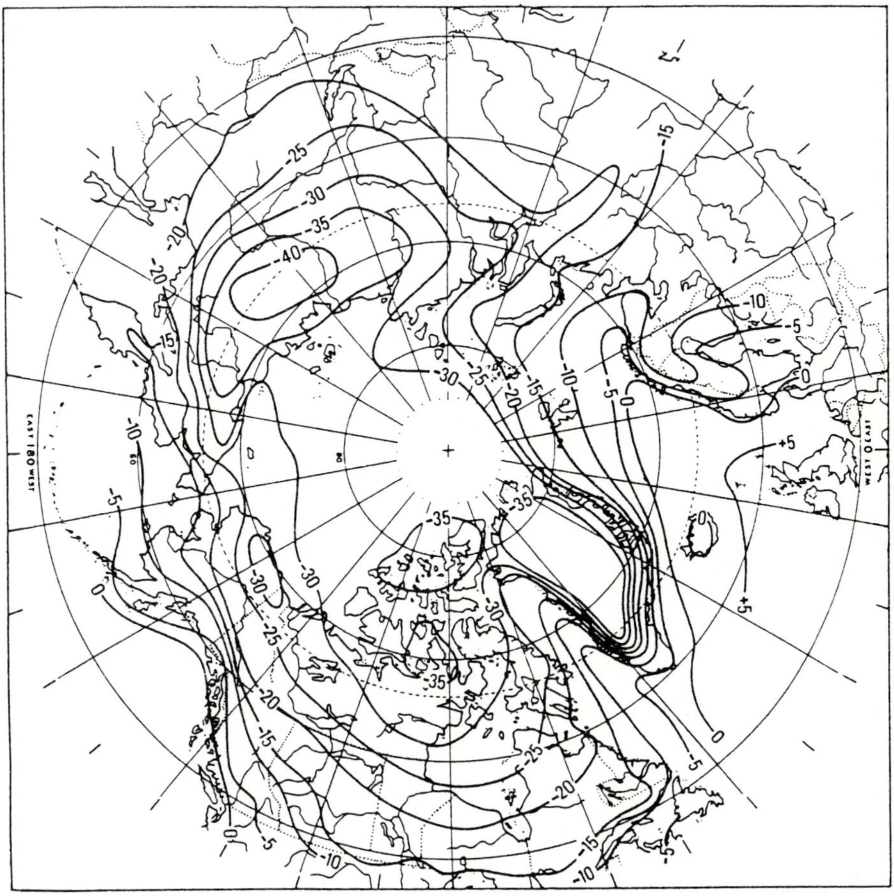

Fig. 1.3. Arctic winter isotherms for February. (Danks 1981, after McKay et al. 1970)

During the winter, persistant anticyclones are formed over Siberia and northwestern Canada, causing extreme cold and dry conditions (Stonehouse 1989). February is one of the coldest months (Fig. 1.3) with mean daily temperatures of -30 °C or lower over large parts of the Arctic (McKay et al. 1970) Temperature extremes of -63 °C have been recorded from Yukon, of -66 °C from the icecap of Greenland and of -68 °C at Verkhoyansk and Oymaykon in Siberia.

1.2.1.3 Precipitation

Mean annual precipiation is low in the Arctic. There are, of course, large variations, but in some locations, such as Brøntlundhus in Peary Land, northern Greenland, it is estimated to less than 25 mm (Molgaard 1982). Examples of mean monthly precipitation at selected arctic sites are presented in Fig. 1.4.

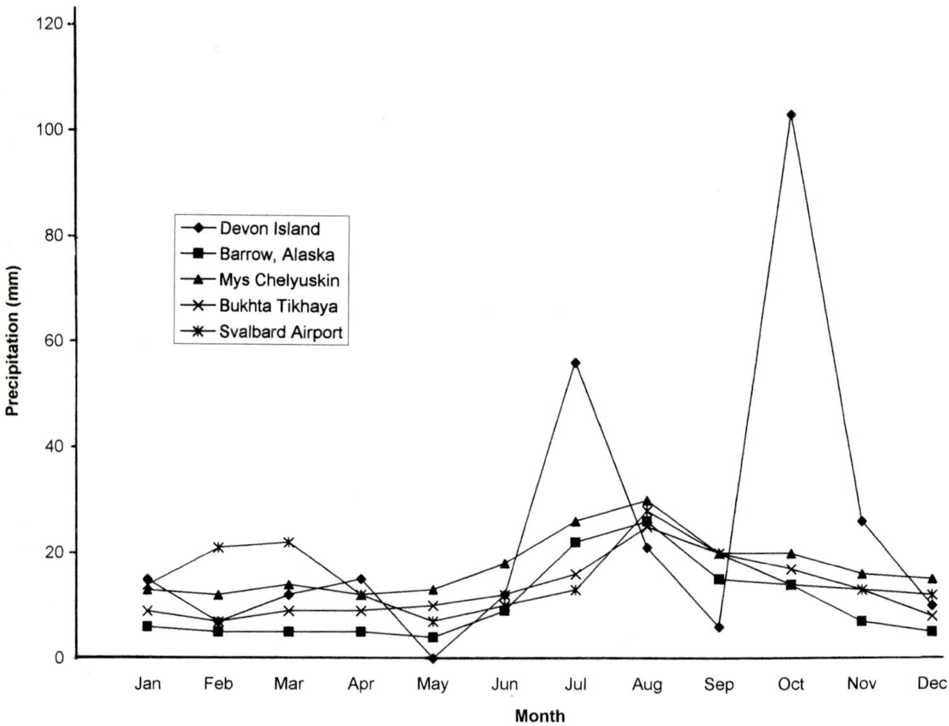

Fig. 1.4. Mean monthly precipitation from selected arctic stations. Devon Island, Canadian Arctic Archipelago, raised beach for 1972-1973 (Bliss 1975); Barrow, Alaska for 1941-1970 (Bunnell et al. 1975); Mys Chelyuskin (Cape Cheluskin), Siberia (Alexandrova 1988); Buhkta Tikhaya, Franz Joseph Land (Alexandrova 1988); Svalbard Airport for 1976-1989 (Hanssen-Bauer et al. 1990)

In most of the stations, precipitation is lowest during the winter, and increases slightly in summer. The high values for July and October from Devon Island may be arbitrary, since data from only 2 years are available.

There are several difficulties involved in recording precipitation in polar areas, and some of the available figures should be considered with certain reservations. According to Maxwell (1992), actual precipitation may be double or more the amounts recorded in standard precipitation gauges on account of the wind effects that are frequently encountered in the Arctic. Due to the wind, snowfall is difficult to measure, and the snow is whipped up from the surface, clogging the gauges and giving false readings (Stonehouse 1989). Better values are obtained by calculating precipitation from snow accumulation. Eastern arctic areas receive substantially more precipitation than the western part of the hemisphere (McKay et al. 1970). Along the Arctic Circle the annual precipitation is 200 mm in Canada, compared with 400 mm in Siberia and 600 mm in the European part of Russia.

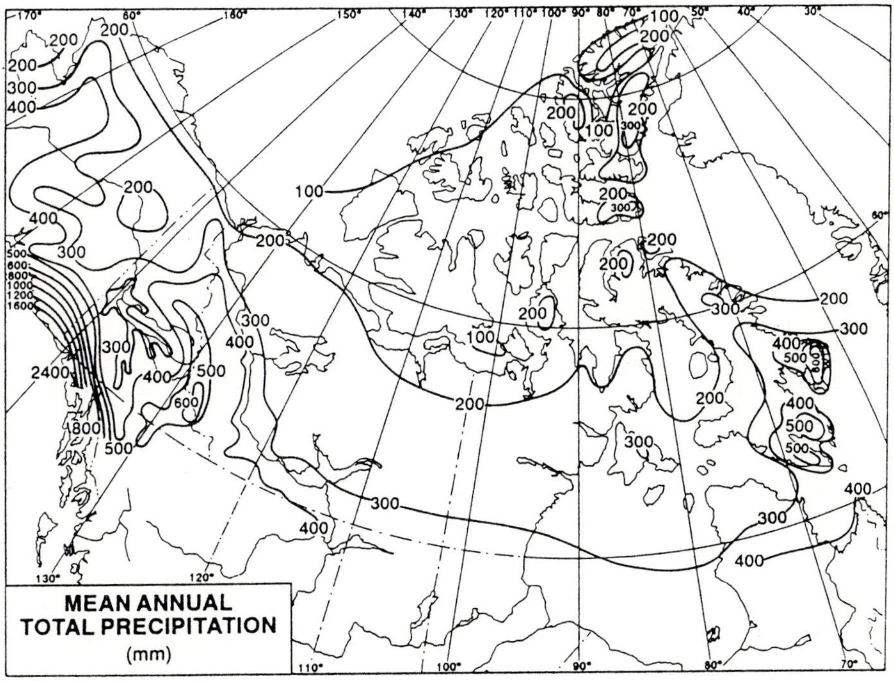

Fig. 1.5. Mean annual precipitation over northern North America. (Maxwell 1992)

Large variations in rainfall occur from year to year and the pattern of precipitation is strongly influenced by the topography of the landscape. Mountainous areas close to open seas may recieve high amounts of precipitation, reflecting the westward movements of moist air masses. Thus, high precipitation results from the influence of the Alaska Range, while the interior parts of northern North America receive considerably smaller amounts (Fig. 1.5). According to Maxwell (1992), the mountain ranges of Ellesmere, Devon and Baffin islands, together with Greenland, act as a barrier to moist air from the North Atlantic. The leeward slopes also receive more precipitation than inland areas.

The climate of Svalbard is relatively mild because of the North Atlantic Drift (Hanssen-Bauer et al. 1990). Monthly mean air temperatures in the 3 coldest months are usually in the range -12 to -16 °C, and mean annual precipitation ranges from 180 to 440 mm at different meteorological stations. High precipitation in the Barents Sea area is explained by the penetration of an eastern branch of the North Atlantic Drift into northern waters (Alexandrova 1988). The maritime conditions cause frequent fogs and dense cloud cover over Franz Joseph Land, which receives about 300 mm of precipitation per year. In comparison, precipitation amounts to no more than 100-230 mm in arctic areas of Siberia and adjoining islands.

1.2.1.4 Snow Cover

At higher latitudes, precipitation falls as snow during most of the year. Due to open waters, more precipitation falls during the summer (Fig. 1.4), and for this reason annual rainfall is approximately equivalent to snowfall (McKay et al. 1970; Danks 1981).

Snow is important in providing a protective cover for plants and animals. Although snowfall may be fairly uniform, the snow is rapidly redistributed by the wind. Accumulation takes place in depressions and on the lee side of ridges, while wind-swept ridge crests are almost free of snow (Danks 1981). The thickness of the snow cover affects the time when the ground becomes free of snow in the spring. This is of great importance for living organisms, since the duration of the summer season is affected. In the polar desert of Alexandra Island in the Franz Joseph Land archipelago, Alexandrova (1988) found that the most luxurious development of vegetation occurs in localities with a thin snow cover. In spite of the severe winter conditions, the early disappearance of the snow is of primary importance. The most favourable conditions for development of vegetation on Alexandra Island occur where the snow forms a layer of 25 cm and disappears by 15-20 June.

Fig. 1.6A, B. A Coastal antarctic scenery, illustrating the dominating cover of ice. **B** Inland nunataks protruding through the ice cap in Dronning Maud Land. (Photos L. Sømme)

1.2.2 The Antarctic

The climate of the Antarctic reflects the special geography of the continent and surrounding islands (e.g. Walton 1984; Dudeney 1987; Stonehouse 1989).

In contrast to the Arctic, which is an ocean surrounded by land, the Antarctic region is a continent surrounded by the sea. South of the Antarctic Convergence, the region can be divided into three zones of different climatic severity and consequently different biodiversity (Fig. 2.7).

The Antarctic Continent covers an area of 14 x 106 km^2, which is almost twice the size of Australia. The mean elevation of the Antarctic Continent is 2000 m, about one quarter is higher than 3000 m, and this is clearly of great importance for meteorological conditions (Walton 1984). The highest point, situated in the Ellsworth Mountains, rises to 5140 m, while the elevation at the South Pole is 2800 m.

Less than 3% of the Antarctic Continent is ever free of permanent snow and ice (Fig. 1.6). Walton (1984) distinguished three types of bare ground. Some areas, such as the the dry valleys of Victoria Land, are practically free of snow and precipitation. Several areas, e.g. along the coast, normally covered by snow during the winter, become free of snow in the summer. Finally, exposed areas of rocks, often in steep mountain slopes of inland nunataks and coastal mountains may be free of snow and ice. While the dry valleys are almost devoid of living organisms, the remaining snow-free habitats offer marginal living conditions for terrestrial plants and invertebrates. Free water is usually available for at least part of the summer. In the maritime and subantartic islands, snow-free ground has larger extension in the summer, but variation among the islands is large. Snow-free areas are the basis for higher biodiversity compared with the continent.

1.2.2.1 Winds

As pointed out by Dudeney (1987), the weather of Antarctica is dominated by a simple circulation system. The temperature difference between the cold continent and relatively warm surrounding ocean creates cyclones that travel clockwise around the continent. In contrast to the Arctic, anticyclone blocking is rare in the Antarctic, and a persistent circumpolar band of low pressures is centred around 60° to 65°S.

Surface winds over the Antarctic Continent are closely related to the direction of the slope of the terrain and associated with persistent temperature inversions (Walton 1984; Dudeney 1987). In principle, surface winds blow from the interior towards the coast in all parts of the continent. Katabatic winds are formed by a combination of surface winds and low pressures. At the coast, gale force katabatic winds, which may persist for weeks, are probably formed by the increasing steepness of the inland escarpment.

1.2.2.2 Temperature

Short cold summers followed by long and very cold winters are typical of the temperature regime of the Antarctic Continent. Due to the extensive formation of sea ice, low winter temperatures dominate both the sea and inland areas. Continental minimum temperatures typically reach -40 to -50 °C (Dudeney 1987). The annual mean temperature at the South Pole is -50 °C, and a record low of -89.5 °C has been recorded at the Vostoc Station on the continental plateau. Even during the summer, air temperatures rarely exceed 0 °C in the continent. Local temperature increase due to insolation is of major importance for the presence of biological microhabitats.

1.2.2.3 Precipitation

Radiative heating of dark rock surfaces causes local melting sufficient to sustain plant and invertebrate life. Apart from this, the Antarctic Continent is extremely arid (Walton 1984). Not only is there no available unfrozen water, but precipitation is very low. According to Dudeney (1987), accumulation of snow on the continent is not easily measured, but an approximation of the amount of precipitation can be obtained from the rate of ice loss into the ocean through the flow of glaciers.

Due to the westerly winds and its high mountains, the Antarctic Peninsula receives more precipitation than the rest of the continent (Walton 1984). Similarly, precipitation is much higher on most of the maritime and subantarctic islands than on the Antarctic Continent. Thus, mean annual precipitation at King Edward Point of South Georgia is approximately 1400 mm. In contrast, Signy Island in the South Orkney Island group receives little precipitation due to the shadow effect of the high mountains of neighbouring Coronation Island.

1.2.3 Microclimate

Corbet (1972) recognized that the warmest climates in the Arctic are found in the shallow air layers above the soil surface and in fresh water. According to Danks (1981), the ground temperature is greatly affected by seasonal and daily changes in solar altitude. At higher latitudes, nocturnal cooling of the ground surface is reduced due to insolation during the night. In this way, ground temperatures suitable for plant growth are increased. The most dramatic changes in microclimate occur with the melting of snow during the brief spring and at the appearance of new snow in the autumn (Bunnell et al. 1975).

During the summer, solar radiation is absorbed by the vegetation and soil surface. Measurements at Barrow, Alaska, (71°N) (Bunnell et al. 1975) and at Devon Island, Canada, (76°N) (Bliss 1975) clearly illustrate that temperatures are highest within the plant boundary layer. Microclimatic recordings at Peary Land, northern Greenland (83°N) demonstrated that temperatures in the scattered and sparse vegetation often exceed ambient temperatures by more than 20 °C under sunny conditions (Molgaard 1982). The response of the arctic plants to insolation differs between different species. Higher temperatures were recorded in *Dryas integrifolia, Saxifraga caespitosa* and *Saxifraga oppositifolia* than in other plants, such as *Carex* sp. and mosses. Diel temperature fluctuations were marked and were higher in slopes exposed to the low sun than on level ground. Under the midnight sun, temperature fluctuations on north-facing slopes were opposite to those of south-facing slopes.

At Alexandra Island in Franz Joseph Land, Alexandrova (1988) showed that the mean daily temperature at the ground surface was 1-4 °C higher than the mean daily air temperature from the end of May to the end of August. The surface maximum temperature, which is considered of enormous importance for the hastening of plant development, exceeded mean daily surface temperatures by up to 8 °C. Furthermore, temperatures were increased 5 cm below the soil surface and below crustose lichens and moss-lichen turf.

Polar microclimates are greatly affected by wind (Danks 1981; Walton 1984). Even light winds may cause considerable loss of heat and the disturbance of humidity profiles. Protection against heat loss is offered by the vegetation and by the litter layer which is occupied by numerous invertebrate animals. In areas with low and discontinuous vegetation, plants and animals are particularly vulnerable.

Soil water content depends on local conditions of drainage (Danks 1981). Although water-logging of the soils above permafrost offsets the aridity in many arctic areas, well-drained sites become very dry (Fig. 1.7A). For this reason, invertebrates may experience periods of aridity that they are able to survive only through behavioural and physiological adaptations.

Walton (1984) has reviewed the literature on microclimatic studies in the Antarctic. The difficulties of accurate measurements and the use of different kinds of instrumentation are discussed. Measurements in the Antarctic have mainly been restricted to radiation and temperature, and there are few data on wind profiles and humidity on a microclimatic scale. Studies on radiation are important to understand the conditions of photosynthetic activity under changing insolation and albedo. Series of temperature recordings are available from many parts of the maritime and continental Antarctic. Some of the series are restricted to the summer, when most scientists visit the Antarctic, but long-term studies with automatic equipment have also been conducted in several locations, such as Signy Island in the South Orkney Islands (Fig. 1.7B).

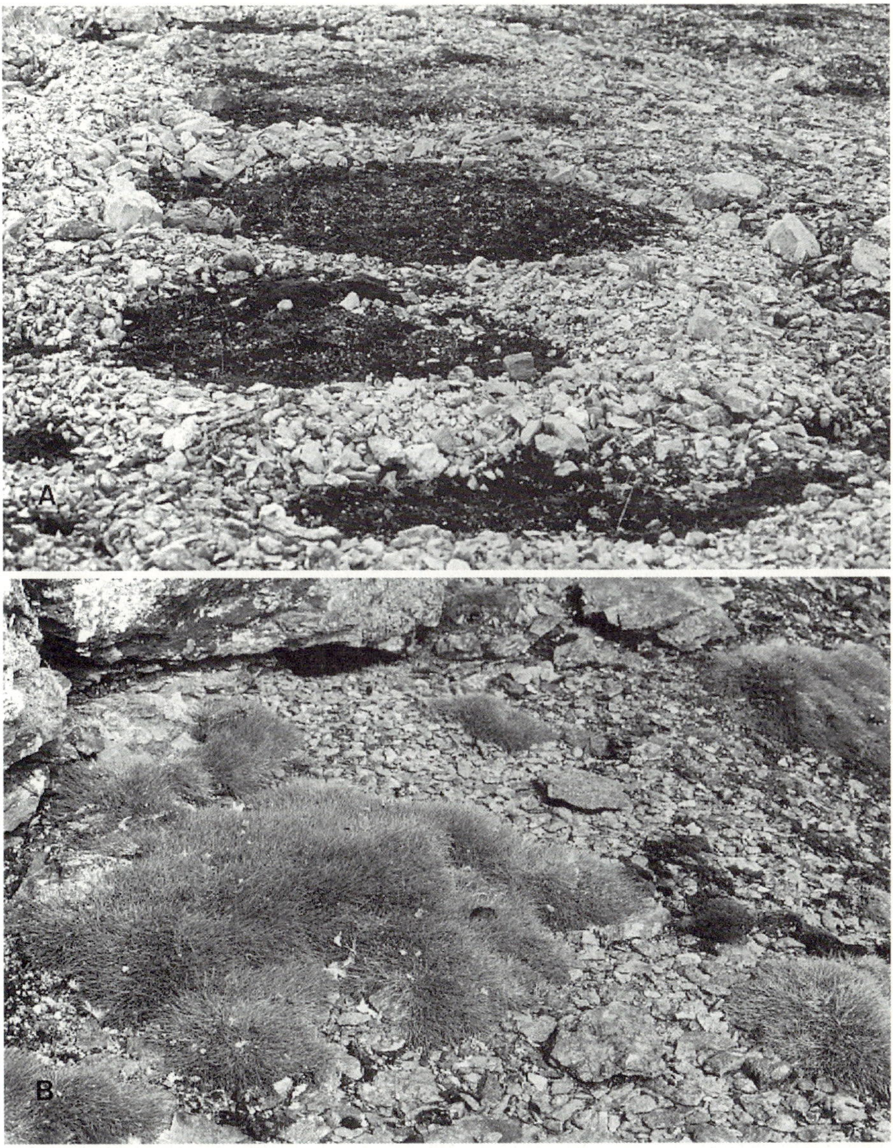

Fig. 1.7A, B. A Sorted rock circle formations from a semi-polar desert at Svalbard. The arrangement of the stones is caused by freezing and thawing. The *central dark* areas are overgrown with bryophytes, lichens and cyanobacteria. They are wet in spring but dry up during the summer. **B** Tufts of the antarctic grass *Deschampsia antarcticus* growing in some of the most favourable microhabitats of Signy Island. (Photos L. Sømme)

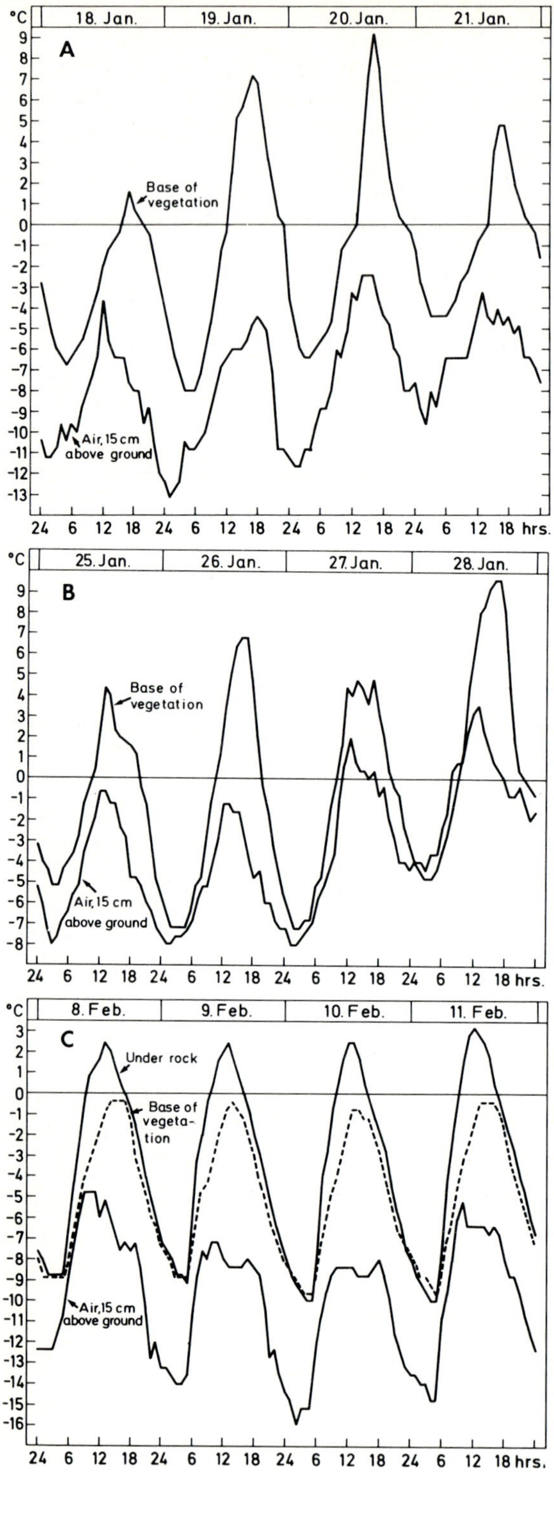

Fig. 1.8A - C. Temperature recordings from microhabitats of Collembola and mites in *Prasiola* vegetation in Mühlig-Hofmannfjella, Dronning Maud Land. Corresponding *shaded* air temperatures about 15 cm above the ground. **A** In crevices of a boulder, **B, C** In gravel. (Sømme 1986a)

As pointed out by Kennedy (1993), solar insolation during summer, combined with thermal blanketing by snow during the winter, creates temperature regimes significantly warmer than macroclimatic data imply. Characteristic of the microclimatic summer conditions are the rapid temperature fluctuations of the soil surface. At Hallett Station in Victoria Land, Pryor (1962) found that the fluctuations were considerably reduced 2 cm below the surface at the time of thawing. No reflection of diurnal temperature fluctuations was observed at a depth of 10 cm, which was close to the surface of the permafrost. As reviewed by Walton (1984), maximum temperatures at the surface of the vegetation may exceed 20 to 30 °C during periods of high radiation.

Recordings from the microhabitats of Collembola at 1600 m above sea level in the Mühlig-Hofmannfjella (72°S, 5°E), Dronning Maud Land exemplify the harsh environment of continental nunataks (Sømme 1986a). Temperatures were recorded at 1-2 cm in loose gravel covered by *Prasiola algae* and in crevices of boulders, harbouring springtails and prostigmatid mites (Fig. 1.8). In January, temperatures rose above 0 °C for 8 to 10 h on most days, and reached a maximum of 8 to 9 °C in sunny weather. Lower temperatures were recorded in cloudy weather and when the vegetation was covered by drifting snow. During the night, temperatures fell several degrees below freezing. In comparison, air temperatures remained below freezing all the time, except for short periods, due to convective heating from surrounding rocks. From the early part of February, marked changes took place, and the temperature at the base of the vegetation remained continuously below 0 °C day and night. The ground did not thaw, although small rocks were heated a few degrees above zero in direct sunshine. The example shows how the microclimate makes it possible for springtails and mites to grow and develop, although only for a few hours per day in the middle of the short summer season. The autumn arrives very early. This particular habitat represents the limit of life in this part of the continent, and beyond the nunataks there is nothing but the ice cap.

1.3 The Limitation of Water in Antarctic Terrestrial Ecosystems

With its enormous ice cap, the Antarctic Continent has been calculated to act as a sink to 90 % of the world's fresh water. In spite of this, the continent is predominantly arid (Walton 1984; Kennedy 1993), since free water is unavailable in most areas and during most of the year. As pointed out by Kennedy (1993), extreme cold has traditionally been considered as the main restriction to the distribution and survival of living organisms in the Antarctic. In his opinion, however, the underavailability of moisture, as a general rule, is of primary importance in the limitation of antarctic terrestrial life. This is not to say that temperature is unimportant, since growth and development are only possible above certain thresholds.

In addition to being frozen most of the year, water in snow-free areas of the Antarctic Continent may not be available to living organisms due to the high salt content of antarctic soils. Not all water, however, is biologically unavailable. Even in the arid continent there are oases, such as the Burger and Schirmacher oases, where wind-blown debris produces increased absorption of solar radiation, resulting in the presence of fresh water pools. The presence of liquid water in rivers, streams and pools, however, is restricted in time. Niches suitable for living organisms may exist for only a few weeks. Most of the soils in snow-free areas are of the ahumic type and characterized by fast drainage. In conclusion, water is a limiting factor, and the nature of biological communities in antarctic habitats must be expected to vary according to local moisture status (Kennedy 1993).

1.3.1 Latitudinal Variations

From his studies in South Victoria Land, Janetschek (1970) pointed out the importance of moisture and temperature in antarctic ecosystems. A gradient of temperature and moisture conditions exists from a "bryo- system" along the coast to the "chalikosystem" on inland nunataks (Fig. 1.9).

Near the coast, melting cycles occur in the upper permafrost layer, and with melting of the covering ice-cemented layer, a constant supply of water is available from the soil during the summer. Under these conditions, vegetation consisting of mosses, algae and lichens is supported. The chalikosystem is composed of barren gravel with "invisible" vegetation of endopedic microphytes (bacteria, cyanobacteria, algae, microfungi, etc.). Moisture is available from local melt water streams and where the ice--cemented layer is shallow enough to be reached by diel melting cycles. With increasing latitude, altitude and local aridity, conditions for terrestrial life disappear.

In Victoria Land, the gradient is interrupted in some areas by the dry valleys (Fig. 1.9). Water formerly present in the upper surface layers has been removed by evaporation over the past 60 000 years. Although summer temperatures exceed 0 °C in the soil and gravel, Janetschek (1970) found little life, and the dry valley area cannot support terrestrial arthropods. He concluded that the thermal regime of the habitats is not in itself limiting, but that life is inseparately linked with the environmental water balance. Later, it was shown that limited living conditions for invertebrates do occur, since Freckman and Virginia (1991) found small populations of nematodes in a number of soil samples from the McMurdo Dry Valley.

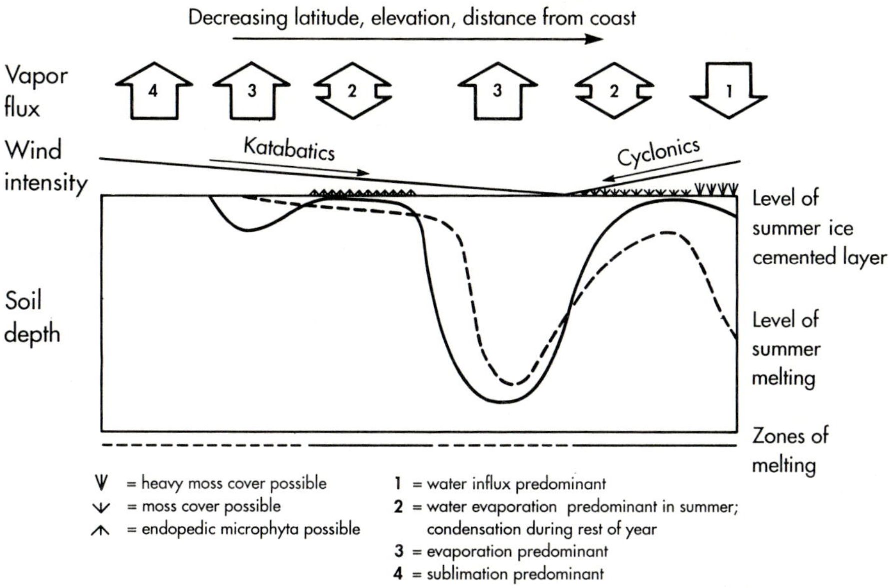

Fig. 1.9. Janetschek's model for the relationship between moisture availability and the distribution of life in Victoria Land. See text for further explanation. (Kennedy 1993)

As pointed out by Kennedy (1993), through its inclusion of the active soil layer, Janetschek's model incorporates an indirect dependence upon temperature for the limitation effect. Influence of temperature is through its modulation of water's physical state, while temperature in itself exerts little direct effect on the distribution of species, as witnessed by the paucity of life in the dry valleys. Although the model was devised using data collected 30 years ago, Kennedy (1993) found that it remains an accurate large-scale description for the distribution of antarctic terrestrial life. According to Janetschek (1970), free water can only exist where there is an overlap between the active soil layer, where melting cycles occur, and the upper ice-cemented soil layer, from which water may be derived. Such overlap occurs at the coast, where a relatively constant supply of moisture is available from the soil in the summer, and many kilometres inland where the shallow ice-cemented layer is reached by diurnal melting cycles (Fig. 1.9). In Victoria Land, there is an area of complete aridity between these zones due to position of the dry valleys between the coast and inland mountains.

1.3.2 Microclimatic Variations

Variations on a smaller scale also confirm the dependence of many antarctic terrestrial organisms on moisture. Wet moss microhabitats support more nematodes than dry lichen-encrusted microhabitats (Tilbrook 1967; Spaull 1973), and rotifers and tardigrades show similar dependence. In the Taylor Valley of southern Victoria Land, Schwarz et al. (1992) showed that the vegetation of algae, cyanobacteria and mosses is restricted to distinct areas of optimum growth, correlated with the flushing of melt water from the Canada Glacier. Simlarly, the presence of rotifers, tardigrades and nematodes is closely associated with the vegetation (Schwarz et al. 1993). The animals were not active prior to the melting period in the spring, and pre-flush counts were taken as an indication of the potential inoculum which would be active when free water become available. Post-melt counts showed substantial increases in the numbers of rotifers and nematodes (Chap. 2.2.2), while development of tardigrades appeared to be slower.

The limiting effect of water is also demonstrated by the vertical distribution of living organisms (Kennedy 1993). Schwarz et al. (1993) found highest numbers of invertebrates in the upper 5 mm of the samples, and Davey and Clarke (1991) have shown that most of the algal biomass in fellfield soil cores is concentrated just below the surface. The microorganisms do not migrate to the top of the surface unless free water is available, and their distribution may reflect a desiccation-avoidance strategy. The occurrence of chasmoendolithic communities also offers an interesting example of microhabitat distribution (Kennedy 1993). These communities comprise bacteria, cyanobacteria, fungi and green algae just beneath the surface of sandstone rocks. In the dry valleys of Antarctica, these organisms do not appear on the surface and are protected against desiccation by the higher relative humidity of endolithic air spaces.

Several studies have also shown that seasonal changes in the populations of bacteria, protozoans, cyanobacteria and green algae are controlled by the availability of free moisture (Kennedy 1993). The optimal growth period of many organisms corresponds to the spring thaw rather that to the period of maximum warmth. The ability of many organisms to enter a state of anhydrobiosis (Crowe et al. 1992) makes it possible for them to survive during shorter or longer periods of drought. Anhydrobiosis is commonly found among antarctic rotifers, tardigrades and nematodes (Chap. 5). In the Antarctic, dehydration resistance is likely to be even more important to survival than the cold resistance of many organisms (Kennedy 1993).

1.4 Conclusions

Deserts and semi-deserts occupy large areas of temperate and subtropical regions. Their distribution is determined by several factors, such as continentality and the formation of persistent anticyclones in certain latitudinal belts. Cold arid environments are found in the Arctic and the Antarctic, and the aridity of polar regions is also based on major global weather systems. Annual precipitation of polar deserts may be as low or lower than that of hot arid and semi-arid environments.

The general climate of an area gives only a superficial impression of the conditions under which plants and animals actually live. Study of the microclimate is important to the understanding of how living organisms are able to survive and develop under the stress of arid environments. The use of microclimatic niches, as part of their adaptation to arid and cold environments, is of vital importance to invertebrates in both hot and cold deserts. In hot deserts, the temperature conditions beneath rocks and within the shadow of vegetation offer protection against excessive heat. The situation is the opposite in polar deserts, where the temperature of the vegetation and upper soil layer may be increased by several centigrades.

The aridity of the Antarctic Continent, and the limitation of water in time and space, illustrate similarities in the environmental conditions of hot and cold deserts. Although the Antarctic Continent represents the most extreme conditions, lack of moisture for shorter or longer periods is apparently of great importance in other antarctic locations, and in the Arctic as well. For this reason, arctic and antarctic invertebrates have to be adapted not only to cold, but for several species adaptation to arid conditions is also required. Invertebrates of temperate and subtropical deserts avoid desiccation both through their behaviour and through physiological adaptations. Species in continental deserts and in mountainous regions also have to survive periods of low temperatures of varying intensity.

2 The Fauna of Hot and Cold Arid Deserts

2.1 The Diversity of Hot Desert Invertebrates

2.1.1 Introduction

The purpose of this chapter is to give an overview of the diversity of the invertebrate fauna of hot deserts. The intention is also to provide background knowledge on the organisms where adaptations to the environment will be discussed in later chapters. The present chapter is not to be considered as a complete review of the enormous literature in the field, but will, through selected examples, give some idea of the magnitude and variability of the invertebrate fauna of hot deserts.

As pointed out previously (Chap. 1.1), hot deserts occupy more that one-quarter of the land surface area of the world. Due to low precipitation and high temperatures, deserts are inhospitable environments. Consequently, primary production is greatly restricted compared with that in many other habitats, and this affects the availability of food resources for herbivores, carnivores and other members of the food webs. Accordingly, deserts may be expected to have low biodiversity, and it has generally been assumed that desert ecosystems are relatively simple and mainly regulated by abiotic factors.

More recent evidence, however, has shown that deserts possess relatively complex biological systems. As pointed out by Polis (1991a), deserts support a surprisingly diverse fauna and flora, including a large number of taxa. Even some taxa considered to be mesic, such as algae, mosses and isopods, occur in deserts. The arid-adapted crab *Holthusiana transversa* which belongs to the fresh-water crabs, is widely distributed in arid and semi-arid regions of Australia (Greenaway 1984).

Due to the distribution of deserts on different continents, large differences in fauna and flora must be expected. The geological history of deserts varies greatly. Their biomes may be relicts from wetter geological periods or composed of species invading from surrounding geographical areas. The invertebrate fauna of many deserts is poorly known, and in most areas estimates of numbers of species are not possible to compile. Consequently, comparisons of the fauna from different deserts are not available at the species level, but within higher taxa similar groups are usually dominating in different parts of the world. Another problem in making comparisons is the way in which deserts and their limits are defined by different authors.

Based on the large literature on desert invertebrates, information on the fauna of different habitats and the composition of different trophic levels have been reviewed by several authors (e.g. Crawford 1981, 1991a; Wallwork 1982; Cloudsley-Thompson 1984, 1991; Polis 1991a,b). By this approach ecological aspects are included, while an impression of the diversity is presented as well. Data from locations where more thorough investigations have been conducted demonstrate that hundreds or thousands of species may be present (Polis 1991a). Three examples of faunal composition of terrestrial arthropods from some of the better-studied areas are presented in Table 2.1. Two of the locations are in California, where investigations have been going on for years (Polis 1991a,b), while the third location is from the Repetek Sandy Desert Biosphere Reserve of the Karakum Desert in Turkmen.

Table 2.1. Examples of species diversity of terrestrial arthropods from hot deserts

Group	Coachella Vally, California [a]	Deep Canyon Reserve California [a]	Karakum Desert [b]
Isopoda	1	3	1
Myriapoda	1	> 2	2
Arachnida	55	97	> 100
Insecta	ca. 2500 [c]	>2540	>4000 [a]

[a] Polis(1991a, b); [b] Walter and Box (1983)
[c] Estimated figures

2.1.2 Soil Fauna: Microarthropods and Nematodes

Soil and litter communities of warm deserts are dominated by nematodes and microarthropods (Crawford 1981; El-Kifl and Ghabbour 1984; Zak and Freckman 1991). Very little is known about the presence of enchytraeid worms in desert habitats (Crawford 1981), while earthworms are known from the fringes of the Sahara Desert and moist localities in the oases and highlands (El-Kifl and Ghabbour 1984).

According to Zak and Freckman (1991), the microflora and microfauna are important components in the regulation of decomposition rates of organic material in arid areas. Their activities are greatly influenced by moisture and temperature, and interactions between the different organisms are important in breakdown processes. Decomposition takes place in surface litter from plants and in litter that is buried by the action of rain and wind, or through the activities of animals. Several arthropods that are permanently active on the surface transport organic material to subterranean sites where decomposition may proceed more rapidly (Wallwork 1982). Harvester ants and termites are particularly important in this respect, but other insects, millipedes, woodlice and mollusks, contribute as well. High densities of soil invertebrates, such as microarthropods and nematodes, are found at sites were plant litter accumulates, and also in the soil surrounding plant roots (Zak and Freckman 1991).

2.1.2.1 Microarthropods

In the Mojave Desert of California, with a mean yearly precipitation of 104 mm, Wallwork (1972) found that 90 % of the microarthropods in dry litter and mineral soil consisted of cryptostigmatid, astigmatid and prostigmatid mites. The remainder of the arthropod fauna comprised Collembola and various insects. The occurrence of microarthropods in the surface litter varied greatly with the season. Highest densities corresponded to the period of maximum rainfall in December and January. The mean density of microarthropods was approximately 1600 specimens per m^2, which is considerably lower than the densities of about 100 000 individuals per m^2 generally recorded from temperate grassland and woodland soils.

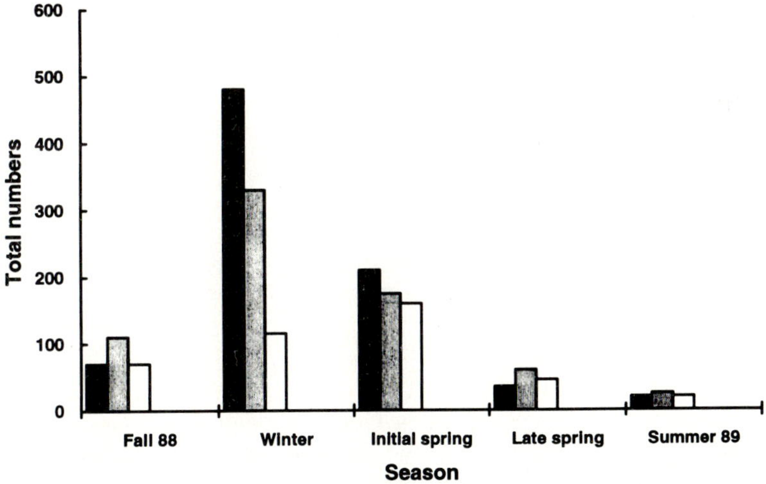

Fig. 2.1. Seasonal patterns of total numbers of mites in soil and litter from three saltbush microhabitats in a Chilean desert site. *Filled columns: Atriplex repanda; stippled columns: Atriplex semibaccata; striated columns: Atriplex nummularia.* (Cepeda-Pizarro et al. 1992)

The patterns of microarthropod densities in deserts are not strictly associated with available moisture since artificial input of water may have no effect on population size (Zak and Freckman 1991). As suggested by Wallwork et al. (1984), desert mites may have evolved a reproductive pattern that is synchronized to the predictable season of precipitation at which microclimate conditions for microbial growth may be optimal. In soil and litter from the saltbush *Atriplex* spp. in the coastal desert of northern Chile, Cepeda-Pizarro et al. (1992) recorded species from 13 prostigmatid (Actenidida) and 7 oribatid families. Highest population densities were found during the winter (Fig. 2.1), at the time of highest precipitation (annual mean of 100-150 mm).

The composition of the microarthropod fauna with regard to trophic levels varies with the season (Zak and Freckman 1991). In the Mojave Desert, fungivores and detrivores are most numerous all year around, while the proportion of general predators is highest during the spring and summer (Franco et al. 1979; Fig. 2.2).

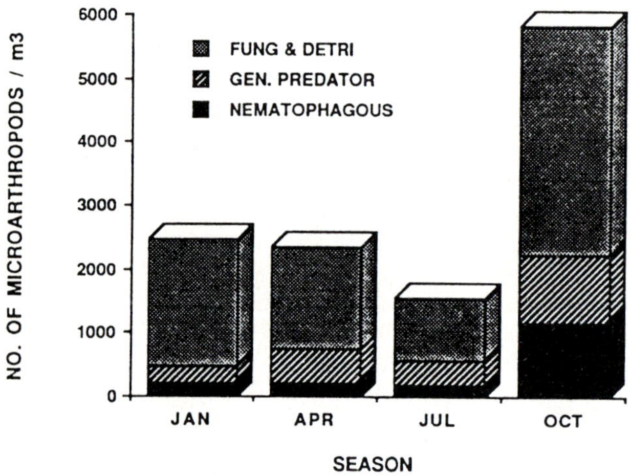

Fig. 2.2. Seasonal patterns in trophic relationships of microarthropods from soil and litter beneath shrubs in the northern Mojave Desert. *Fung & Detri* fungivore-detrivore. (Zak and Freckman 1991; data from Franco et al. 1979)

Nematophagous species have a peak in October, corresponding to high moisture levels. Experimental studies in the Chihuahuan Desert showed that microarthropod densities changed with the degree of composition of litter from *Lepidioum lasiocarpum*, a common Chihuahuan Desert annual. The prostigmatid tydeid and anoteid mites were the earliest invaders, followed by mesostigmatid and other prostigmatid mites. Collembolans and predatory gammasid mites appeared in the litter when more than 40% of the mass had been lost. Removal of the microfauna resulted in a significant decrease in litter decomposition rates. Tydeid and cryptostigmatid mites are dominant in litter beneath perennial plants in the Mojave Desert, where total densities of microarthropods may be close to 6×10^3 m^{-2} (Franco et al. 1979).

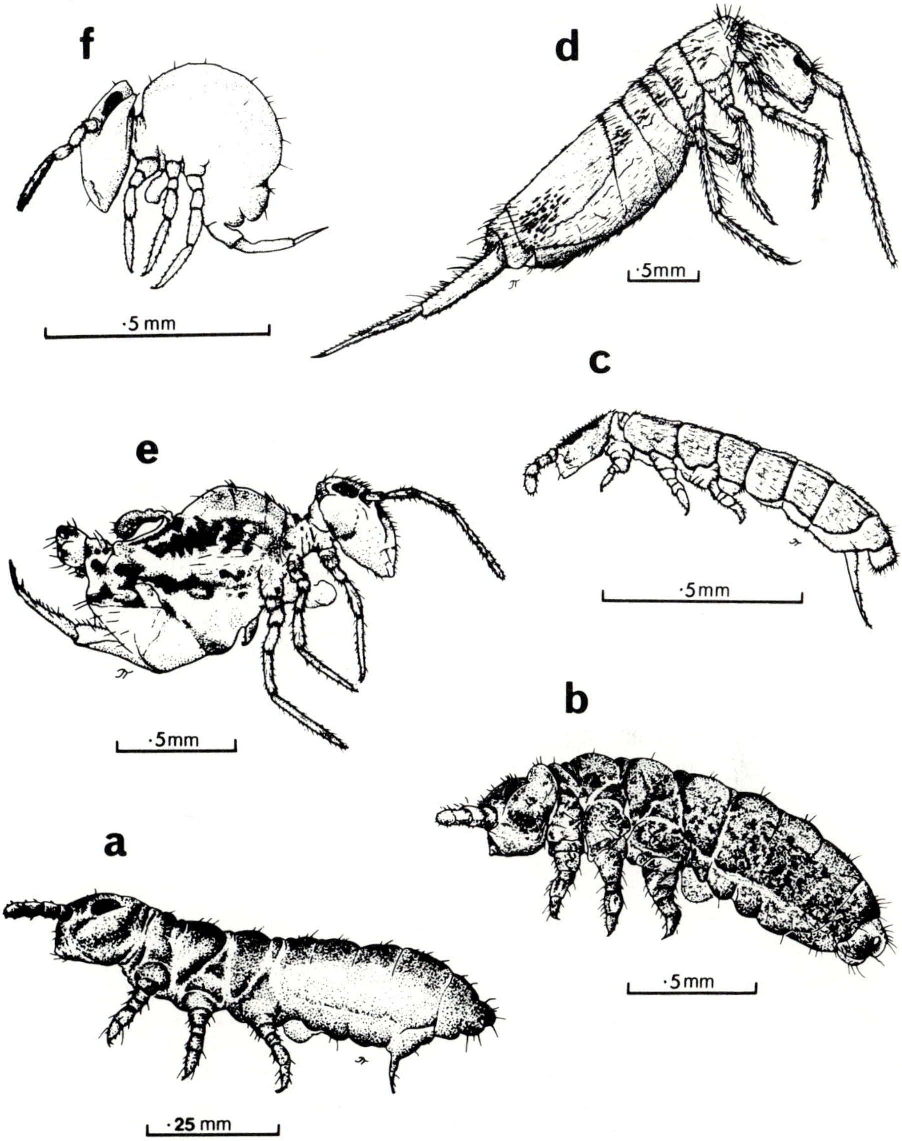

Fig. 2.3a-f. Representative Collembola from the Australian arid zone. **a** *Xenylla* sp. (Hypogastruridae) **b** *Setanodosa* sp. (Neanuridae). **c** *Folsomia* sp. (Isotomidae). **d** *Acanthocyrtus* sp. (Entomobryidae). **e** *Corynephoria absoloni* (Sminthuridae). **f** *Sphaeridia* sp. (Sminthuridae). (Greenslade 1982)

Collembola are important in the decomposition of organic material. They occur in soils throughout the world, including soils of hot deserts (Wallwork 1982). The collembolans occupy several trophic levels, feeding on bacteria, fungi, nematodes and other organisms. Several species are detritivores, ingesting particles of decaying plant material. In hot deserts the number of species normally declines with increasing aridity. Thus, Greenslade and Greenslade (1984) showed that the number of collembolan species (40) at the dry inland locality of Koonamore Station in South Australia was less than half the number (97) at Western Kangaroo Island on the coast. The taxonomic composition of Collembola at Koonamore is very similar to that at Khartoum in Sudan (Greenslade 1981). According to Greenslade (1982), 47 genera of Collembola are recorded from the arid zone of Australia (Fig. 2.3). On an evolutionary scale, many of the species may have originated in humid climates in localized xeric environments such as the bark of trees or coastal sand dunes.

2.1.2.2 Nematodes

In spite of their fragile appearance, nematodes are among the most numerous soil animals in hot deserts (Freckman 1986). Densities of 1.24×10^6 m^{-2} at 0-30 cm depth have been recorded in the Mojave Desert of Nevada. Highest densities are found in the upper 10 cm of the soil and the densities decrease with distance from desert shrubs (Fig. 2.4). While a large proportion of nematodes from humid environments are plant parasites, desert nematode communities are dominated by bacteriophages, fungivores and omnivore-predaceous groups (Freckman and Mankau 1977; Zak and Freckman 1991).

Nematodes and microarthropods may affect decomposition rates of organic material in the desert dunefields of New Mexico (Elkins and Whitford 1982). Other studies have cast doubt on the importance of nematodes in decomposition of surface litter, since nematodes are inactive in anhydrobiosis for much of their life time (Zak and Freckman 1991). Below the surface, part of the nematode population may be involved in soil decomposition processes. In the Chihuahuan Desert, nematodes have been recorded to a depth of 11-12 m in association with the deep-rooting woody legume mesquite *Prosopis glandulosa* (Freckman and Virginia 1989). A large proportion of these nematodes were plant feeders but other trophic groups were also present, suggesting that nematodes in general may affect plant growth and natural cycling in such desert ecosystems.

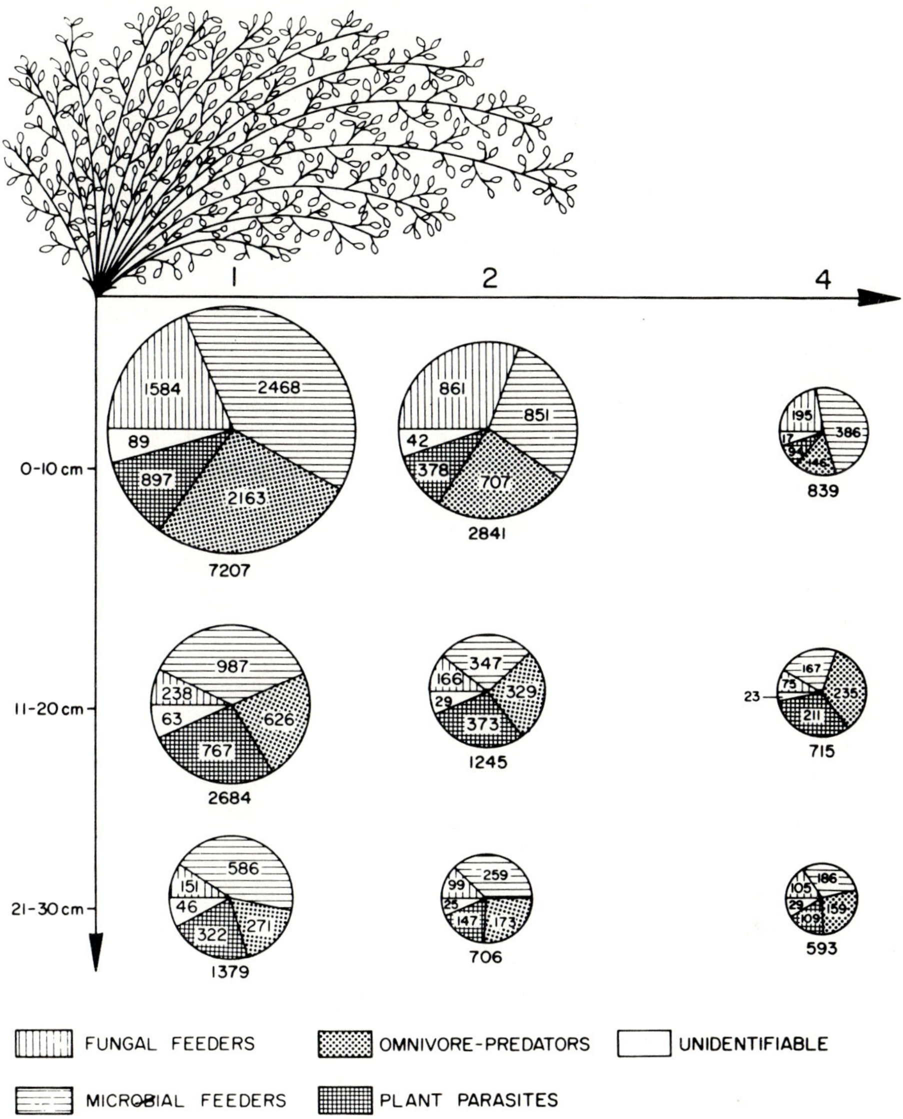

Fig. 2.4. Spatial distribution of nematode trophic groups associated with desert shrubs at Rock Valley, Nevada. Numbers per 500 cm^{-3} soil from four desert shrubs. (Freckman and Mankau 1977)

2.1.3 Ants and Termites

The success of ants and termites in desert regions is based on their preadapted ability to construct underground nests and on their social life (Cloudsley-Thompson 1991). Termites also construct mounds in arid areas, but tend to live entirely underground in the hottest deserts. Most desert termites have little resistance to water loss, and will forage on the soil surface only during humid periods. Ants are generally well adapted to arid environments but retreat from the surface to their nests when temperatures become too high.

The nests of ants and termites may extend several metres into the ground. Social life in nests constructed from suitable substrates provides the colony with a stable microclimate (Crawford 1981). Within the nests, food can be stored, and reproduction takes place relatively unaffected by outside conditions. Termite mounds are renowned for their ventilation and internal system of channels, as well as for their orientation to absorb heat only when air temperatures are low (Crawford 1981; Cloudsley-Thompson 1991).

Both ants and termites are major components in desert ecosystems, as reviewed by MacKay (1991). Their biomass greatly exceeds that of other desert invertebrates, and even those of birds and mammals in some deserts. Ants are important predators, and seed harvester ants are particularly abundant in arid habitats. Termites are important in the decomposition of organic matter, nutrient cycling and the effects of their tunnels on soil properties and chemistry (MacKay 1991).

2.1.3.1 Ants

The diversity of ant species is amazingly high in arid regions, although present knowledge of their distribution and occurrence in most deserts is far from complete. Collections from the Chihuahuan Desert revealed 50 species, and 59 species were found in a single desert canyon in California (MacKay 1991). Ant colonies are found in deserts with annual precipitation of less than 50 mm, but Marsh (1986) showed that the species richness increased with precipitation in the Namib Desert (Fig. 2.5).

Ant species richness is especially high in arid regions of Australia. The number of species is not completely known but appears to be as high as in humid regions (Greenslade and Greenslade 1984). Many species are predators, foraging on the soil surface, and taking a wide variety of prey, including other ants. In Australia, harvester ants are more efficient seed consumers than mammals (Morton 1985). According to Abushama (1984), there are 66 species of ants in the Sahara Desert, but most of them live in rather humid areas and only 10 species are xerophiles.

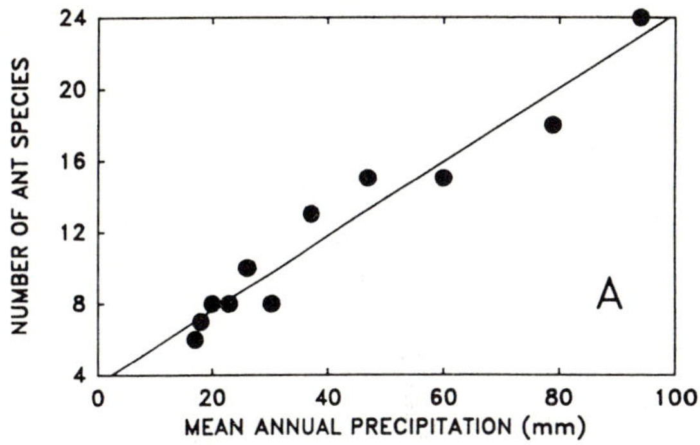

Fig. 2.5. The effect of mean annual precipitation on desert ant species richness along a climatic gradient in the Namib Desert. (Marsh 1986)

Arid areas in South America, separated by the Andes Mountain Range, differ in ant species richness (Medel and Vásquez 1994). Sixteen species of harvester ants have been found in the Argentinian Monte Desert, but only three in the Chilean coastal arid zone. The difference may reflect a higher degree of isolation on the western side of the mountains, while an invasion of species from mesic tropical and subtropical areas may have taken place in the Monte Desert after its formation. The presence of the cosmopolitan genus *Solenopsis* in the coastal deserts of Chile suggests that they also lived in the western margin of South America prior to the uplift of the Andes Mountain Range.

2.1.3.2 Termites

Termites are found in deserts all over the world. According to Crawford (1981), 32 species are known from Indian desert ecosystems and 18 from the Arabian Peninsula. Mound-building termites live in various parts of Africa and Australia, while species in the drier parts of Central Asia are strictly subterranean. Some of the species of the Sahara Desert are active below the surface, probably feeding on subfossil trunks from the ancient flora of the humid Pleistocene period (Abushama 1984). In North America, four species are found in the Chihuahuan Desert, six genera in the Sonoran Desert and nine species in the Coachella Valley of California (MacKay 1991). The termite density increases with aridity in Central Australia (Holt 1987). According to Cloudsley-Thompson (1989), many Australian species have not yet been described. There could well be more than 90 species of *Tumitotermes* alone, of which only 30 have been named so far.

2.1.4 Herbivores

As pointed out in a review by Wisdom (1991), herbivorous insects of the desert are faced with considerable environmental obstacles. Firstly, they must be directly adapted to high temperatures and low humidities or else avoid excessive stress through their behaviour. Secondly, they are indirectly affected by the response of their host plants to heat, lack of water and occasional deficiency of minerals such as nitrogen and phosphate.

Another group of herbivores are the desert snails, of which *Sphinctero-chila boissiere (=zonata)* plays an important part in the food chain in the Negev Desert (Shachak et al. 1976a). The snails spend the summer in their burrows in an inactive state, which is broken by rain in the autumn. Several species of desert snails are known from South and North America, as well as from Australia and southern Africa (Crawford 1981).

The fauna of herbivorous insects in desert regions comprises numerous species of different groups. It is not possible, and it would be outside the scope of the present book, to present a comprehensive list of all groups from warm deserts around the world. The following paragraphs, however, give some examples of the occurrence and variety of insect herbivores in hot deserts.

The fauna of herbivorous insects reflects those orders commonly found in other regions (Wisdom 1991). As an example, 14 insect orders are present in the extreme parts of the Sahara Desert (Abushama 1984). Species of Hemiptera, Homoptera, Lepidoptera and Orthoptera are dominating groups of herbivores in most deserts (Wisdom 1991). Within the order of Lepidoptera, several families of moths are often represented. Grasshoppers and locusts are some of the most abundant and dominating insects, and Wisdom (1991) reported 77 species of grasshoppers from arid areas in New Mexico. Representatives of Coleoptera, Diptera and Thysanoptera are also commonly found in desert ecosystems. Among the beetles, plant feeders such as Bruchidae, Chrysomelidae, Curculionidae and Meloidae are represented.

Roots of desert plants represent a large biomass but relatively little is known about the insects feeding on these resources. According to Crawford (1981), coleopteran larvae are certainly associated with roots, including species of plant-feeding families such as Curculionidae, Scarabidae, Elateridae and Tenebrionidae. Similarly, larvae of several families of Lepidoptera feed on roots and tubers, and larvae of anthomyiid flies are most likely to be found on bulbs of desert plants.

Ephemeral plants in the desert grow and bloom quickly in response to rain, die and leave their seeds. Although their appearance is of short duration, annual plants are visited by an array of insects. According to Crawford (1981), insects such as Orthoptera, Hemiptera, Homoptera and Thysanoptera feed on living plant tissues. Among the Coleoptera, both larvae and adults may be plant consumers, but usually only the larvae of Lepidoptera and Diptera. Other species forage on nectar and pollen, the most important groups being solitary bees, wasps and flies (Crawford 1981; Cloudsley-Thompson 1991).

There are indications that the occurrence of insects in the desert is synchronized with the appearance of their annual host plants (Crawford 1981). Examples are larvae of Noctuidae, Geometridae and Pyralidae in Australia, which respond rapidly to sporadic rains by increased rates of development. Insects on ephemeral vegetation in desert dunefields typically involve species with short developmental times such as mirid and lygaeid bugs, aphids, chrysomelid beetles and many lepidopteran larvae (Crawford 1991a).

The literature on herbivores of perennial desert shrubs, including both xerophilic shrubs and succulents, has been reviewed by Crawford (1981). Since perennials persist for several years, they provide more complex habitats which may result in the establishment of specialized niches by their consumers. The diversity is illustrated by the insect fauna of the mesquite tree *Prosopis glandulosa* in the Sonoran and Chihuahuan deserts of North America (Wisdom 1991). Over its geographical range, this tree supports approximately 200 species of herbivorous insects which consume the leaves, fruits and stems.

An array of insects is found on perennial plants, and the following paragraphs are cited directly from a list of categories compiled by Crawford (1981):

Defoliators. Especially common are larvae of Lepidoptera, Diptera and Coleoptera, adult Coleoptera such as chrysomelid beetles, and all feeding stages of acridid Orthoptera. In addition to insects, certain Gastropoda are common.

Flower Feeders. Examples are meloid beetles and thrips.

Fruit and Seed Eaters. Internal feeders consists of lepidopteran larvae and bruchid, cerambycid and curculionid beetle larvae. External feeders include nymphs and adults of Hemiptera, and also larvae of Lepidoptera.

Pollen and Nectar Feeders. This category includes adult bees, wasps, beetles, flies, butterflies and moths, many of which are effective pollinators as well.

Sap Feeders. Most frequent are mites, and sucking insects such as a variety of hemipterans, homopterans and thrips.

Gall Formers. These include cynipid and chalcidoid wasps, tephritid and cecidomyid flies, psyllid homopterans, as well as several families of Lepidoptera which tunnel in galls.

Borers and Girdlers. Cerambycid and buprestid beetles feed on the cambrium and bark of many woody plants. Some of these insects also bore into the stems.

2.1.5 Detritivorous Macroarthropods

In addition to microarthropods and nematodes, large numbers of macroarthropods are associated with soil and litter. Desert snails are also members of the soil and litter community. Both herbivores and detritivores are common in such habitats, although many species must be considered as omnivores. Ants and termites (Chap. 2.1.3) belong to the soil fauna, and a number of predators (Chap. 2.1.7) prey on its array of invertebrates. In most deserts, individual detritivores are far more numerous than surface-active carnivores and herbivores (Crawford 1991b).

Detritivorous macroarthropods make use of their habitats in several ways (Crawford 1981, 1991b). Under favourable conditions they appear on the surface in large numbers, feeding on available resources. The surface activity of arthropods such as tenebrionid beetles, isopods and millipedes, is regulated by seasonal events or diel rhythms. At other times, the fauna moves to more sheltered underground locations where it may feed on buried plant litter. Crevices or spaces beneath rocks, fragments of plants and other objects are used by many species to escape from the surface, and many species dig burrows in the soil (Crawford 1981).

Typical representatives of detritivores include Isopoda, which are widely distributed in warm deserts all over the world (Crawford 1981; Cloudsley-Thompson 1991). Lewis (1984) listed 14 species from the Sahara Desert. *Hemilepistus reaumuri* is a common species in deserts of North Africa and the Middle East and remarkably large numbers are found in the Negev Desert (Shachak 1980). The various species feed on dry leaves from *Hammada* sp. and other plants (Shachak et al. 1976b). Diplopoda of the order Spirostreptida are numerous in the Chihuahuan Desert (Crawford 1981, 1991b) and large species are also found in African deserts.

Among the insects, Thysanura are common sand dwellers in the Sahara Desert (Abushama 1984), but contribute little to the biomass due to their small size. More important are Dictyoptera, of which there are several species in Africa and Asia. The sand cockroach *Heterogamia syriaca* is the most abundant species in the Sahara (El-Kifl and Ghabbour 1984). This cockroach represent some of the highest biomass of a single non-social species, while *Arenivaga* spp. of the Mojave Desert in North America are less numerous (Crawford 1991b). Among the Orthoptera, crickets are omnivores and live in burrows in the soil (Abushama 1984). Grasshoppers such as *Ammobaenetes phrixocnemoides* from the Chihuahuan Desert contribute to decomposition of plant material (Crawford 1991b).

Tenebrionid beetles are among the best-studied desert arthropods, in particular with regard to their morphological, physiological and behavioural adaptations (Crawford 1981; Cloudsley-Thompson 1991). According to Cloudsley-Thompson and Chadwick (1964), they are the insects best adapted to desert life. Their numbers increase as conditions become more and more adverse to other forms of life. They are particularly abundant in drier parts of the Namib Desert, where almost 200 species have been recorded (Koch 1962). The long-legged *Onymacris* spp. are found on sand dunes and along dry river courses in southern Africa, while the genus *Adesmia* is widespread in North Africa and Central Asia (Crawford 1981).

Polis (1991b) listed 12 species of tenebrionids from the Coachella Valley in California, and Thomas (1983) showed that some species prefer gravelly soil and others sandy soil in the Mojave Deserts. Densities of tenebrionid beetles in the Mojave Desert may exceed 100 000 per ha, and even larger population sizes have been reported from Khazakhistan (Crawford 1991b).

Tenebrionids of the genera *Pimelia*, *Prionotheca*, *Ocnera* and *Adesmia* are common in the Sahara Desert (Abushama 1984). According to Mikhail (1993), four tenebrionid species live in the dry river (wadi) beds of the almost rainless desert of southern Egypt. Aldryhim et al. (1993) trapped 26 species in a wadi of central Saudi Arabia. More than 300 species of Tenebrionidae are known from Saudi Arabia and 95% of them are found in arid and semi-arid environments.

2.1.6 Necrophagous and Coprophagous Arthropods

A number of necrophagous insects are attracted to carcasses of reptiles, birds and mammals, and constitute an important part of the desert fauna. As an example, Coe (1978) made observations on the decomposition of carcasses of elephants in Kenya. The soft tissues were consumed by larvae of calliphorid flies, while dermestid beetles attacked the skin. Termites were active in the removal of cartilage and ligaments from the carcass.

Coprophagous insects feed and develop in the dung from mammals. Different species of flies may breed immediately in the dung, while scarabaeid beetles bury it, and termites and trogine beetles feed on it after a certain amount of drying has occurred (Crawford 1981).

2.1.7 Predaceous Arthropods

General reviews on the biology and ecology of predaceous arthropods from hot deserts have been presented by several authors, e.g. Crawford (1981), Cloudsley-Thompson (1991) and Polis and Yamashita (1991).

Arthropod predators appear to be a particularly important component of the desert invertebrate fauna (Crawford 1991b; Polis and Yamashita 1991).

They comprise a high proportion of the total number of species compared with other trophic groups of arthropods. As pointed out by Crawford (1981), desert predators are often long-lived and ready to catch their prey during sporadic favourable conditions, and may have to starve for long periods in between.

Predaceous arthropods include a number of insects from different orders and families, generally reflecting the fauna of temperate and moist tropical communities. Further details about the diversity of species have been presented by Polis and Yamashita (1991). Carnivorous beetles, wasps and flies are among the most abundant species, in addition to parasitic wasps and flies.

Several groups of arachnids such as mites, spiders, scorpions and solpugids are important predators in deserts. Pseudoscorpions and harvestmen are less common (Cloudsley-Thompson 1984). Data on the diversity of spiders, scorpions and solpugids from several deserts in North America, Africa and Asia have been compiled by Polis and Yamashita (1991). On the average, the deserts included in this survey harbour about 7 species of scorpions, 54 species of spiders and 10 species of solpugids. The number of species in North American deserts ranges from 2-13 scorpions, 30-100 spiders and 6-28 solpugids. In comparison, Cloudsley-Thompson (1984) reported more than 25 species of scorpions from the Sahara Desert, but the fauna of spiders from this area is poorly known.

Among desert spiders, non-web-building species are dominating, and include predators that hunt actively as well as those that sit and wait for their prey (Polis and Yamashita 1991). Several families of web builders are also present in most deserts.

Desert scorpions are typical inhabitants of warm regions and are found on all continents with the exception of Antarctica (Polis 1990). Their densities vary considerably with vegetation, texture of soil and the presence of suitable rocks for their retreat (Cloudsley-Thompson 1991). In North America, species of the family Vaejoridae dominate among the scorpions, while scorpions of the family Scorpionidae are most common in Africa and Australia (Polis and Yamashita 1991). Buthidae are present in all warm deserts and in North America some genera occur in densities of more than 5000 per ha.

Large desert centipedes of the order Scolopendromorpha are mainly found in arid regions of Africa and North America (Crawford 1981). Five orders of Diplopoda with a total of 12 species are present in the Sahara Desert (Lewis 1984).

Different groups of predaceous prostigmatid and mesostigmatid mites are part of the soil and litter communities (Zak and Freckman 1991), and densities vary with season. The giant velvet mites *Dinothrombium* spp. (Fig. 2.6) appear in the Sahara Desert after the rains that trigger the emergence of termites, upon which they feed (Cloudsley-Thompson 1991).

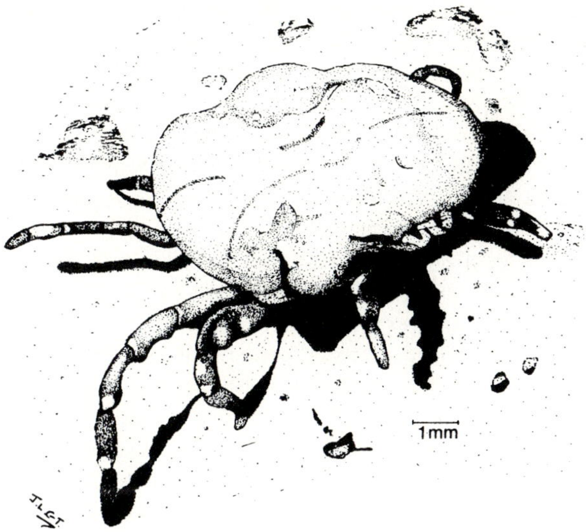

Fig. 2.6. A giant velvet mite (*Dinothrombium* sp.) from the Sahara Desert. (Cloudsley-Thompson 1991)

2.2 The Fauna of Polar Regions

2.2.1 Introduction

Compared with temperate regions, the fauna of the Arctic and the Antarctic, including polar deserts, is very limited. Assuming that half the fauna is known, Kevan and Danks (1986) suggested that the total fauna of arctic terrestrial and fresh water arthropods corresponds to only 25% of the British fauna. The antarctic fauna of terrestrial arthropods appears to be better known (Block 1984a), but the number of species in the Antarctic is much lower than in the Arctic. The terrestrial communities of the Antarctic contain an impoverished flora and fauna, relatively simple in terms of number of component species. According to Wallwork (1973), the distribution of microarthropods suggests that a relic fauna is present, and that this has been enriched by penetrating species, particularly in the subantarctic islands. Presently, these islands contain elements of a southern cold temperate fauna.

There are several reasons why the invertebrate faunas of polar regions are limited. Due to the geographical isolation of the Antarctic Continent and its offshore islands, invasion by new species is restricted. As pointed out by Block (1984a), the prevailing winds are not favourable for dispersal of small invertebrates to the Antarctic, but air currents together with bird transport may be important for colonization in this region. The subantarctic islands may have been colonized in this way after they appeared, particularly after the Pleistocene. The fauna of the isolated, volcanic Bouvetøya cannot include relic elements since the island emerged above sea level only one or two million years ago (Sømme 1986b). The present fauna, including three species of Collembola and four species of Acari, must therefore have arrived by air.

In the Arctic, geographical isolation may greatly affect the colonization of some islands, such as Greenland and Svalbard. The arctic fauna is relatively young and invaded the northern areas after the last glaciation. As suggested by Danks (1981), the post-glacial colonization of arctic areas may still not be complete. Many invertebrate species may not yet have arrived. The reduction in number of species may also be due to lack of resources. An example is the limited diversity of Lepidoptera at northern latitudes because of reduced plant species diversity (Ryan 1981). Subsequently, there is a reduction in the diversity of parasites of Lepidoptera.

In addition to geographical isolation and post-glacial invasion, the extreme climatic conditions of polar areas make it impossible for most invertebrates to live there. Those that survive are able to tolerate the long and cold winters, combined with low summer temperatures. Most ecological studies on polar invertebrates have been focussed on adaptations to low temperatures. However, low precipitation on the Antarctic Continent and over large arctic areas classifies them both as 'polar deserts', and it is likely that moisture is an environmental factor of great importance (Chap. 1.4). In some polar habitats, food and water may be temporarily available during snow melt and rain. Tardigardes, nematodes and rotifers that enter anhydrobiosis during periods of dryness are well adapted to such conditions. More studies on the tolerance to desiccation of polar invertebrates are necessary for us to understand the role of drought in the distribution and diversity of the fauna.

From a terrestrial point of view, it is convenient to divide the south polar region into three zones based on differences in climate, vegetation and fauna (Holdgate 1964; Wallwork 1973). An outer subantarctic zone comprises the islands in the region of the Antarctic Convergence between 46°S and 55°S (Fig. 2.7). The maritime antarctic zone includes the Antarctic Peninsula, some of the islands of the Scotia Ridge and Bouvetøya. Finally, the continental antarctic zone covers the main antarctic landmasses. In accordance with the climatic gradient from north towards higher southern latitudes, species of terrestrial plants and invertebrates decrease dramatically in numbers.

The theories of continental drift and the breakup of Gondwanaland suggest that Antarctica was formerly linked to South America and Australia (Stonehouse 1989). At the time of the separation, some 200 million years ago, the climate was more favourable and the Antarctic Continent had a rich fauna and flora. Concurrent with a southern drift of the continent, the glaciation of the Antarctic was completed about 10 millions years ago.

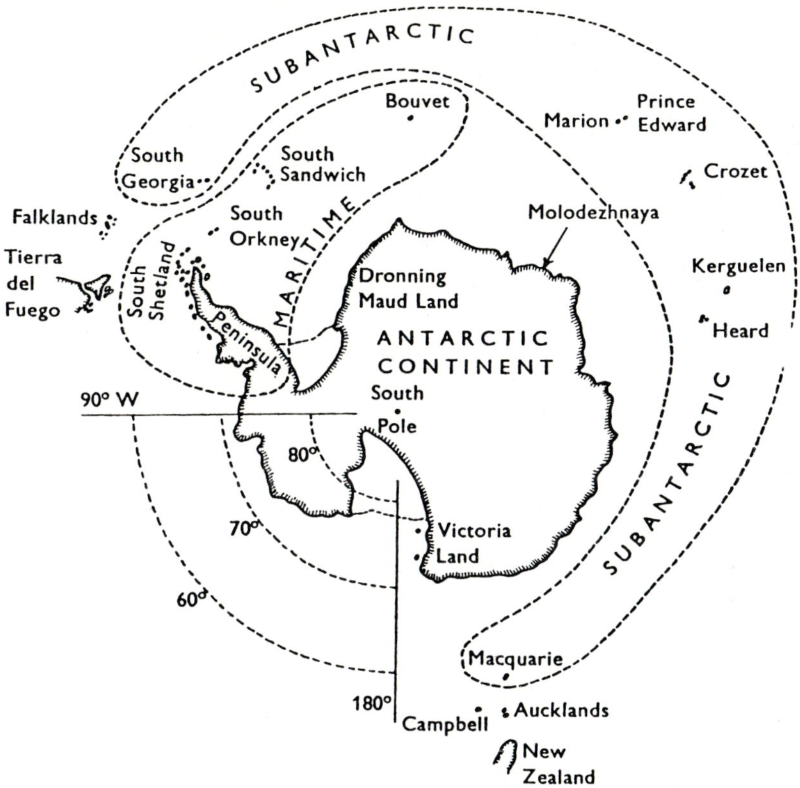

Fig. 2.7. Map of the south polar region to show the distribution of the various zones. (Wallwork 1973)

2.2.2 Nematodes

From a biogeographical point of view, nematodes offer an interesting parallel to antarctic terrestrial arthropods since the distributions of many species are restricted to one of the zones. A review of the literature by Maslen (1979) shows that 22 species, of which 12 are endemic, have been recorded from the subantarctic islands, 40 species (34 endemic) from the maritime antarctic zone and 10 species (7 endemic) from the continental zone. It is uncertain if this high degree of endemism is related to differences in geological history or to differences in ecological conditions.

Recently, Wharton and Brown (1989) reported 7 species of nematodes from the McMurdo Sound region, which is the most southern location recorded for these animals. In samples of vegetation of cyanobacteria and bryophytes in the Taylor Valley, southern Victoria Land, Schwarz et al. (1993) found densities of nematodes in the range $18 \times 10^3\ m^{-2}$ to $430 \times 10^3\ m^{-2}$.

Nematodes are also present in the dry valleys of southern Victoria Land. In the McMurdo Dry Valley, Freckman and Virginia (1991) recorded densities of nematodes up to 4000 per kg dry soil, which is comparable to densities in hot desert soils. Higher densities occurred in soils with higher moisture contents from melting ice and snow. Only three genera were found in the McMurdo Dry Valley, comprising bacterial feeding and omnivore-predaceous nematodes. By manipulation of microclimatic conditions in the field, Freckman and Virginia (1991) showed that the densitites of the bacterial feeding nematode *Scottnema* sp. increased in response to increase in both temperature and amount of water. The greatest effect was noted in plots given a combination of increased temperature plus water plus carbon in the form of sucrose. In contrast, the larger omnivore-predator *Eudorylaimus* sp. did not increase in numbers during 1 year, suggesting that other factors, such as nutrition are limiting.

At Signy Island in the maritime Antarctic, four different groups of nematodes have been recorded with regard to feeding habits; omnivores, predators, microbial feeders and fungal feeders (Spaull 1973), but no plant feeders. Estimates of nematode biomass in soil and vegetation ranged from 0.16 to 8.25 g m^{-2} (Spaull 1973), while population densities in two research sites were 0.77×10^6 m^{-2} and 0.45×10^6 m^{-2}, respectively (Maslen 1981).

Even higher numbers are reported from the Arctic, where densities of nematodes in the tundra ranges from 1×10^6 m^{-2} to 5×10^6 m^{-2} (Chernov 1985). Most of the arctic nematodes feed on algae and bacteria, but omnivores, predators and fungal feeders are present as well. Species diversity is high, and Chernov et al. (1977) reported 53 species of soil-dwelling nematodes from Cape Cheluskin at the Taimyr Peninsula.

2.2.3 Tardigrades

Several species of tardigrades have been reported from Antarctica and reviews of the previous literature have been presented by Jennings (1976a) and Block (1984a). Many of the tardigrades have a worldwide distribution, and only a few species are restricted to the Antarctic. Most of the species have been collected from the maritime antarctic zone, where a total of 23 species has been recorded (Dastych 1984; Usher and Dastych 1987). Although most tardigrades are found in well-vegetated environments, several species were collected from fellfields and apparently unvegetated gravel. On the Antarctic Continent, tardigrades have been reported from bryosystems and chalikosystems in the Syowa Station area (Sudzuki 1979) and in the Mühlig-Hofmannfjella (Sømme and Meier 1995). Schwarz et al. (1993) found densities of tardigrades up to 123×10^3 m^{-2} in bryophytes in the Taylor Valley, southern Victoria Land.

At Signy Island, densities of tardigrades varied from 11 000 m^{-2} in moss turf to $14\,130 \times 10^6$ m^{-2} in the foliose *Prasiola* algae (Jennings 1976a).

2.2.4 Rotifers

A review of the antarctic fauna of Rotifera has been presented by Dartnall and Hollowday (1985). Although a list of 45 species of aquatic rotifers from Signy Island is given, the fauna of other areas is relatively poorly known. Most species are aquatic, but several have also been reported from terrestrial habitats in the maritime and continental antarctic zones (Jennings 1976b; Sudzuki 1979). Since rotifers are difficult to identify, the number of species is uncertain. Most of them belong to the class Bdelloida, but monogonotid species are also present in the Antarctic. Terrestrial rotifers are mainly found in wetter communities, where they live in the interstitial water of soil and mosses. Like nematodes and tardigrades, rotifers may enter cryptobiosis under stressful periods of drought.

In moss communities, growing in flushes of melt water from the Canada Glacier in Taylor Valley, southern Victoria Land, Schwarz et al. (1993) foung high densities of rotifers. Up to 112 x 10^3 m^{-2} specimens were revived from samples of vegetation soaked in water prior to the spring melting period, and up to 548 x 10^3 m^{-2} were recorded after melting. The most abundant rotifers were bdelloids, while monogonotid species appeared to be more common at the end of the season. Due to their fast revival from anhydrobiosis and their parthenogentic reproduction, bdelloid rotifers may increase rapidly in numbers when environmental conditions are favourable (Ricci 1987; Schwarz et al. 1993).

2.2.5 Other Invertebrates

Earthworms have been recorded from several subantarctic islands, including South Georgia (Block 1984a). A few species of enchytraeids are also found in the maritime antarctic zone (Block and Christensen 1985). Oligochaetes are in general dependent on moisture, and therefore restricted to moist habitats. Similarly, gastropod land snails prefer moist environments. In the Antarctic they are restricted to the sub-antarctic zone (Block 1984a).

Enchytraeids are widely distributed in the Arctic (Nurminen 1973; Chernov 1985; Alexandrova 1988). They are mainly associated with wet habitats, and it is not known if any stages are tolerant to desiccation. The terrestrial enchytraeid species are predominantly similar throughout the Holarctic Region (Nurminen 1973). At Spitsbergen, Svalbard, a total of 28 species has been recorded (Birkemoe and Dózsa-Farkas 1994).

Table 2.2. Number of species of terrestrial arthropodes reported from Arctic North America, Svalbard and the Antarctic

Group	Arctic Canada and Alaska[a]	Canadian Artic Islands[a]	Queen Elizabeth Islands[a]	Svalbard[b]	Sub-Antarctic[c]	Maritime Antarctic[c]	Continental Antarctic[c]
Aranea (spiders)	117	30	18	15	14	0	0
Acari (mites)	354	179	117	70	128	32	29
Myriapoda	3	0	0	0	3	0	0
Collembola (springtails)	110	61	55	47	37	8	10
Insecta	1648	585	361	185	173	27	38
Total	2232	855	551	317	355	67	77

[a]Danks (1990), [b] Sømme (1993), [c]Block (1984a)

2.2.6 Terrestrial Arthropods

2.2.6.1 Antarctica

Mites. More than 150 species of terrestrial mites are known from the antarctic zones (Block 1984a). Most of them are found in the sub-antarctic islands (Table 2.2) and the numbers of species decrease towards the south. When all terrestrial free-living, parasitic, fresh water and marine mites are included, Pugh (1993) listed a total of 528 species of Acari from Antarctica and the Southern Ocean.

Dry terrestrial habitats are dominated by prostigmatid and cryptostigmatid mites (Fig. 2.8), as well as by some predatory mesostigmatid species. The fauna of the maritime and continental zones contains a high degree of endemism (Wallwork 1973; Block 1984a). Of the 23 species of prostigmatid mites inhabiting the polar deserts of the Antarctic Continent, only *Nanorchestes antarcticus* is found further north. This species is also the most southern terrestrial arthropod in the world, and is recorded from 85°32'S.

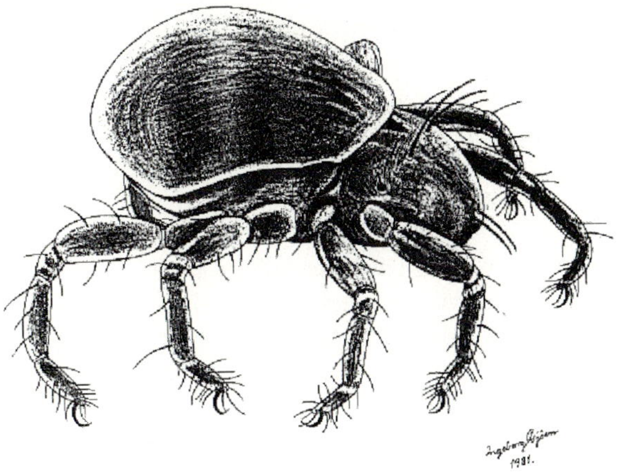

Fig. 2.8. *Alaskozetes antarcticus*, a common cryptostigmatid mite from the maritime antarctic zone. (Drawing from SEM photo)

Spiders. In addition to mites, the arachnid fauna of the Antarctic includes 14 species of spiders. They are restricted to the subantarctic islands.

Springtails. The total fauna of antarctic Collembola numbers almost 50 species, of which 70% are endemic (Wallwork 1973; Block 1984a). As a group, springtails are widespread and numerous. Several subantarctic species are also distributed further north. The numbers of collembolan species are harply reduced in the maritime and continental zones, where

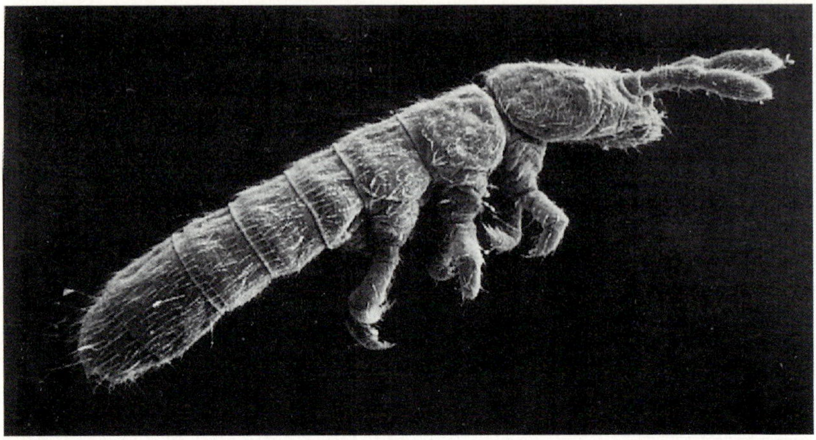

Fig. 2.9. *Cryptopygus antarcticus*, one of the most common species of springtails in the Antarctic. (SEM photo, Biological Institute, University of Oslo)

8 and 10 species have been recorded, respectively (Table 2.2). Collembola inhabit a variety of habitats, in particular vegetation of mosses, lichens and algae. One of the most widespread species, *Cryptopygus antarcticus* (Fig. 2.9) has been recorded in densities of more than 115 x 10^3 m^{-2} in mosses at Signy Island (Block 1984a). In the continental antarctic zone, collembolans are found in inland habitats where access to moisture is restricted (Janetschek 1970; Sømme 1986a).

Insects. There are no free-living insects in the continental antarctic zone, but 38 species of Phthiraptera live on birds and mammals (Table 2.3). In addition, one species of flea, *Glaciopsyllus antarcticus*, has been recorded, and it is suggested that its larvae may overwinter in empty nest sites (Murray et al. 1967).

Compared with the Arctic, the lack of free-living insects in the Antarctic is striking. This is at least partly due to the severe climatic conditions of the Antarctic Continent. Both low temperatures and lack of moisture contribute to the hostile environment. Most of the present terrestrial arthropods of the Antarctic Continent are probably relics from a warmer period, and geographical isolation greatly restricts the invasion of new species.

Geographical isolation may also explain the poverty of the insect fauna in the maritime and subantarctic islands. Two species of midges, *Belgica* spp., are the only free-living insects of the maritime antarctic zone. Although the climate is relatively mild, the fauna of the subantarctic island include fewer species than far more extreme arctic locations (Table 2.3).

Table 2.3. Diversity of insects in the Arctic and the Antarctic

Order	Arctic North America[a]	Canadian Arctic Islands[a]	Queen Elizabeth Islands[a]	Greenland	Svalbard[b]	Sub-Antarctic[c]	Maritime Antarctic[c]	Continental Antarctic[c]
Ephemeroptera	22	3	0	1	0	0	0	0
Odonata	6	0	0	0	0	0	0	0
Plecoptera	20	3	0	0	0	0	0	0
Orthoptera	6	0	0	2	0	0	0	0
Psocoptera	0	0	0	0	0	3	0	0
Phthiraptera	52	44	32	47	37	67	25	38
Hemiptera	66	18	6	17	8	4	0	0
Thysanoptera	5	3	3	5	0	2	0	0
Neuroptera	2	0	0	3	0	0	0	0
Coleoptera	203	30	13	44	12	40	0	0
Diptera	787	307	199	275	97	44	2	0
Siphonaptera	20	6	4	?	2	6	?	1
Lepidoptera	162	62	23	41	5	3	0	0
Trichoptera	25	3	1	7	1	0	0	0
Hymenoptera	272	106	80	87	23	4	0	0
Total	1648	585	361	529	185	173	27	39

[a]Danks (1990), [b]Sømme (1993), [c]Block (1984a).

2.2.6.2 Arctic

In many geographical areas the arctic fauna is still poorly investigated. The list of North American arctic terrestrial arthropods compiled by Danks (1981), and with later additions (Danks 1990), is the most complete survey. By including named species as well as those reported as taxonimically different but unidentified, Danks (1990) concluded that a minimum of 2232 species of terrestrial arthropods are present in the Arctic Canada and Alaska (Table 2.2). The figure includes spiders, mites, chilopods, springtails and insects north of the arctic timberline. By contrast, more then 22 000 species of insects alone are reported from Great Britain.

With increasing latitude the number of arctic terrestrial arthropods is greatly reduced. Thus, when the Canadian arctic islands are considered separately, the total number of species amounts to only 855. In the Canadian arctic archipelago the Queen Elizabeth Islands represent an even more extreme area, and the number of terrestrial arthropods is reduced to 551 (Table 2.2).

Compared with the fauna of the Canadian arctic islands, and the minimum expected species of the Queen Elizabeth Islands, the terrestrial arthropod fauna of Svalbard is rather impoverished (Table 2.2). Although situated far to the north (77 to 81°N), the climate of Svalbard is relatively mild due to the warm North Antlantic Stream (the Gulf Stream). The low number of 317 species, including potential unidentified species, is probably due to the isolated geographical position of the island (Sømme 1993). Only 560 insect species have been reported from Greenland (Böcher 1992), and this is a rather low figure considering the position of the island between 60 and 81°N.

The composition of the terrestrial arthropod fauna of the Arctic differs in many respects from that of temperate and tropical regions. Some taxa are better represented in the north, while others have lower diversity or are lacking. The difference in diversity suggests that some groups have been more successful than others in colonizing the hostile arctic environments.

Springtails. Among the more successful taxa, Collembola appears to be well represented and the reduction in species with northern latitudes is less than in other groups (Table 2.2). The springtails occupy an important part of the soil fauna, and appear in enormous numbers of specimens, e.g. 52×10^3 m^{-2} from the polar desert plateau at Devon Island (Addison 1977). At Spitsbergen, densities reached 38×10^3 m^{-2} in a dry lichen heath, 244×10^3 m^{-2} in wet mosses and 268×10^3 m^{-2} on grassland (Bengtsson et al. 1974).

Even within short distances, the densities of springtails varies greatly with different microhabitats, as demonstrated by Chernov et al. (1977) in the polar desert of Cape Cheluskin at the Taimyr Peninsula. In areas with only 5% vegetation coverage, a density of 9×10^3 m^{-2} was found in apparently naked soil, while densities increased to 70×10^3 m^{-2} and 188×10^3 m^{-2} in stripes and cushions of moss within nearby cracks of soil. With a vegetation coverage of 20%, densities ranged from $19 \times 10^3\ m^{-2}$ in bare soil to $366 \times 10^3\ m^{-2}$ in moss cushions. The latter figure is higher than densities recorded from an adjoining lichen-moss community.

Arachnida. Similarly, mites are a dominant group of the soil fauna, represented with a large number of species and large populations. Compared with the maritime and continental Antarctic, the biodiversity of mites in the Arctic is much higher (Table 2.2). As an example, Hammer (1944) reported 55 species of oribatid mites from Greenland and suggested that a large part of them had survived glaciation on ice-free nunataks. Along the arctic south-north gradient, spiders are the taxon with proportionally the greatest reduction in species.

Insects. The number of insect species is also greatly reduced and several orders are absent at the more extreme northern latitudes. (Table 2.3). Among the typically fresh-water insects, Ephemeroptera, Odonata and Plecoptera are represented with several species north of the arctic timberline, but numbers decline rapidly further north. Trichoptera is represented by several species in arctic North America, but by only one species on the Queen Elizabeth Islands and Svalbard. Orders such as Orthoptera, Thysanoptera and Neuroptera are also poorly represented.

Most striking are the low numbers of Coleoptera in the Arctic. Beetles comprise the largest of all insect orders, with an estimated number of 350 000 species on a world basis. The total fauna of Coleoptera in arctic North America is estimated at ca. 200 species, while only 12 or 13 species are known from the most northern latitudes (Table 2.3). Beetles represent approximately 39% of all known insects in the world, but they represent only 3% of the species on the Queen Elizabeth Islands (Danks 1990).

Lepidoptera are relatively well represented in the Arctic, although the number of species declines towards higher latitudes. Hymenoptera are found in even higher numbers, and most of the species are parasitoids.

As in the Antarctic, Phthiraptera is a dominant order in the Arctic. These insects live a protected life in the feathers of birds or the fur of mammals. Close to the warm skin of the host, they do not experience the hostile conditions of the surrounding environment. The number of species recorded from arctic areas reflects the numbers of host animals.

The most dominating insect order in the Arctic is Diptera (Table 2.3). Almost 50% of the insects recorded from arctic North America and 61% of the insects from Queen Elizabeth Islands belong to this order (Danks 1990). In comparison, only 16% of the insects of the world are Diptera. At Svalbard, Diptera is represented with 97 species (Sømme 1993). As in other arctic areas, midges of the family Chironomidae are dominant. Approximately half of the flies at Svalbard and on the Queen Elizabeth Islands belong to this family. Their success is at least partly related to the fact that the larvae live in fresh water and moist soil, protected from the more extreme climatic fluctuations at the surface. During periods of hatching, adult midges appear in large swarms on calm days. In some habitats, craneflies (Tipulidae) also attain enormous populations (Brodo 1990), and 16 species are known from the Canadian arctic islands. Other dominant families are Sciaridae, Muscidae and Anthomyidae (Danks 1990).

2.2.6.3 Trophic levels

From an ecological point of view, all trophic levels are represented among arctic terrestrial arthropods (Danks 1990). Detritus-eating species, including a number of dipterous larvae, mites and springtails, are especially significant. These groups are found in more protected habitats such as moist soil and fresh water. The numbers of aquatic species increase in proportion of the entire fauna from 10% in temperate areas to almost 60% in severe arctic sites (Danks 1990) (Fig 2.10).

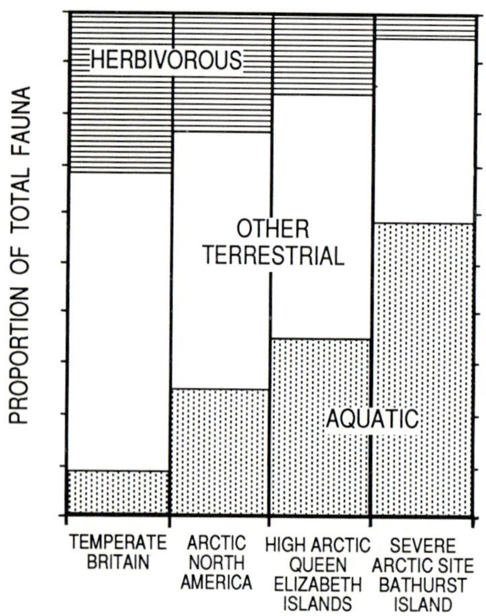

Fig. 2.10. Relative representation of herbivorous, non-herbivorous terrestrial, and aquatic species of insects in a temperate area and in arctic areas of different severity. (Danks 1990)

On the other hand, the numbers of herbivorous arthropods decline from 30 % in temperate areas to less than 5 % at high arctic latitudes. This, of course, is also related to a decline in the number of plant species. The arctic fauna includes sucking and chewing species, as well as species living in galls and mines in the plant tissue. As pointed out by Kevan and Danks (1986), plants provide shelter for many insects, including species that do not depend directly on them as food resources. Within the leaves and flowers, temperatures reach several centigrade above ambient levels. A substantial number of arctic plants are insect-pollinated, the Diptera being the most important pollinators.

In general, a relatively large proportion of terrestrial arctic arthropods comprises predators and parasitoids. Predators are represented by spiders, mites, flies and beetles, and parasitoids by wasps and tachinid flies.

As an adaptation to avoid the extreme environment, it appears that a large proportion of arctic species is restricted to moist habitats where desiccation is not a problem of survival. The species that experience arid conditions are herbivores and predators which are active on the surface during the summer, as well as species inhabiting soils that temporarily dry out. All of them spend most of their lives in an inactive state while overwintering. They are exposed to more or less severe temperature conditions, and different strategies of cold hardiness have evolved among them (Chap. 9).

2.3 Comparison of Hot and Cold Arid Environments

Although the invertebrate fauna of most temperate and subtropical deserts are far from completely investigated, high diversity appears to be present in most regions. Numerous species are found at all trophic levels, including the soil fauna, surface-active species, predators and plant feeders. The high diversity may have resulted from the invasion of species from adjoining areas, gradually adapting to the adverse conditions of the desert. In some deserts, relict faunas from earlier geological periods with more humid climates may be present.

Compared with hot deserts, species diversity of arid polar areas is much poorer. Due to geographical isolation, the fauna of the Antarctic is very limited. Arctic areas were invaded by invertebrates from the south relatively recently, following the termination of the last period of glaciation. More than 2200 species of terrestrial arthropods are known from Arctic Canada and Alaska, but a large proportion of these species are not found in arid regions. With increasing latitudes, aridity and cold become more severe and species richness decreases. Geographical isolation of some of the arctic islands, such as Greenland and the archipelago of Svalbard, also accounts for low biodiversity.

In addition to quantitative differences, there are distinct qualitative differences in the fauna of hot and polar deserts. Spiders are poorly represented in the Arctic but are dominating predators in warm environments. Other important groups of predators from hot deserts such as scorpions and solpugids are entirely lacking in polar areas. Microarthropods and nematodes, however, are well represented in both environments, and appear in higher densities in cold than in hot arid areas.

With regard to insects, most orders are poorly represented or lacking in polar regions. In contrast, insects are extremely abundant in hot deserts, reflecting the fauna of adjoining temperate and wet tropical areas. As an example, beetles are abundant in hot deserts but have low diversity in cold habitats. While Diptera are not a prominent part of the hot desert insect fauna, they are abundant in the Arctic. Most species, such as chironomid midges, have larval stages in water and moist soil.

The survey of these widely separated biomes shows how different taxa have been able to cope with adverse environmental conditions. Life in hot arid environments requires special morphological, physiological and behavioural adaptations. Although temperatures are lower, several polar species are also exposed to desiccation. In addition, they have to survive cold summers and extremely low winter temperatures. The combination of drought and cold may explain the poverty of the polar fauna as compared with the fauna of hot deserts. Both cold hardiness and resistance to desiccation are thoroughly treated in the following chapters.

In one respect the species of terrestrial invertebrates from hot and cold arid environments do show similar response: when conditions are adverse, they disappear from the surface and live a hidden life. In hot deserts many species seek shelter when conditions are adverse, and use periods of rain or other favourable conditions for their opportunistic activities. Similarly, many polar species live a hidden life in the soil or in water. During the winter they shelter deep below the snow, to appear only in the spring to continue development and reproduction during the short summer season.

3 Tolerance to Arid Conditions in Terrestrial Arthropods

The large literature on the adaptation of terrestrial arthropods to arid environments includes a number of excellent books and review articles that summarize the most important results within the field (e.g. Edney 1977; Crawford 1981; Hadley 1985, 1994; Cloudsley-Thompson 1991.) In the present chapter, emphasis is placed on developments during later years. Most authors have paid little attention to adaptations of alpine and polar species to arid environments, and for this reason a comparison of species from colder areas with those from hot and arid subtropical and temperate habitats is presented.

As pointed out by Edney (1977), the term resistance to desiccation is used in two ways. Firstly, it means tolerance of larger water losses and low water contents. Secondly, it refers to the restriction of water loss in one way or another. Both of these abilities would be important adaptations to arid environments. In most desert arthropods the ability to conserve water is apparently best developed.

The rate of water loss varies considerably between different species of terrestrial arthropods, as is apparent from a series of examples given by Edney (1977). When details of the sources are not known, water loss can be expressed simply as percent fresh weight per time unit (e.g. % fw h^{-1}), but is frequently related to surface area (e.g. $\mu g\ cm^{-2}\ h^{-1}$) when transpiration is mainly cuticular. According to Hadley (1994), cuticular water loss is most commonly expressed as $\mu g\ cm^{-2}\ h^{-1}\ mmHg^{-1}$. In this expression, the water loss per surface area and time is divided by the vapour pressure difference, which is the driving force to move water across the cuticle. The use of this unit facilitates comparisons between different species, since changes in vapour pressure at different experimental temperatures are at least partly compensated. Hadley (1994) listed close to one hundred examples of permeability in species of terrestrial arthropods from the literature, but warned that comparisons must be made with caution. Not only do the experimental conditions of different authors differ, but biological status, such as size, age, sex and water content, of experimental animals also differs.

3.1 Microarthropods

Most Collembola are highly sensitive to desiccation (e.g. Christiansen 1964; Verhoef and Witteveen 1980), although large variations in cuticular permeability have been recorded (Table 3.1). Whilst drying, many Collembola behave as porous hygroscopic solids (Vannier 1982; Block et al. 1990; Harrisson et al. 1990). The characteristic components of the process include an initial decrease in the rate of drying, followed by a constant rate period, and finally a falling rate which is exponential. The purely physical and passive manner in which several Collembola dry, and isotomids in particular, may partly explain their high rates of water loss (Harrisson et al. 1991).

Similarly, large variations in susceptibility probably occur among mites, although less information is available. Among oribatid mites, several species exhibit high tolerance of desiccation (Worland and Block 1986).

Different experimental procedures have been used to measure sensitivity to desiccation in Collembola, and it is often impossible to convert the data of different authors to comparable units (Block et al. 1990; Harrisson et al. 1990, 1991). The rate of water loss depends on temperature, the vapour saturation deficit and the water content of the organisms. According to Harrisson et al. (1990), better comparisons are obtained by converting the results to conductance, which involves standardizing rates of water loss on surface area. The conductance k (mm min^{-1}) is calculated from the average evaporative flux $\overline{x}E$ (µg mm^{-2} min^{-1}) divided by the evaporative driving force or vapour density deficit ΔD (g m^{-3}):

$$k = \frac{\overline{x}E}{\Delta D}.$$

Based on data compiled from the literature, Harrisson et al. (1990) compared the cuticular conductance of 16 species of Collembola from a range of habitats (Table 3.1). The drier the habitat, the lower is the conductance value. Extremely low values of conductance were recorded in Seira domestica living in house dust and walls, and in species from dry forest litter (e.g. Vannier 1978; Verhoef and Witteveen 1980.). Medium values were represented by species from moist litter and soil (e.g. Eisenbeis 1982), while highest values were found in four species of cave dwellers living under constant high humidities (Vannier and Thibaud 1978).

Table 3.1. Cuticular permeabilities to water vapour of 16 species of Collembola from a range of habitats, under external dying conditions of constant temperature (20-21 °C) and dry air (0% RH). JI and JII refer to respective first and second juvenile instars. For references see text and Harrisson et al. (1990)

Species	Conductance (mm min^{-1})	Habitat
Seira domestica	0.5	House dust, old walls libraries; France
Allacma fusca adult	0.8	Lower forest vegetation canopy; France
Allacma fusca JII	1.1	Mostly top of the forest floor and lower vegetation canopy; France
Orchesella cincta	4.0-24.2	Drier litterlayer of mixed birch-pine forest; The Netherlands
Entomobrya nivalis	5.9	Mixed birch-pine forest; The Netherlands
Allacma fusca JI	6.8	Forest litter layer in contact with soil; France
Orchesella villosa	14.4-27.2	Drier litter layer of forest: Germany
Tomocerus flavescens	35.1	Moist areas between drier litter layer of beach forest; Germany
Tomocerus longicornis	35.1	Same as *T. flavescens*
Tomocerus minor	40.5-98.1	Below moist litter layer of forset; Europe
Tetrodontophora bielanensis	61.1-76.1	Large species in soils of humid forests; France
Parisotoma octooculata	72.0	Beneath moist moss clumps and guano enriched, mineral soils; maritime Antarctic
Tomocerus problematicus	73.3-95.2	Obligate cave dweller in French Pyrenees
Podura aquatica	86.1	Water surfaces of ponds and ditches; England
Tritomurus falcifer	101.2	Mountain caves; Garonne in France
Plutomurus sp.	118.5	Cave dweller; Mount Fuji; Japan
Trichomurus scutellatus	153.8	Cave dweller; Yugoslavia
Tomocerus catalanus	159.0	Cave dweller; Catalan region of Spain

3.1.1 Alpine Species

While temperature studies have been the object of most investigations on adaptations to high altitudes, relatively few studies have been concerned with the tolerance of alpine microarthropods to desiccation. With a growing awareness that moisture may be just as important as temperature under alpine conditions, it is likely that more attention will be given to studies on adaptations to arid conditions in the future.

At high altitudes, some species of microarthropods live in dark foliose lichens, such as *Alectoria* and *Umbelicaria* spp. (Leinaas and Sømme 1984). With regard to heat and drought, desert-like conditions are experienced in periods of low precipitation. Solar radiation is readily absorbed by the black surface of the lichens, and temperatures may rise to extreme levels. Water losses make the lichens dry and crisp. Under such conditions some species of microarthropods probably migrate to less hostile environments, and only a few species with high resistance to desiccation can survive in the dry lichens themselves.

A characteristic collembolan fauna is also known from lichens and mosses on granite boulders at 600 m in northern Austria (Bauer 1993; Bauer and Christian 1993). Most species depend on vegetation bridges connected to the surrounding vegetation but some, such as *Xenylla boer*, are more resistant to desiccation and inhabit the boulders all year round.

The rock-inhabiting Collembola *Xenylla maritima* and *Anurophorus laricis* from Finse, Norway, survived up to 3 and 14 days, respectively, at 20 °C and 50% RH (Leinaas and Sømme 1984). Although the most tolerant species under laboratory conditions, *X. maritima* preferred the more protected north facing sides of the boulders, while *A. laricis* appeared in equal numbers on both north and south facing walls. In comparison, *Tetracanthella wahlgreni*, which lives partly on the boulders and partly in vegetation at their base, died within 3-4 h at 20 °C and 50 % RH. Similarly, the lichen-feeding species *Vertagopus westerlundi* and *Vertagopus sarekensis* with their main distribution in the low and high alpine zones of Norway, respectively, differed greatly in resistance to desiccation (Leinaas and Fjellberg 1985). Newly hatched individuals were considerably more susceptible than adults, but showed similar interspecific difference. In *V. sarekenis* the eggs represent the most resistant stage, while in *V westerlundi* the eggs are more resistant than newly hatched juveniles but more sensitive than adults. The more susceptible *V. sarekensis* lives both on boulders and in the ground, and may escape drought more easily than *V. westerlundi*, living in bark and lichens. The difference in resistance is reflected in the cuticular structure of the two species, for *V. westerlundi* it appears to be more specialized with reduced areas of transpiring cuticle.

The protective effect of the vegetation is even more pronounced in the paramo of the Venezuelan Andes (Garay 1981). At 4200 m the senescent leaves surrounding the stems of the giant rosette plants *Espeletia* spp. may harbour up to 130 000 specimens of microarthropods. The density of springtails and mites in the protected soil directly below the rosette plants is two to three times higher than the density in the adjoining open, dry grassland.

In conclusion, the survival of collembolans under arid alpine conditions may depend on their resistance to desiccation, based on reduced cuticular permeability, as well as on the ability to seek protected microhabitats with more favourable moisture conditions.

3.1.2 Polar Species

From his studies in South Victoria Land, Janetschek (1970) concluded that moisture plays a major role in the limitation of terrestrial life (Chap. 1.4). The springtail *Gomphocephalus hodgoni* is characteristic of coastal bryophyte systems, where liquid water is available during the short summer. In the inland chalikosystem, species like *Antarcticinella monoculata* have a spotwise distribution, based on the diel occurrence of melt water. No terrestrial arthropods occur in the Dry Valley of Victoria Land, due to lack of moisture.

The importance of the availability, uptake and loss of water for the survival of antarctic microarthropods has been emphasized by Block (1985a). In the fauna of Signy Island there is a close relation between the habitat occupied by different species and their resistance to dehydration. Among the seven species listed by Worland and Block (1986), the cryptostigmatid mites *Alaskozetes antarcticus* and *Halozetes belgicae* had the lowest rates of water loss under experimental conditions. Both species feed on the surfaces of rocks and lichens, and are apparently able to tolerate exposure to solar radiation. During desiccation at 10 °C and 5 % RH, a small proportion of *A. antarcticus* survived losses of 60 % of their initial fresh weight. The collembolans *Parisotoma octooculata* and *Archisotoma brucei* are the most susceptible species, being confined to moist habitats. The mites *Stereotydeus villosus* (Prostigmata), *Gammasellus racovitzai* (Mesostigmata) and the springtail *Cryptopygus antarcticus* represent intermediate tolerance.

The difference in rates of water loss between *Parisotoma octooculata* and *Cryptopygus antarcticus* is partly explained by the lower conductance values of the latter (Block et al. 1990). In addition, *P. octooculata* contains less water than *C. antarcticus*. High conductance values (Table 3.1, Fig. 3.1) and low water content emphasize the necessity for *P. octooculata* to live within moist habitats in the maritime Antarctic. The greater risk of dying through desiccation clearly influences its distribution. Where both species co-exist, *C. antarcticus* is more abundant and occupies a wider range of habitats than *P. octooculata*. The latter prefers damp mineral soils, particularly those which have been enriched with nutrients from sea bird colonies, while *C. antarcticus* occurs in most organic material, with or without mosses. The wider distribution of *C. antarcticus* may be related to its higher water content, lower susceptibility to dehydration and, consequently, lower conductance (Worland and Block 1986; Block et al. 1990).

In some species of springtails, compensation for rate of water loss with decreasing temperatures has been observed. While *Parisotoma octooculata* has high conductance when measured at 20 °C (Table 3.1), and would have been classified as a hygrophilic species, Harrisson et al. (1990) showed that the cuticular conductance changes with temperature and consequently with the vapour density deficit (Fig. 3.1).

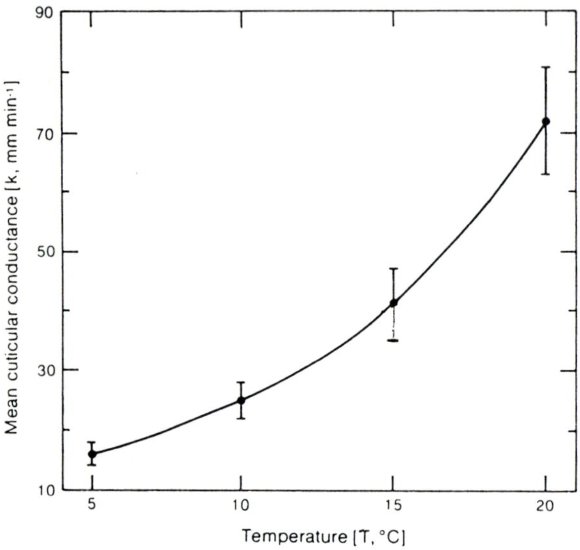

Fig. 3.1. The relation between the mean cuticular conductance to water vapour (k) and the drying temperature (T), whilst drying laboratory specimens of *Parisotoma octooculata* to their lower field hydration value at constant drying temperatures of 5, 10, 15 and 20 °C. Mean $k \pm$ SE, n=5. (Harrisson et al. 1990)

When drying *P. octooculata* at constant temperatures and 0% RH, the conductance decreased from 72 mm min^{-1} at 20 °C to 15.8 mm min^{-1} at 5 °C. Accordingly, *P. octooculata* becomes mesophilic at low temperatures and, by extrapolation of the results, may be considered as almost xerophilic at -15 °C. Although rates of water loss in this way will be considerably reduced by lowered temperatures, *P. octooculata* still remains very susceptible to drying if exposed to appreciable drying forces for any length of time, without having access to liquid water (Harrisson et al. 1990)

Cryptopygus sverdrupi, inhabiting nunataks at elevations up to 1600 m in Dronning Maud Land, exhibits even higher resistance to desiccation (Fig. 3.2) (Sømme 1986a), although values for conductance are not available. *C. sverdrupi* has a patchy distribution in connection with the nitrophilous *Prasiola* algae near occasional occurrence of melt water around bird colonies.

In February, at the end of the austral summer at 72°S, *C. sverdrupi* tends to seek shelter underneath small rocks when environmental temperatures fall below 0 °C (Fig. 1.7). Low moisture content prevails in the barren ground, and *C. sverdrupi* probably seeks the microhabitat with highest moisture content in preparation for overwintering.

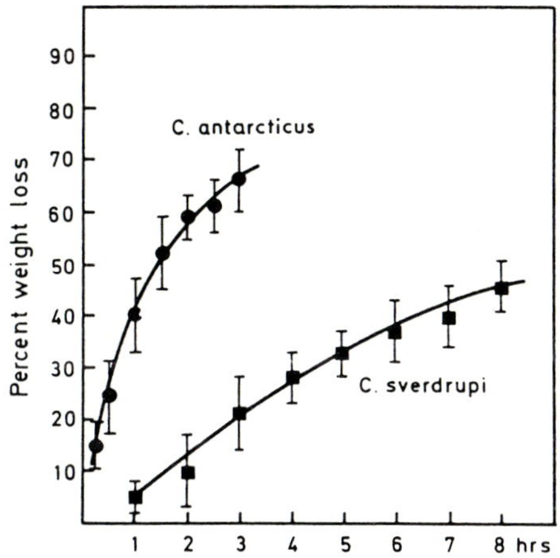

Fig. 3.2. Weight loss plots of *Cryptopygus sverdrupi* and *Cryptopygus antarcticus* at 5 % RH and 15 °C. (Sømme 1986c)

3.2 Coleoptera

Tolerance of desiccation and physiological mechanisms behind the adaptation to arid environments have been studied in a number of species of beetles (e.g. Edney 1977; Zachariassen et al. 1987; Cloudsley-Thompson 1991; Hadley 1994). Due to their large size and frequent occurrence in lowland and alpine deserts, beetles are often chosen as experimental animals. Their taxonomy is also better known than that of most other groups of terrestrial arthropods, and this makes it easier for ecologists and physiologists to identify the species they are dealing with.

In dry habitats, plant debris underneath shrubs, fallen plants and the undersides of rocks offer sanctuaries against desiccation. Thus, two strategies of survival are possible for beetles and other terrestrial arthropods; either avoidance of the most xeric conditions or the evolution of greater resistance to desiccation. For species restricted to moist microhabitats within an arid area, there are obviously many constraints. Activities such as feeding and reproduction may be limited to periods with higher precipitation, or may possibly take place during the night. In arid alpine environments, diurnal temperatures of protected microhabitats are lower than those of the surroundings, and this will slow down growth and development in beetles and other invertebrates.

3.2.1 Alpine Species

3.2.1.1 Tenebrionids

Among beetles, both lowland and alpine tenebrionids are known to be extremely tolerant of dehydration. On Tenerife, several species live from 2000-2500 m in dry environments around the base of Pico del Teide (Ottesen and Sømme 1987). *Pimelia radula ascendens*, *Hegeter lateralis* and *Hegeter tenuipunctatus* (Fig. 3.3). All species survived more than 40 days at 5% RH and 20 °C. Their low rates of water loss place them among the most resistant to cuticular transpiration of the lowland species listed by Edney (1977).

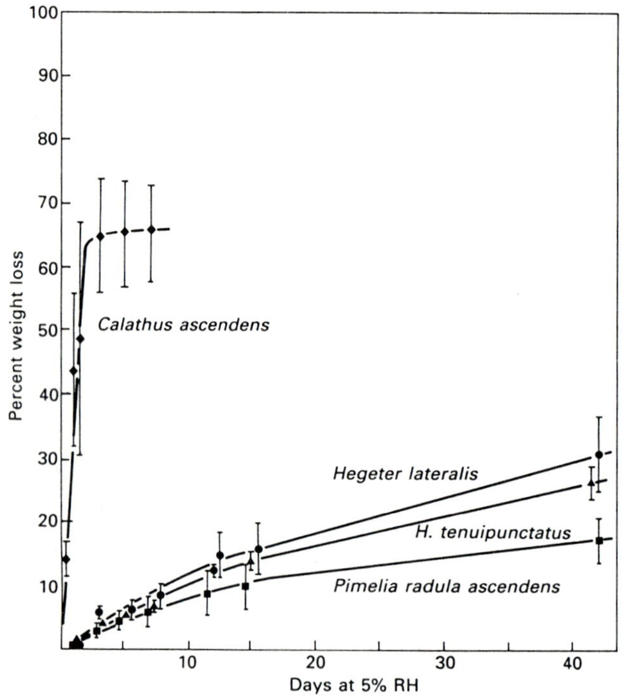

Fig. 3.3. The effect of desiccation at 5 % RH and 21 °C on weight loss in three tenebrionid beetles and the carabid beetle *Calathus ascendens* from 2000-2500 m at Pico del Teide, Tenerife. (Ottesen and Sømme 1987)

In comparison, rates of water loss were higher in five species of tenebrionids from the High Atlas Mountains, collected at 2600-3200 m (Gehrken and Sømme 1994). At 5% RH and 21 °C, *Pimelia thomsoni*, *Pimelia* sp. and *Pachyla telvetensis* all survived 23 days on the average, while *Opatrum rhaticum* and *Crypticus* sp. survived for 12.6 and 10.8 days, respectively. The different tenebrionid species tolerated water losses of up to 51 to 63% of their total body water before they succumbed.

Rates of water loss in beetles normally increase with body weight (Zachariassen et al. 1987, 1988). In this respect, considerable differences exist between species within families and even higher differences may be found between families. When the rate of water loss is plotted against fresh body weight in a double $\log_{10}$ plot (Fig. 3.4), it appears that tenebrionid beetles from the High Atlas Mountains (Gehrken and Sømme 1994) have higher rates of water loss than species from arid lowland areas in East Africa (Zachariassen et al. 1988), while species from Pico del Teide have even lower values.

3.2.1.2 Carabids

Compared with tenebrionids, considerably higher rates of water loss are found in carabid beetles exposed to dry air. At 5% RH and 21 °C, *Calathus ascendens* from Pico del Teide lost 1.61% fresh weight h^{-1} (Fig. 3.3) (Ottesen and Sømme 1987). At this rate of water loss the beetles died within 12 h. Similarly, the rate of water loss in the carabid *Harpalus dissitus* (0.42% fw h^{-1}) from the High Atlas Mountains was much higher than in tenebrionids from the same area. Combined with a lower tolerance of loss of water in itself, *H. dissitus* only survived for 3 days.

Among insects from the Venezulean Andes at 4200 m, two *Colpodes* spp. had rates of water loss of 1.40% fw h^{-1} and 0.88% fw h^{-1}, respectively (Sømme 1986c). Two unidentified Harpalini species from 4500 m at Mount Chimborazo, Ecuador lost water at rates of 1.23 and 1.09% fw h^{-1} (Sømme, unpubl.), and a small unidentified Bembidiini species from 4800 m at Mount Chimborazo at the high rate of 5.10% fw.h^{-1}. When these values are converted to rate of water loss in mg h^{-1} (Fig. 3.4), it appears that all carabid species from the Andes Mountains have rates higher than those of carabid beetles from arid lowland areas in East Africa (Zachariassen et al. 1987, 1988).

Although these alpine carabid beetles live in relatively dry mountainous areas, they are able to avoid dehydration in their natural environment. Both *Calathus ascendens* and *Harpalus dissitus* are commonly found in plant debris below shrubs and cushion plants, where some moisture is retained. The two *Colpodes* spp. seek shelter underneath rocks and fallen *Espeletia* plants, and the three carabid species from Mount Chimborazo are found under rocks as well (Chap. 6.2.5).

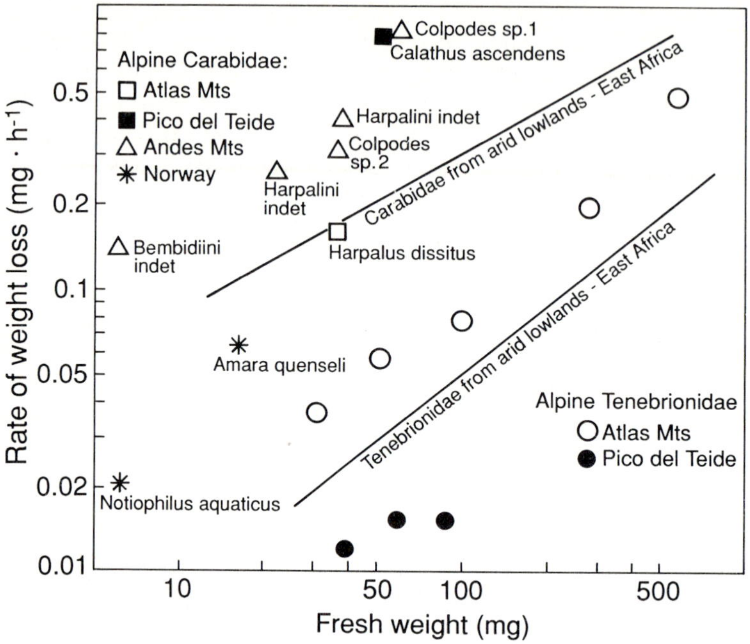

Fig. 3.4. Rates of water loss in carabid and tenebrionid beetles from arid alpine environments, plotted in a double $\log_{10}$ plot against body weight. Regression lines represent similar data for seven carabid and eight tenebrionid species from lowland East Africa, according to Zachariassen et al. (1987,1988). All measurements were made at 20-21 °C in an atmosphere of ca. 5 % RH. See text for further explanation and details regarding the different alpine species

Rates of water loss values within the same range were also reported by Andersen et al. (1986) in the carabid *Notiophilus aquaticus* (0.35% fw h^{-1}), *Amara quenseli* (0.89% fw h^{-1}). Both species occur in relatively dry habitats above the timber line in the Norwegian mountains. *Notiophilus aquaticus* and *Amara quenseli* have rates of water loss lower than those of beetles from arid African environments (Fig. 3.4), illustrating that water-conserving capacity is not a feature exclusively found in hot desert species.

In conclusion, it appears that many alpine carabid beetles have relatively high rates of water loss, but are able to survive in arid areas by seeking sheltered microhabitats.

3.2.2 Polar Species

The proportion of beetles to other insects in polar areas is generally lower than in other geographical zones (Table 2.3). A total of 185 species of beetles is known from the North American Arctic, 13 species from the high

North American Arctic (Danks 1990), and 9 species from Svalbard (Sømme 1993). Forty species have been reported from the subantarctic islands, while no beetles are found in the maritime and continental Antarctic (Block 1984a). The small number of species in the arctic and antarctic islands may be due to geographical isolation, but also suggests that beetles are less adapted than many other insects to overcome environmental constraints. Cold is obviously an environmental factor of great importance, but moisture will also be limiting under periods of drought (Chap. 1.4). Most polar beetles are found in relatively hygric environments, and few studies of tolerance to desiccation have been performed.

South Georgia (54°S, 37°W) receives relatively large amounts of precipitation (Smith and Walton 1975) and appears as a green and lush island when viewed from the seaside. The continuous vegetation cover, however, reaches elevations of only 100-200 m. At higher elevations the vegetation is sparse and patchily distributed (Fig. 3.5).

Fig. 3.5. Mosses growing along a melt water creek at South Georgia form a contrast to the surrounding arid and rocky ground. (Photo L. Sømme)

The upper flora is dominated by mosses growing along creeks leading from snowfields and glaciers. Most of the ground is covered by loose rocks and gravel, which is the site of lichens, green algae and cyanobacteria. Eight species of Coleoptera, representing the world's most southern beetles, have been recorded from South Georgia (Block 1984a). Two representatives of the family Perimylopidae, *Perimylops antarcticus* and *Hydromedion sparsutum* (Fig. 3.6) have a distribution along a gradient from moist lowland grassland to alpine fellfields at 300 m or higher (Block and Sømme 1983; Ring et al. 1990).

Fig. 3.6. *Hydromedion sparsutum* (Coleoptera; Perimylopidae) from South Georgia. (Photo L. Sømme)

The larvae of *Perimylops antarcticus* feed on microphytes and mosses, suggesting an adaptation to more extreme environments where other plants sources are scarce (Ring et al. 1990). They are often observed feeding on the surface of dry gravel. Adults of *Perimylops antarcticus* are found among rocks and gravel, while larvae and adults of *Hydromedion sparsutum* seek more protected sites close to patches of mosses.

In accordance with their choice of microhabitat, the lowest rate of water loss at 15 °C and 5% RH was found in larvae of *Perimylops antarcticus* (Ring et al. 1990). The mean rate in specimens from fellfields was 0.32% fw h^{-1} in larvae, and 1.55% fw h^{-1} in adults. In *Hydromedium sparsutum* rates of water loss were 1.01 in larvae and 1.40% fw h^{-1} in adults. These rates of water loss are within the range reported for many carabid beetles from dry lowland and alpine areas (Fig. 3.4). In addition to low rates of water loss, larvae of *Perimylops antarcticus* had higher water contents and tolerated greater weight losses than did adults of their own species or adults and larvae of *Hydromedion sparsutum* (Ring et al. 1990).

According to Chown (1993), the maximum water losses tolerated by six adult weevil species from the subantarctic Marion Island (47°S, 38°E) are within the range of beetles from South Georgia. The cool climate of Marion Island is also characterized by relatively high precipitation. Among weevils, there were considerable differences in the time to reach the maximum tolerated water loss, and in the rates of water loss % fw h^{-1}). *Bothrometopus randi*, *Mesembriorrhinus parvulus* and *Phaliroeus eatoni*, collected from rock faces - a habitat that may remain dry for considerable periods of time at Marion Island - had the lowest rates of water loss and highest resistance to desiccation. A second group of three weevil species, *Antarctonesiotes elongatus*, *Ectemnorhinus marioni* and *E. similis*, had higher rates of water loss. Both *Ectemnorhinus* spp. were collected from moist lowland habitats. *Antarctonesiotes elongatus*, however, lives in a fellfield habitat. Due to its small size, it may easily find sheltered conditions, and hides rapidly as a response to a decrease in moisture on sunny days.

The larvae of *Phaliroeus eatoni*, *Bothrometopus randi* and *Ectemnorhinus similis* have mean rates of water losses in the range of 0.56 to 0.69% fw h^{-1} at 10 °C and 5% RH (Chown and van Drimmelen 1992). All larvae rapidly increased their water content when given access to free water. Although they lost from 24 to 31% of their initial water content, haemolymph osmolality remained low, suggesting that they have some capacity of osmoregulation. The larvae of *Phaliroeus eatoni* inhabit the supralittoral zone, while *Bothrometopus randi* and *Ectemnorhinus similis* are associated with moss cushions. Both *Phaliroeus eatoni* and *Bothrometopus randi* are rock-face dwellers and cease foraging when their substratum dries out during sunny periods. As the substratum becomes drier, the larvae were observed to track the receding water film on the rock surfaces, before seeking shelter in crevices or moss cushions.

Adaptations to desiccation in both larvae and adults have been demonstrated through these experiments, in spite of great differences between habitats of beetles from the subantarctic islands. As in larvae of *Perimylops antarcticus* (Ring et al. 1990), the most tolerant weevil species also had the highest water contents (Chown 1993), and this may be of adaptive value in species from relatively dry habitats. Due to microclimatic conditions, some tolerance to desiccation is important, even in an area with a moist macroclimate since the habitats of the beetles may temporarily dry out.

3.2.3 Tropical and Temperate Lowland Species

The water balance of beetles from dry lowland habitats has been the subject of a great number of investigations (Edney 1977; Cloudsley-Thompson 1991; Hadley 1994). Among others, arid areas in Africa support a rich fauna of Coleoptera (Zachariassen et al. 1987). Beetles of the family

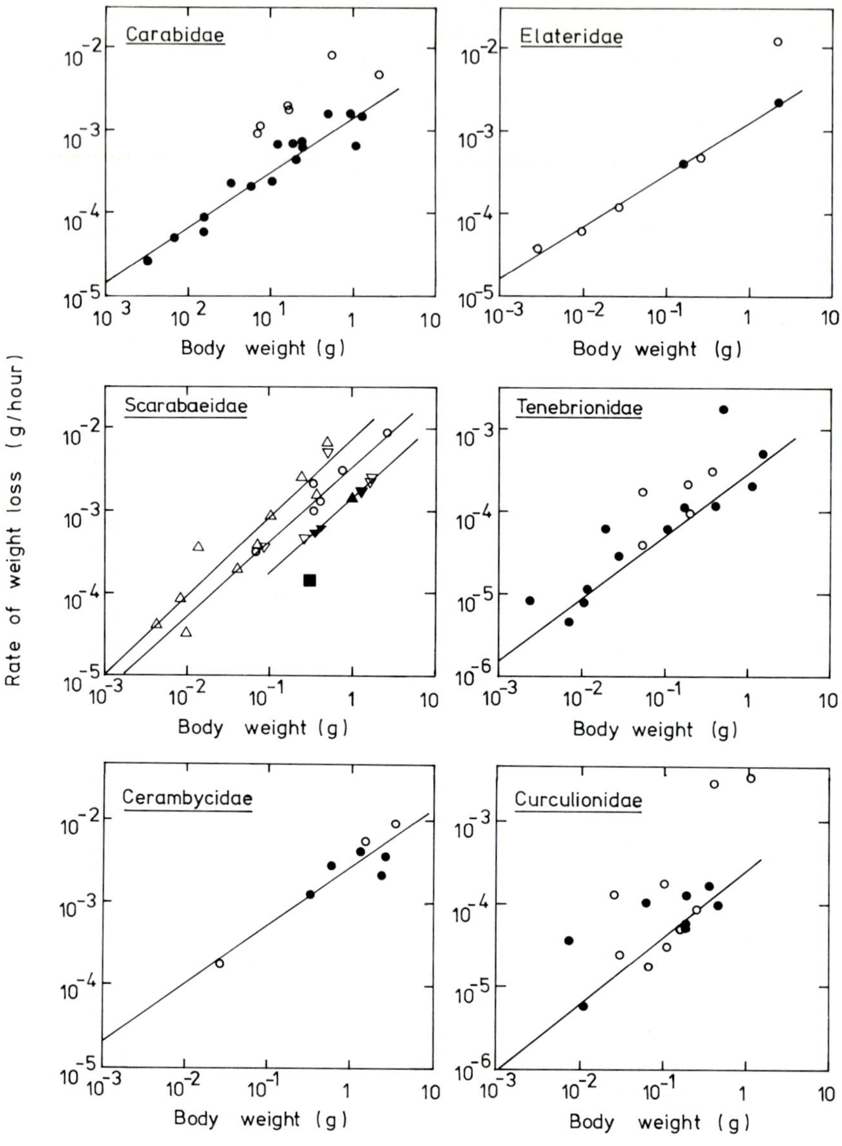

Fig. 3.7. Double $\log_{10}$ plot of rates of water loss in East African beetles of different families as a function of body weight. All measurements were made at 20 °C and ca. 5 % RH. *Each point* represents 1-20 individuals of one species. For the *Scarabaeidae*, subfamilies *Aphodiinae* and *Scarabaeinae* are represented by ▲ and Δ , *Dynastinae* and *Rutelinae* by o, *Cetoniinae* and *Trichiinae* by ▼ and ∇ , and *Troginae* by ■ . *Closed symbols* represent beetles from xeric areas, whereas *open symbols* represent beetles from mesic and hygric areas. (Zachariassen et al. 1988)

Tenebrionidae appear to be particularly well adapted, but species of Carabidae and members of other families are represented as well. Carabids are common in xeric areas of East Africa, and Andersen et al. (1986) reported that they have considerably higher rates of water loss than tenebrionids from the same area. On the other hand, carabids from these areas have lower rates of water loss than most carabids from temperate regions.

While most experiments on beetles and other terrestrial arthropods have been carried out at low RH, rates of water loss under natural conditions are greatly affected by different levels of humidity. The differences in water loss at different RH were clearly demonstrated by Hadley et al. (1990) in larvae of the tiger beetle *Cicindela marutha*. At 25 °C the larvae lost 3.09% fw h^{-1} at 53% RH and 0.57% fw h^{-1} at 97.5% RH. At higher temperatures, higher rates of water loss were observed. The larvae of *C. marutha* avoid unfavourable conditions by retreating into their burrows (Chap. 6.1.3).

Comparison of beetles from xeric and mesic/hygric habitats in East Africa show great differences with regard to water loss at 20 °C and 5% RH (Andersen et al. 1986; Zachariassen et al. 1987, 1988). Within different families of Coleoptera, beetles from humid areas tend to have higher rates of water loss than do beetles from arid areas (Fig. 3.7). The differences are most pronounced among carabid beetles, but are also apparent among tenebrionids and curculionid beetles. For species of Elateridae and Cerambycidae, Zachariassen et al. (1988) found only small differences. Within the family Scarabidae, however, there are substantial differences in rates of water loss between different subfamilies (Fig. 3.7).

As reviewed by Edney (1977), many desert arthropods may tolerate water losses in excess of 30% of their body weight. For an insect with a water content of 70%, this corresponds to a loss of ca. 43% of total body water. According to Zachariassen et al. (1988), East African carabids, scarabids and tenebrionids have water contents of 65-70%, and are injured when it drops below 50%. This corresponds to a loss of 25-35% of total body water, which is less than the tolerance of tenebrionids from the High Atlas Mountains (Gehrken and Sømme 1994). Curculionids from East Africa, however, may tolerate a loss of 45% of total body water content.

3.3 Other Terrestrial Arthropods

3.3.1 From Lowland Deserts

Rates of water loss and tolerance of desiccation have been studied in a number of other desert arthropods, and reviews of the results have been presented by several authors (e.g. Ahearn 1970a; Edney 1977; Crawford 1981; Polis 1990; Cloudsley-Thompson 1991; Hadley 1994).

According to Crawford (1981) there is little evidence that desert arthropods tolerate lower water levels in their tissues than do species from mesic environments. Crawford (1981) used the term desiccation resistance in the sense that many desert taxa are superior in terms of restricting body-water loss when stressed by prolonged exposure to dry air.

Among typical desert arthropods, the North American cockroach *Arenivaga investigata* is well equipped to resist water loss (Crawford 1981). The sand cockroach *Heterogamia syriaca* travels upwards the sand dunes in Egypt at dusk, which coincides with temperature depression and the dew formation.

A wide range of values for rates of water loss have been reported in scorpions from xeric and mesic habitats (Cloudsley-Thompson 1991; Hadley 1994). Very low rates of water loss at 33 °C and 0% RH are found in the African scorpions *Leuiurus quinquestriatus* and *Androctonus australis*, with values of 0.03 and 0.032% fw h^{-1}, respectively. Under similar conditions the Sonoran Desert scorpion *Hadrurus hirsutus* lost 0.042 % fw h^{-1} (Cloudsley-Thompson 1967). In contrast, the rate of water loss in the East African scorpion *Lychas burdoi* was 0.25% fw h^{-1} at 20 °C and 5% RH (Bjerke et al. 1989). These results indicate that some species are more dependent than others in finding sheltered habitats to avoid desiccation.

Exceptionally poor desiccation resistance is also found in the adult whip scorpion *Mastigoproctus giganteus* (Uropygi), which readily loses water (Crawford and Cloudsley-Thompson 1971). The loss increases in proportion to the water vapour saturation deficit, suggesting abrasion of the epicuticular wax layer. In unabraded juvenile specimens the water loss is below 0.4% fw h^{-1} (Ahearn 1970b).

3.3.2 Alpine Millipedes

Millipedes (Diplopoda) lack a waterproofing epicuticular wax layer and may lose water quickly by transpiration (Cloudsley-Thompson 1991). Their water balance depends partly on the possibility of regaining lost water from the surroundings. Nevertheless, resistance to desiccation is high in some desert millipeds, e.g. in *Orthoporus ornatus*, where Crawford (1972) found rates of 2.7-7.9 $\mu g\ cm^{-2}\ h^{-1}\ mmHg^{-1}$.

Meyer and Eisenbeis (1985) also pointed out that resistance to desiccation in millipeds is correlated with the distribution of the species and the microclimatic conditions of their habitats. They found great differences in the resistance to desiccation in five species along an altitudinal gradient in the Ötztal Alps. Water loss rates expressed in relation to body surface area was relatively high at 25 °C and 0% RH in all species. The lowest rate, with a value of 45.9 $\mu g\ cm^{-2}\ h^{-1}\ mmHg^{-1}$, was found in *Trimerophorella nivicomes* from windswept mountain ridges at 2500 m. Highest rates were found in *Haasea fonticulorum* (236.9) and *Ochogona caroli* (147.0) from an *Alnus viridis* community at 2000 m and *Mastigona mutabilis* (284.8) from an oak forest at 670 m.

Surprisingly, rates of water loss in *Leptoiulus saltuvagus* (55.5) from the *Alnus* forest and *Enantiulus nanus* (98.1) from the oak forest were also relatively low. *Ochogona caroli* and *Haasea fonticulorum* co-exist in alder litter, but differences in tolerance to water loss may facilitate ecological isolation within the same habitat.

The millipedes from the Ötztal Alps survive water losses of 21 to 44% of their initial body water (Meier and Eisenbeis 1985). They are also able to absorb moisture from the surroundings by the eversion of rectal tissue. Cuticular permeability was highest at atmospheric vapour saturation, and decreased greatly as ambient RH fell from 99 to 0%.

3.4 Conclusions

Tolerance to arid conditions has been studied in a large number of terrestrial arthropods from hot deserts. Species of widely different taxa are able to sustain prolonged exposure to low atmospheric humidities. This ability is mainly based on a greater resistance to the loss of body water and less on a higher tolerance of low water content. Less attention has been paid to alpine and polar species, but the information that is available shows great variation in their ability to survive in arid environments.

In microarthropods there is generally a close relation between the moisture conditions of their habitats and the resistance to dehydration. Rate of water loss is high in collembolans from humid caves and extremely low in species from habitats that are dried out by solar radiation. Alpine and polar species associated with lichens on rocks and vegetation are better adapted to arid conditions than species from wetter habitats. Most species actively seek protected microhabitats to avoid periods of severe drought.

Rates of water loss in beetles from arid alpine areas may be similar to or lower than in lowland species. Tenebrionid beetles are more resistant to dehydration than carabids. The latter group depends more on shelter in moist plant debris or under rocks. Beetles are not a dominant insect group in the Antarctic, but species from the subantarctic islands are temporarily exposed to arid conditions. They exhibit different resistance to water loss and avoid desiccation stress through their behaviour.

From the examples cited in the present chapter, it appears that there is great variation in rates of water loss among alpine and polar terrestrial arthropods. It also appears that behavioural avoidance of arid conditions is an important part of their adaptation. Behavioural adaptations in species from polar deserts are further discussed in Chap. 6.2.

4 The Physiology of Tolerance to Desiccation

4.1 Introduction

Several physiological mechanisms are involved in the maintenance of water balance in terrestrial arthropods. Combined with the behavioural avoidance of the most stressful exposures to drought, physiological adaptations to reduce water losses are a requirement for life in arid environments.

Transpiration through the cuticle is an important component of evaporative water loss in terrestrial arthropods (Fig. 4.1).

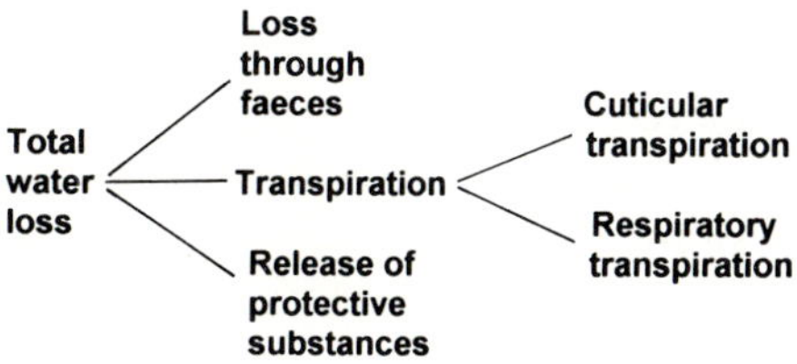

Fig. 4.1. Different routes of water loss in insects and other terrestrial arthropods

In species from arid environments losses are reduced through physiological and chemical modifications of the cuticle. Evaporative water loss through the spiracles is also important and cannot be avoided during the exchange of gases through respiration. In most insects water losses are reduced by closing the spiracles in between bursts of gas exchange. The release of protective substances from glands through cuticular openings or the excretion of haemolymph through the joints of certain beetles also contributes to the total water loss.

In addition to evaporative water loss, water is lost through the faeces (Fig. 4.1). Absorption that takes place in the rectum is of great importance for the water balance of insects from arid environments. Further adaptation to desiccation is seen in the osmoregulation of many species, protecting the cells from osmotic stress.

Eisenbeis (1989) considered moisture to be of selective importance in forming the body shape and size of soil arthropods. In species from mesic and semi-humid habitats, there is a tendency for shorter appendages and larger trunks. The reduced surface area of the appendages, and a larger proportion of surface area restricted to the globular or cylindrical trunk, reduce cuticular transpiration. The principle of surface minimization does not apply to species where better protection against cuticular transpiration has evolved. According to Remmert (1981), humid habitats favour populations of small individuals, while arid conditions cause increased rates of water loss and mortality in small animals.

4.2 Evaporative Water Losses

Due to different methods used by different authors, the comparison of evaporative water losses between species is sometimes difficult to make. Harrisson et al. (1990) and Block et al. (1990) suggested that comparison would be easier if insect physiologists expressed rates of water loss in relation to surface area. In fact, this has been a common practice of several authors, and values for cuticular water permeability for a large number of terrestrial arthropods have been listed by Edney (1977) and by Hadley (1994). Cloudsley-Thompson (1991) also stated that the major source of water loss from desert arthropods is probably through cuticular transpiration and less in respiration, except in the case of large insects in flight.

4.2.1 Beetles

In many desert beetles, the elytra of the abdomen are fused, forming a subelytral cavity (Fig. 4.2).

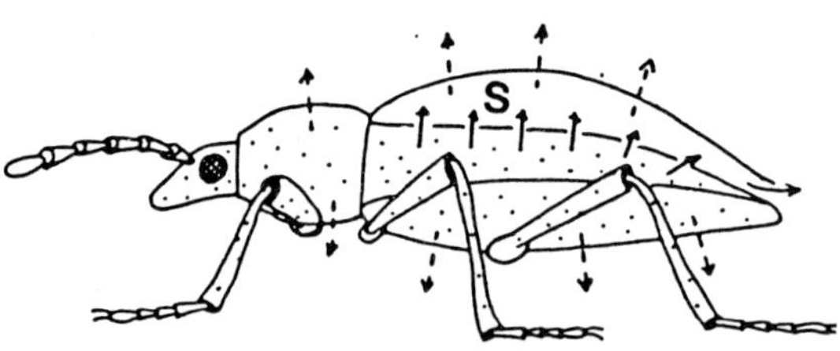

Fig. 4.2. Paths of transpiratory water loss in a dry habitat beetle. *Broken arrow* Transcuticular water loss; *solid arrows* respiratory water loss; *S* subelytral cavity. (Zachariassen and Maloiy 1989)

Respiratory water loss from the abdominal spiracles goes through this cavity, and in addition water is lost from the thoracic spiracles. The subelytral cavity acts as a buffer to reduce water loss from the abdominal spiracles (Cloudsley-Thompson 1964; Ahearn and Hadley 1969). In the East African tenebrionid *Phrynocolus petrosus*, Zachariassen (1991a) found that the water loss was doubled when the subelytral spiracles were surrounded by dry air under experimental conditions. In this species, the passage from the subelytral cavity to the ambient air is closed most of the time, leaving only short periods during which water vapour can diffuse out of the chamber. During closure, the subelytral air remains nearly saturated with humidity, and the exchange of respiratory gases through the abdominal spiracles will result in only a small water loss.

In contrast to the results of Zachariassen (1991a), Ahearn (1970a) found that cuticular transpiration was the greatest source of water loss in the desert tenebrionids *Eloedes armata*, *Centrioptera muricata* and *Cryptoglossa verrucosa* from Arizona. At temperatures from 25 to 40 °C, losses through the spiracles were negligible, while a marked increase in respiratory transpiration was observed from 40 to 45 °C. The results indicate a temperature breakdown in spiracular water regulation due to increased respiratory activities.

Ahearn (1970a) also found higher cuticular transpiration in freshly killed beetles than in live ones. Apparently, the maintenance of cuticular water permeability of living insects at a low level is partly due to active, energy-requiring processes within the cuticula (Hadley 1984). This is supported by evidence suggesting that the apical membrane of the epidermal cells offers resistance to the outward movement of water (Edney 1977). When the animals die, cuticular water permeability will increase, and this should be considered in experiments where dead insects are used to differentiate between cuticular and respiratory transpiration. As suggested by Zachariassen and Maloiy (1989), the difference in experimental techniques may explain the difference between their own results and those of Ahearn (1970a).

Recent studies have also shown that a greater part of evaporative water loss may be through respiration even in resting insects. Cooper (1983) concluded that respiratory water loss makes up 36% of total evaporative water loss in *Eloedes armata*. By the use of a tritiated water technique, Nicolson et al. (1984) found that 45% of the evaporative water loss probably is through the pronotal spiracles in the Namib desert tenebrionid *Onymacris plana*. By isolating the different compartments in *Phrynoculus petrosus*, Zachariassen (1991a) demonstrated the distribution of evaporative water losses in this beetle. Water loss from the subelytral chamber makes up less than 10% of the total evaporative water loss, while loss over the pronotal spiracles amounts to as much as 70%. The results suggests that transcuticular water loss makes up only about 20% of the total evaporative losses.

Studies on the relation between the rates of water loss and metabolic rates also support the idea that respiratory water loss is of major importance to a number of beetles from dry habitats (Zachariassen et al. 1987; Zachariassen and Maloiy 1989). A double $\log_{10}$ plot of these two factors revealed that data from carabid and tenebrionid beetles from xeric habitats in Kenya fell along one straight line (Fig. 4.3).

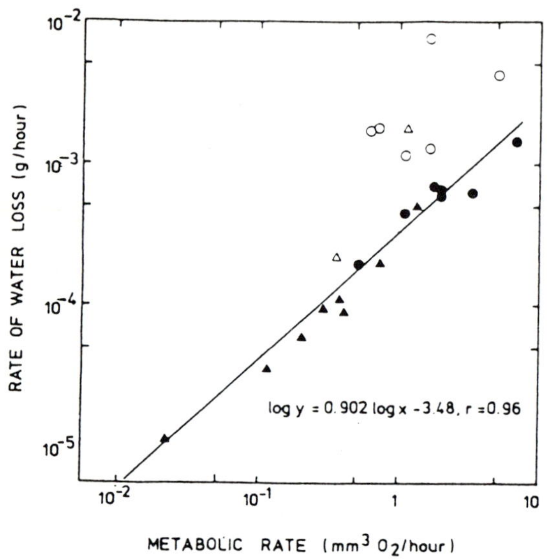

Fig. 4.3. Rates of transpiratory water loss of carabid beetles (*circles*) and tenebrionids beetles (*triangles*) plotted as a function of metabolic rate. Scale on both axes is logarithmic. *Closed symbols* represent beetles from dry habitats and *open symbols* from mesic or hygric habitats. The *solid line* is the calculated linear regression line for beetles from dry habitats. (Zachariassen et al. 1987)

Although the carabids and tenebrionids have very different rates of water loss, the rates varied as a linear function of the metabolic rate. Apparently, water loss in both groups is closely associated with respiration. In contrast, beetles from mesic and hygric habitats had higher rates of evaporative water loss (Fig. 4.3), suggesting that a larger proportion is lost through the cuticle. Similarly, Gehrken and Sømme (1994) found a close relationship between rates of water loss and metabolic rates in tenebrionids, a carabid beetle and a curculionid beetle from the High Atlas Mountains, while data for the chrysomelid *Timarcha mairei* fell outside the regression line.

4.2.2 Scorpions

According to Hadley (1990), the cuticle of some scorpions is virtually impermeable to water, on account of its chemical composition and its abundance of cuticular lipids. Rates of water loss among species from xeric habitats are as low as 0.025 mg cm^{-2} h^{-1}. In spite of this, integumental water loss accounts for most of the water lost at normal environmental temperatures.

Direct measurements of respiratory transpiration in scorpions are lacking, but low levels of metabolism are typical (Hadley 1990). Accordingly, the rates of respiratory transpiration are also low. Scorpions are equipped with book lungs located within cavities which open to the outside through narrow openings. The respiratory membranes are moist, and some water loss cannot be avoided during uptake of oxygen. In the protected book lungs the area of transpiration is reduced, and diffusion is slowed down within the chambers. Some species have muscles that regulate the size of the openings, and they may be closed to limit water losses in temporarily stressful situations.

4.3 The Cuticula of Terrestrial Arthropods

4.3.1 Structure

The integument of arthropods is composed of an epithelial layer of cells and the cuticula, and serves several functions. This outer armour-like exoskeleton is one of the reasons for the success of this phylum in conquering terrestrial environments. Orginating from aquatic organisms, insects, arachnids and myriapods have adapted to a large variety of terrestrial habitats. No other animal phylum exhibits such a diversity of species, ranging from the tiniest mites to giant tropical beetles and moths. Presently, terrestrial arthropods are found almost everywhere under conditions where life may exist, including the harsh environments of hot and cold deserts.

The extensive distribution and diversity of terrestrial arthropods can, of course, not be explained by their cuticle alone, but also by a variety of other biological factors. The cuticle, however, is important in providing a barrier against the surroundings, protecting the organisms from mechanical injuries and making it possible for arthropods to occupy habitats where more soft-bodied animals cannot live. The complex structure of the cuticle includes stratification in horizontal layers of different strength and flexibility. While the exocuticle is hardened by sclerotization, the endocuticle remains soft and less rigid. In many species, most of the body is covered by the rigid exocuticle, while this layer is missing in intersegmental membranes and in the joints of legs, mouth-parts and antennae. The soft endocuticle of these body areas permits the animal to perform complex movements. The soft-bodied larvae of many arthropods have a thin flexible endocuticle, permitting movements by change in the local hydrostatic pressure.

The more rigid parts of the cuticle offer almost unlimited possibilities for the attachment of muscles. The diversity and complexity of insect muscles forms the basis for their rapid movements, including the ability to run, jump and fly. From an evolutionary point of view, great movability makes it possible to invade new habitats and develop new life patterns. The cuticle not only covers the outer part of the body, but also lines the tracheal tubes and sections of the gut, and is in this way associated with physiological functions, such as respiration and digestion. Pigments and physical structure of the cuticle determine the body colouration of insects and other terrestrial arthropods.

4.3.2 Restriction of Water Loss

As pointed out by Hadley (1984, 1985), no cuticular function is likely to be more important than its role in restricting water loss. This feature is essential for the evolution of terrestrial arthropods, and for the survival under arid conditions at the individual level.

The importance of the cuticle in providing a barrier to the loss of water has been reviewed by several authors (e.g. Wigglesworth 1972; Edney 1977; Hadley 1994). The main barrier to water loss is found in the thin epicuticula which covers the outside of the other cuticular layers. Although only a few micrometres in thickness, at most, the epicuticula has a multilayered structure penetrated by microscopic wax canals. Recently, Hadley (1994) proposed a model on how the different components of the epicuticular lipids contribute to waterproofing.

A series of early investigations confirmed the role of epicuticular lipids in determining the properties of the cuticle in relation to water transport. Transpiration of the cuticle varies considerably between species and is affected by temperature and the water vapour saturation deficit. When temperatures are increased, a critical level is reached, above which there is an abrupt rise in the rates of water loss. The transition temperature is higher than temperatures normally encountered in the natural environments of the insects, but tends to be higher in species adapted to arid conditions. The phenomenon has been explained by rearrangement of the surface wax molecules, but serious questions have been raised about the validity of these theories (Hadley 1984, 1994).

Edney (1977) has summarized the evidence that illustrates the importance of the superficial epicuticular lipid layers in providing a barrier to transpiration. A large increase in transpiration results from mechanical abrasion, such as contact with inert dust. Following ecdysis, the new cuticle is highly permeable, but the acquisition of impermeability coincides with the deposition of surface lipids. Lipids extracted from the epicuticle have melting points close to the transitional temperatures of the insect, explaining the abrupt increase in transpiration. Experimentally, such lipids have been applied to inert membranes, making them impermeable to water.

As pointed out by Hadley (1984, 1985), epicuticular lipids are not the only waterproofing barrier of the integument. Although the epicuticle represents the principle barrier to transpiration, lipids are also located in the exo- and endocuticula, providing additional protection against water loss. Furthermore, other components of the cuticle, such as chitin and protein molecules, may also contribute to its high impermeability.

Under experimental conditions, the rate of water loss from terrestrial arthropods is usually not constant but decreases with time (Edney 1977). The decrease may be ascribed to the loss of water from the cuticle itself, changing its permeability by a closer packing of the molecules. This hypothesis is contradicted by the fact that the epicuticle and the exocuticle of most arthropods contain little water (Hadley 1994) and cannot account for the initial high water losses. Another possible explanation is an increased activity of the animals during the initial phase of the experiments. If the continued decline in water loss represents a true increase in cuticular resistance, an important consequence would be that the evaporative water loss through the cuticle is reduced over time at low environmental humidities.

Finally, the epidermal cells may participate in the control of cuticular water loss (Edney 1977; Hadley 1984). Some water diffuses through the apical membrane of the cells, and it has been suggested that their resistance to water flux may be altered under increasing stress of desiccation. The control provided by the epidermal cells is probably an energy-requiring process. This may explain why water loss rates of dead insects often exceed those of living ones when measured under identical conditions (Hadley 1984; Zachariassen and Maloiy 1989).

4.3.3 Cuticular Lipids

Earlier work on epicuticular lipids has been reviewed by Edney (1977). According to Hadley (1984, 1994), the dominant constituents are straight-chain saturated hydrocarbons. High molecular weight branched hydrocarbons usually accompany these n-alkanes and may even be more abundant. In 50 % of the arthropods studied so far, the epicuticle also contains unsaturated hydrocarbons. True waxes are formed by esters derived from oxygenated derivates of hydrocarbons. Alcohols, sterols and glycerides are also present, but the role of the different components is poorly known. Hydrocarbons and wax esters are generally the most water-repellent molecules among the lipids.

The physical structure and arrangement of the epicuticular wax deposits are also important in controlling water loss (Hadley 1984, 1994). The use of scanning electron microscopes has revealed a wide variety of arrangements. In many insects the surface is covered by hard waxes, forming complicated and irregular patterns. So-called waxblooms projecting outwards from the surface are present in several Lepidoptera, Coleoptera, Hemiptera and other insects. In the black widow spider *Latrodectus hesperus* and the house cricket *Acheta domestica*, the surface is uniformly covered by a thin film of lipids. According to Edney (1977), the cockroach *Periplaneta americana* has a greasy lipid layer, while other insects are covered by harder waxes.

4.3.4 Desert Arthropods

Substantial wax deposits may be observed on the surface of many desert insects and arachnids (Cloudsley-Thompson 1991). Elaborate wax blooms, frequently multicoloured, are particularly prevalent among tenebrionid beetles. In *Cryptoglossa verrucosa* of the Sonoran Desert, an extensive meshwork of wax filaments forms under low relative humidity (Hadley 1984, 1994). Similarly, *Onymacris rugatipennis* from the Namib Desert has a bluish white appearance under arid conditions when covered by a wax bloom, while the beetle is black when the wax is removed. According to McClain et al. (1984), the percentage of species with a wax bloom increases from the coastal fog area of the Namib Desert to the inland hot dry areas. In species inhabiting both regions, the intensity of the bloom increases with hotter and drier climate.

In a detailed study on the dynamics of the wax bloom in *Cauricara phalangium*, McClain et al. (1984) showed that the amount of wax varies with season. This long-legged tenebrionid beetle is endemic to the Namib Desert, and spends the night in shelter among pebbles and ephemeral grass. The beetles are active during the austral winter season when temperatures are relatively low, although maximum temperatures above 35 °C are recorded in all months. When the adult *C. phalangium* emerge in April, they possess a full complement of white wax bloom, and this remains intact during the first 6 weeks. As the season progresses, the amount of wax diminishes, especially in the females. Due to its high reflectivity, the white wax probably protects the beetles against overheating on warm days. This is reflected in the correlation between the most intense wax bloom and the hottest part of the season. Furthermore, the wax bloom results in reduced rate of water loss. At 25 °C and 10 % RH, mean rate of water loss in *C. phalangium* with wax present in the month of April was 0.024 mg cm^{-2} h^{-1}, compared to 0.034 mg cm^{-2} h^{-1} without wax in August. Although the wax bloom may disappear in cold weather and the end of the season, it is renewable. Both in the field and in the laboratory the wax was regenerated at high temperatures and low humidities. McClain et al. (1984) concluded that the wax bloom contributes to the protection of these beetles against high temperatures and the radiant heat load in the Namib Desert.

According to Cloudsley-Thompson (1991), adult whip scorpions of the species *Mastigoproctus giganteus* readily lose water at 26 °C and RHs up to 95 %. It is possible that continuous contact with sand in older specimens results in abrasion of the waterproofing epicuticular wax layer. In intact young specimens, Ahearn (1970b) found a critical temperature of 37.5 °C, above which water is rapidly lost through transpiration.

4.3.5 Evaporative Cooling

There are few studies on regional differences in the cuticular permeability to water in terrestrial arthropods, although differences between sclerotized cuticle and membraneous soft cuticle have been demonstrated in insects and scorpions (Hadley et al. 1989). The transcuticular water loss rate of the desert cicada *Diceroprocta apache* from southwestern North America is exceptionally high, and rates are different from different regions of the body. According to Hadley et al. (1989) pores of 7-8 μm in diameter are found on the dorsal regions of the thorax and abdomen, but are lacking in the lateral thorax. By the use of a transpiration monitor equipped with a tiny stainless steel capsule attached to the cuticle, values of water loss rates were obtained from the different regions.

At 27 °C, the water loss rates were 6.6, 6.9 and 8.7 mg cm^{-2} h^{-1} for dorsal thorax, lateral thorax and dorsal abdomen, respectively. Great differences, however, were found at 41.5 °C, when mean water loss rates were to 37.1, 9.9 and 85.1 mg cm^{-2} h^{-1} for the dorsal thorax, lateral thorax and dorsal abdomen, respectively. Tracing from the dorsal thorax, and to a lesser degree from the dorsal abdomen, showed irregular peaks representing bouts of water extrusion. In contrast, no fluctuations were recorded from the lateral thorax. Death of the animals at the higher temperature resulted in a significant decrease in the highest rates of water loss, demonstrating that the extrusion of water is an active process.

The high rates of water loss are apparently related to the pores on the dorsal thorax and abdomen (Hadley et al. 1989). *Diceroprocta apache* offers a rare example on extrusive use of evaporative cooling in insects. Although seeking cooler microclimates in the shade, the male cicadas often sing at air temperatures exceeding 48 °C. Active extrusion of water begins at 39 °C, and the cicadas are able to maintain a body temperature lower than the ambient temperature. Prolonged evaporative cooling requires a sufficient source of water, and this is obtained from the xylem of the mesquite trees on which the cicadas feed.

4.4 Water Absorption

4.4.1 The Hindgut of Insects

Most insects have the capacity to absorb water from the hindgut of their digestive system. Towards the rear end, the epithelial cells tend to increase in size and are arranged in longitudinal folds (Wigglesworth 1972). Intermediate stages exist between such epithelial pads and a more uniform epithelium. In some species, membranes of the cells facing the lumen are extensively folded and associated with large mitochondria. In the hindgut, a gradual desiccation of the liquid content from the midgut takes place. This is particularly evident in the rectum where the gut content is in contact with the epithelial pads. The faeces leave the anus more or less as dry pellets. Rectal water absorption occurs in a number of insect taxa, and may even be of importance in species where the content remains fluid (Wigglesworth 1972). Species adapted to arid environments often possess special structures for more effective water absorption from the gut or other parts of the body.

4.4.2 The Cryptonephridial System

4.4.2.1 Structure and Function

In a number of insects the distal end of the Malpighian tubules are closely connected to the rectum to form a cryptonephridial system. This is found in lepidopterous larvae, many sawfly larvae and in several taxa of beetles (Saini 1964). Early work in this field has been reviewed by Maddrell (1971). In the mealworm *Tenebrio molitor*, the faecal pellets become so dry that they are able to take up water from an atmosphere of about 90 % RH (Edney 1977). In this way, water is effectively removed from the gut content, and in addition the larvae are able to gain water from uptake of air through the anus. Both drying of the faecal pellets and absorption of water vapour requires the establishment of a strong osmotic gradient, and this is achieved through the cryptonephridial system.

Grimstone et al. (1968) made a detailed study of the cryptonephridial system of *Tenebrio* larvae. In principle, three compartments are present; the rectal lumen surrounded by the perirectal space and the lumen of the Malpighian tubules. The haemocoel surrounding the whole structure may be considered as a fourth compartment. In *Tenebrio molitor* six Malpighian tubules enter the cryptonephridial system in a bundle. Under the perinephric membrane that surrounds the system, they spread out and cover the surface in zig-zag bands.

The most important points of this structure have been summarized by Maddrell (1971). The perinephric membrane is very thick and composed of several layers of flattened cells (Fig. 4.4).

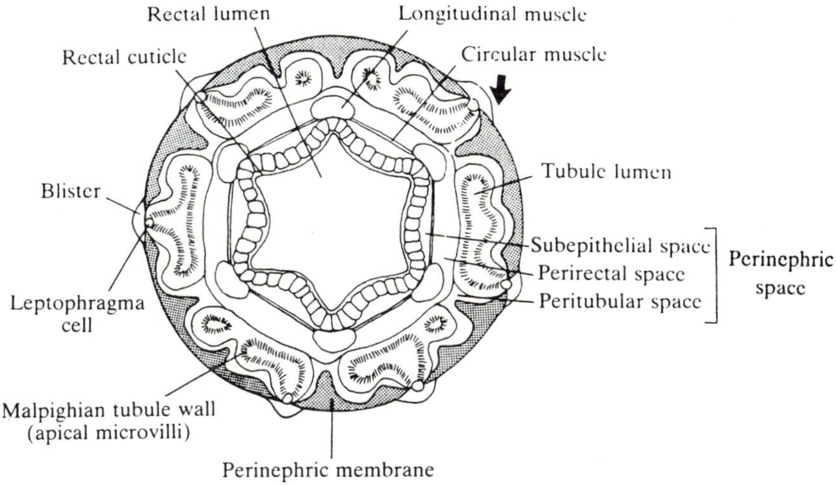

Fig. 4.4. Schematic transverse section of mealworm cryptonephridial complex showing ephitelia and fluid compartment interrelationships. (O'Donnell and Machin 1991)

Each cell consists of little more than two plasma membranes separated by a very thin layer of cytoplasm. At regular points the perinephric membrane is reduced to areas of only two extracellular basement membranes. On the outside these points appear as blisters or 'boursouflures'. The perforations are probably very permeable and represent a close connection between the haemolymph and the underlying Malpighian tubules. At the boursouflures the wall of the Malpighian tubules is also extremely thin. It is made up of a specialized leptophragma cell which has a plane surface against the perinephric basement membrane. On the inside, thin microvilli radiate into the lumen of the Malpighian tubule. The cytoplasm of the leptophragma cells contains many mitochondria, a sign of high metabolic activity.

The rectum, which lies within this structure, is approximately hexagonal in cross section, and is separated from the Malpighian tubules by the perirectal space (Fig. 4.4). The rectal epithelium is made up of a single layer of cells from 10 to 70 μm in thickness. They contain large numbers of microtubules as well as mitochondria and ribosomes distributed throughout the cytoplasma. They lack, however, the regular arrays of mitochondria and folded lateral walls in the blowfly, locust and American cockroach (Maddrell 1971) that are associated with rectal water absorption, suggesting that the rectal cells of *Tenebrio* are less active.

Based on the experiments of Ramsay (1964) and of Grimstone et al. (1968), a general hypothesis for the function of the cryptonephridial system has been developed (Maddrell 1971; Edney 1977). The leptophragma cells of the Malpighian tubules actively secrete K^+ ions, passively followed by Cl^- from the haemolymph into the tubules. A remarkable feature of this process is that the movement of ions is not accompanied by osmotically compensating amounts of water from the haemolymph. The dense walls of the perirectal membrane are probably impermeable to water.

It was originally thought that an active uptake of water took place in the rectum (Maddrell 1971; Edney 1977) parallel to absorption of K^+ into the Malpighian tubules, but this was later contradicted by Machin (1979, 1983). Based on thawing of quick frozen sections of the cryptonephridial complex of *Tenebrio molitor*, progressively decreasing water movement from the rectal lumen to the Malpighian tubules was demonstrated. This suggests that inward water flow takes place without the active involvement of the rectal epithelium. The results were also confirmed by studies of the electrochemical gradient in the mealworm rectal complex (O'Donnell and Machin 1991).

In his original studies Ramsay (1964) found increasing concentrations of solutes from the anterior to the posterior part of the perirectal fluid in *Tenebrio*. Freezing point depressions of 3 to 4 °C were measured in the anterior perirectal fluid, and as high as 6 to 8.5 °C in the posterior fluid. During desiccation, the freezing point depression of the haemolymph never exceeded 1.5 °C. The high freezing point depressions of the perirectal fluid were confirmed by Machin (1979, 1983) (Fig. 4.5).

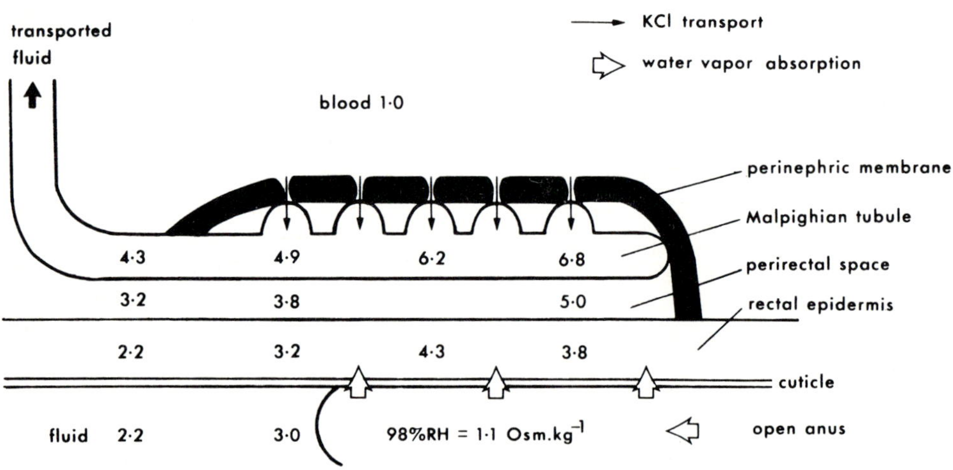

Fig. 4.5. Diagram summarizing the interrelationships of compartments and ion transport with osmotic pressure distribution in mealworm rectal complex. Osmotic pressure given in osmol kg^{-1}. (Machin 1983)

In *Tenebrio*, there is good agreement between the threshold of water absorption and the maximum osmotic pressure generated in the cryptonephridial system.

Although the system apparently evolved for water reabsorption from faeces and excreta, Machin (1983) pointed out that water vapour can be absorbed with equal facility. Studies of frozen sections of the rectum indicate that during vapour absorption the anterior third of the rectum is filled with fluid and the remainder two thirds with air. Significant contributions to the uptake of water occur when humid air is brought into contact with large areas of condensation. When the areas contain fluids of depressed water vapour pressure absorption takes place into the rectal complex. There is good agreement between the threshold of absorption at 88 % RH, which corresponds to 6.7 osmol kg^{-1}, and the maximum osmotic pressure of the system.

By the use of ion-selective microelectrodes, O'Donnell and Machin (1991) were able to study ion activities and electrochemical gradients in both cellular and extracellular compartments of the rectal complex of *Tenebrio*. Values of K^+ in the Malpighian tubules of the posterior rectal complex exceeded 2700 mmol l^{-1}, which approaches the high values necessary to drive the uptake of water from the rectal lumen. Na^+ and H^+ ions are also actively accumulated in the Malpighian tubules, while Cl^- probably enters passively as a consequence of the electric potential across the luminal wall. The results of O'Donnell and Machin (1991) also suggest that cations move from the haemolymph to the perinephric space. The perinephric space is extended around the Malpighian tubules as the peritubular space (Fig. 4.4), and it is suggested that this is the immediate source of ions for transport into the tubules. In this way the perinephric space is considered as a functional compartment, and an explanation of its role within the rectal complex is presented.

Several questions remain to be further investigated. It is assumed that water from the cryptonephridial system is returned to the haemolymph from the Malpighian tubules. The uptake of water creates a forward flow in the tubules, and a gradual dilution takes place. As pointed out by Machin (1979), water and ions are probably transferred from the diluted tubule fluid to the haemolymph by a coupled transport along the free loops of the Malpighian tubules.

4.4.2.2 Desert Beetles

Although not the only solution to the problem of water conservation in arid environments, the cryptonephridial system represents an important adaptation in desert beetles. Rectal water vapour absorption has been demonstrated in the larvae of six desert tenebrionids (Coutchié and Machin 1984). Among these, three *Onymacris* spp. and two *Physosterna* spp. belong to the fauna of the Namib Desert, while *Eloedes armata* inhabits the deserts of the southwestern United States.

Larvae of *Onymacris marginipennis* have a threshold of water vapour absorption at 83 % RH (Coutchié and Machin 1984), and those of *Onymacris plana* at 81 % RH (Machin and O'Donnell 1991), which is lower than in *Tenebrio molitor*. Due to the lower thresholds, uptake rates of water in the desert species are more than double those of mealworms of the same size and under identical ambient RH. The difference is due to a more effective concentration of ions in the Malpighian tubules of the rectal complex (Machin and O'Donnell 1991). Although NaCl is available to *Onymacris* spp. in their desert environment, Na^+ plays a moderate role in the accumulation of ions.

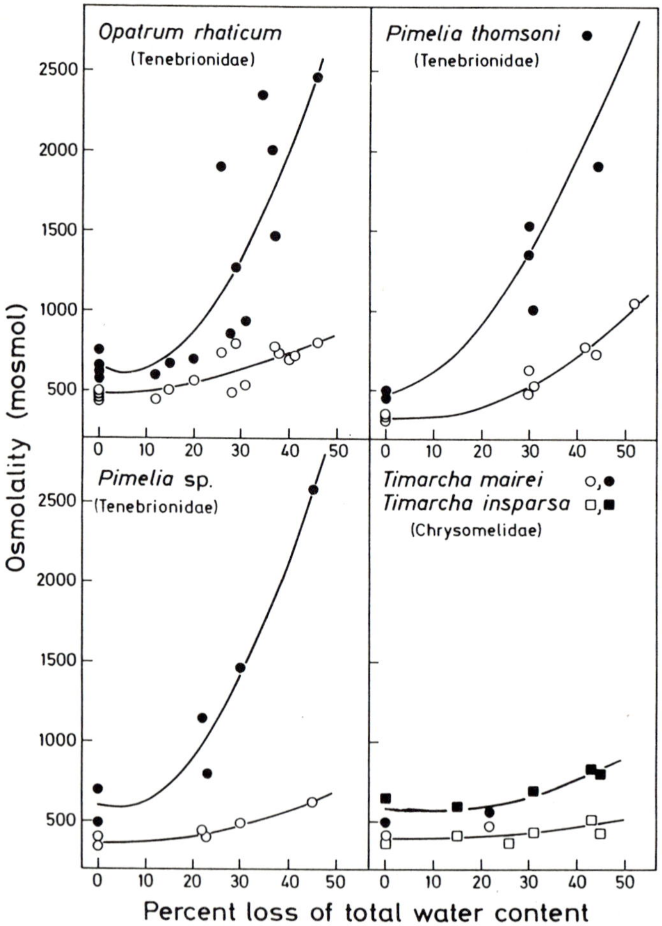

Fig. 4.6. Changes in the osmolality of the haemolymph (*open symbols*) and perirectal fluid (*closed symbols*) with increasing water loss in beetles from the High Atlas Mountains and Sierra Nevada in Spain. (Sømme 1994)

Active uptake of K^+, passively followed by Cl^-, accounts for the greater part of the concentration increase, and the concentrations of KCl reach higher levels than in saturated solutions. Thus, the high osmotic activity of the tubule fluid, necessary to facilitate water vapour uptake at RH of 81-83 %, is explained by supersaturation. Avoidance of crystallization may depend on 'supersaturation proteins' which bind to crystal surfaces (Wright and Machin 1993), representing an analogy with the 'antifreeze proteins' of supercooled insects. The difference between *Tenebrio* and *Onymacris* is also reflected in the morphology of the rectal complex (Machin and O'Donnell 1991). Among other morphometric differences, the functional area of water absorption is larger and the number of boursouflures higher in *Onymacris.*

The presence of a cryptonephridial system has also been demonstrated in adult tenebrionid beetles from the High Atlas Mountains (Gehrken and Sømme 1994). In *Pimelia thomsoni*, *Pimelia* sp. and *Opatrum rhaticum*, the distal ends of six Malpighian tubules are joined in a common trunk before entering the rectum beneath the perinephridic membrane. Difference in osmolality between the perirectal fluid and haemolymph suggests that the cryptonephridial system of these tenebrionids play an important role in osmoregulation (Chap. 4.4.3; Fig. 4.6). A cryptonephridial system comprising four Malpighian tubules, is also present in the crysomelid beetle *Timarcha mairei.* Two pairs of tubules join before entering the perirectal membrane, and branch off again at the surface of the rectum.

4.4.3 Water Absorption in Other Terrestrial Arthropods

Since it was originally described in the mealworm *Tenebrio molitor*, water vapour absorption has been documented in a large number of terrestrial arthropods (Edney 1977; O'Donnell and Machin 1988; Hadley 1994). They include mites and insects, not only from deserts but also from other arid environments. Surprisingly, booklice (Psocoptera) have the capacity for water absorption, although they mainly inhabit moist habitats (Rudolph 1982). That insect eggs may be able to absorb water vapour from the air has recently been demonstrated in the Australian walking stick *Extatosoma tiaratum* (Yoder and Denlinger 1992). Absorption took place at RH as low as 40 %.

4.4.3.1 Non-Desert Arthropods

Although an extremely interesting topic, water vapour absorption in arthropods that do not live in arid habitats falls outside the scope of this book and is mentioned only shortly in the present section. Rectal uptake through a cryptonephridial system, as described above (Chap. 4.2.3), represents only one of several possible mechanisms of water vapour absorption. (O'Donnell and Machin 1988). In booklice (Psocoptera) and biting lice (Mallophaga) absorption takes place within the oral cavity through special structures of the hypopharynx (Rudolph 1982, 1983). The thresholds of absorption are extremely low in biting lice, although uptake rates of water are high, adding up to an efficient mechanism of water conservation.

Oral absorption is also known from mites and ticks (O'Donnell and Machin 1988). The house dust mite *Dermatophagoides farinae* lives in a dry environment where water vapour absorption may be of critical importance for survival. Due to a high surface area:volume ratio, water may easily be lost through the cuticle of such small arthropods. Ticks spend 98 % of their life away from the host (Needham and Teale 1986), and water vapour absorption is an important aspect of their water balance.

Water vapour absorption in the firebrat *Thermobia domestica* may take place at humidities as low as 45 % RH (Noble-Nesbitt 1970). The absorption takes place by means of a special anal sac, which is quite different in construction from the cryptonephridial system of *Tenebrio*.

Terrestrial isopods are generally considered to have poor water conservation capacities, but recent studies have demonstrated high rates of water vapour absorption in the Crinocheta and Diplocheta (Wright and Machin 1990, 1993). Among the species studied were the diplochete *Ligia oceanica* and 12 species of crinochete oniscids from England and Ontario. Uptake takes place in the ventral pleon, where water is probably being absorbed across the pleopodal endopods. The system is based on the secretion of a strong hyperosmotic fluid onto the surface. Rhythmical movements of the pleopods probably circulate air over the uptake fluid. It has also been suggested that cyclic compression of the air within the pleon cavity is required to increase water vapour activity above that of the uptake fluid. *Ligia oceanica* is restricted to littoral habitats and can tolerate submersion, desiccation and high haemolymph osmolality. The Crinocheta are able to colonize mesic-xeric biotopes (Wright and Machin 1993). This involves several adaptations to reduce water losses and to replenish lost water by water vapour absorption. Uptake thresholds in these isopods are relatively high compared to insects, but uptake fluxes are among the highest recorded. Among the crinochete species of this study, *Armadillidium vulgare* had the lowest threshold of 0.866 a_w as well as the lowest water loss flux, while *Oniscus asellus* had the highest water vapour absorption threshold with a value of 0.929 a_w and a high water loss flux. In general, species with low water vapour thresholds also displayed lower cuticular permeability; both adaptations to arid environments. The Crinocheta are primarily night-active animals, and it is suggested that water losses during night foraging are compensated for by seeking moist shelters during the day.

4.4.3.2 The desert cockroach

Among desert arthropods, water vapour absorption in the desert cockroach *Arenivaga investigata* has been studied in great detail (Edney 1977; O'Donnell and Machin 1988; Cloudsley-Thompson 1991). The cockroaches inhabit sand dunes in the Mojave Desert of California, where they feed on dead leaves from desert shrubs. The upper layer of the sand is usually very dry, but by digging 10 to 45 cm down, the desert cockroach finds areas of 80 % RH or more. The night-active nymphs and females are able to take up water vapour from humidities above 82.5 % RH in the sand.

According to O'Donnel and Machin (1988), two bladder-like diverticula of the hypopharynx are protruded during absorption. These bladders are covered by a closely packed mat of microscopic cuticular hairs. Two so-called frontal bodies, situated above the oral cavity, produce a fluid which is applied to the bladders through surface grooves. The fluid is produced by the exudation of a haemolymph ultrafiltrate through a porous cuticular plate on the frontal bodies. The exudation takes place in pulses through the contraction of frontal muscles.

The moist surface of the diverticula is considered to be the site of water vapour absorption. This is confirmed by recording the surface temperatures of the bladders, which are up to 0.2 °C warmer than those of the surrounding mouth parts (O'Donnell and Machin 1988). The rise in temperature is probably a result of the heat released from water changing phase during condensation on the bladders. The temperature difference varies with ambient humidity, reflecting different rates of condensation onto the bladder surface.

The mechanism of water vapour condensation in the desert cockroach is not completely understood. The frontal fluid has an osmolality slightly higher than that of the haemolymph, but much lower than is required to take up water from air of 82 % RH. O'Donnell and Machin (1988) came to the conclusion that the fluid plays an ancillary role in the absorption, and that the reduction of water activity at the absorption site is a property of the cuticular hairs covering the bladder surface. The complicated mechanism is based on cyclical swelling of the hairs in response to periodic alterations in the ionic strength of the frontal fluid. Alteration in iso-electric points of cuticular chitin or proteins could result in modulations of the water affinity of the bladder hairs. It is known from polyelectrolytes in general that swelling takes place as the ionic strength of the surrounding fluid is reduced. The high osmotic pressure of the bladder surface may result from evaporation of saliva, producing a solution, the water activity of which corresponds to that of the atmosphere. This equilibrium is disturbed by the exudation of small quantities of fluid from the frontal bodies. When exposed to a fluid of relatively high ionic concentration, the water affinity of the bladder hairs is lowered. They shrink and release water which may be swallowed by the cockroach.

4.4.3.3 Desert Woodlice

In spite of their high cutaneous transpiration, some isopods are able to survive the arid conditions of deserts (Edney 1977; Cloudsley-Thompson 1991). *Hemilepistus reaumuri* is found in the desert of North Africa and the Middle East, while *Hemilepistus afghaniscus* inhabits the deserts of Central Asia. Their tolerance to arid environments is at least partly based on the fact that desert woodlice dig burrows into the ground. At a depth of 40 cm they escape the extremely high surface temperatures and find moisture conditions close to saturation. Even so, overall water losses are high (Wright and Machin 1990) and their survival is not completely explained.

In *Hemilepistus afghaniscus*, Coenen-Stass (1981) showed that water was lost at RH below 93 %, but that the woodlice gained weight at higher humidities. In *Hemilepistus reaumuri* no weight loss occurred at 98 % RH, but weight increased in a completely saturated atmosphere. The results strongly indicate that water vapour absorption takes place. Based on knowledge of water absorption in other isopods (Wright and Machin 1993), this seems a reasonable explanation for the adaptations of woodlice.

4.5 Osmoregulation

4.5.1 Occurrence in Different Taxa

The maintenance of a relatively constant osmotic pressure of the body fluids is a general necessity for normal physiological and biochemical processes. To avoid harmful and life-threatening effects, the concentration of solutes must remain within relatively narrow limits. In addition to the capacity of desert arthropods to take up water and to avoid the loss of water through behaviour, many species have the ability to regulate the osmotic pressure of their haemolymph. In this way vital processes may be protected, partly by the accumulation of solutes in other body compartments.

Many terrestrial arthropods exhibit a considerable capacity for osmoregulation. Results from a number of insect species support the generalization that this group may be strong haemolymph regulators (Riddle et al. 1976; Edney 1977; Hadley 1994). Both the degree of osmoregulation and tolerance of increased solute concentrations vary between groups of arthropods. Thus, Riddle et al. (1976) demonstrated that the tenebrionid *Eloedes hispilabris* from New Mexico had a high capacity for haemolymph osmoregulation, while the vejovid scorpion *Paruroctonus aquilonalis* tolerated almost a doubling of solute concentrations during desiccation. Myriapods represent an intermediate position. The millipede *Orthoporus ornatus* displayed a pattern of both osmoregulation and tolerance, depending on sex and the duration of desiccation (Riddle et al. 1976).

Osmoregulation in other myriapods seems to be less efficient than in insects (Crawford et al. 1975; 1986). No osmoregulation was demonstrated in the East African scorpion *Lychas burdoi* after evaporative water losses of weight up to 20 % (Bjerke et al. 1989). Some arthropods exhibit still higher tolerance to haemolymph hyperosmolality. For instance, the oniscid *Ligia oceanica*, which inhabits the littoral sone, obtains values of 2000 mosmol kg^{-1}.

4.5.2 Desert Tenebrionids

In the tenebrionid *Onymacris marginipennis*, Coutchié and Crowe (1979) showed that haemolymph osmolality is regulated between 320 and 500 mosmol kg^{-1}, although haemolymph volume decreased as a linear function of decreasing water content. In *Eleodes hispilabris*, haemolymph osmolality only increased from ca. 450 to 500 mosmol kg^{-1}, while one-third of the total body water was lost (Riddle et al. 1976). Similarly, while the haemolymph volume was almost halved, haemolymph osmolality increased from 460 to 520 momsol kg^{-1} in the tenebrionid *Trachyderma philistina* of the Negev Desert (Broza et al. 1976). Naidu and Hattingh (1986) found that while the haemolymph volume of the Namib Desert tenebrionid *Stips stali* dropped from 16.2 to 10.8 µl, osmolality only increased from 428 to 533 mosmol kg^{-1} (Fig. 4.7A,B). If the capacity for osmoregulation was not present, a value of 642 mosmol.kg^{-1} would have been expected in this species.

Osmoregulation also takes place during the rehydration of desiccated beetles. In *Stips stali*, drinking depressed haemolymph osmolality below the original level (Fig. 4.7B) (Naidu and Hattingh 1986), and this was also evident in males of *Eleodes hispilabris* (Riddle et al. 1976). In other tenebrionids (Broza et al. 1976; Naidu and Hattingh 1986) the haemolymph volume also initially increases above normal level during drinking, but later returns to normal values (Fig. 4.7B).

Osmolality found during desiccation and rehydration in the tenebrionids is not the result of simple haemolymph concentration and dilution. The maintenance of haemolymph osmolality during drastic changes in haemolymph volumes suggests that osmotically active substances must be removed from the haemolymph during dehydration and returned to it when water is taken up (Naidu and Hattingh 1986). In accordance with this, Coutchié and Crowe (1979) showed that the concentration of ions, amino acids and trehalose in larvae of *Onymacris marginipennis* is relatively constant, despite changes in the volume of the haemolymph. The quantities of all the major components increase as the volume of haemolymph increases, and this accounts for the observed regulation of haemolymph osmolality.

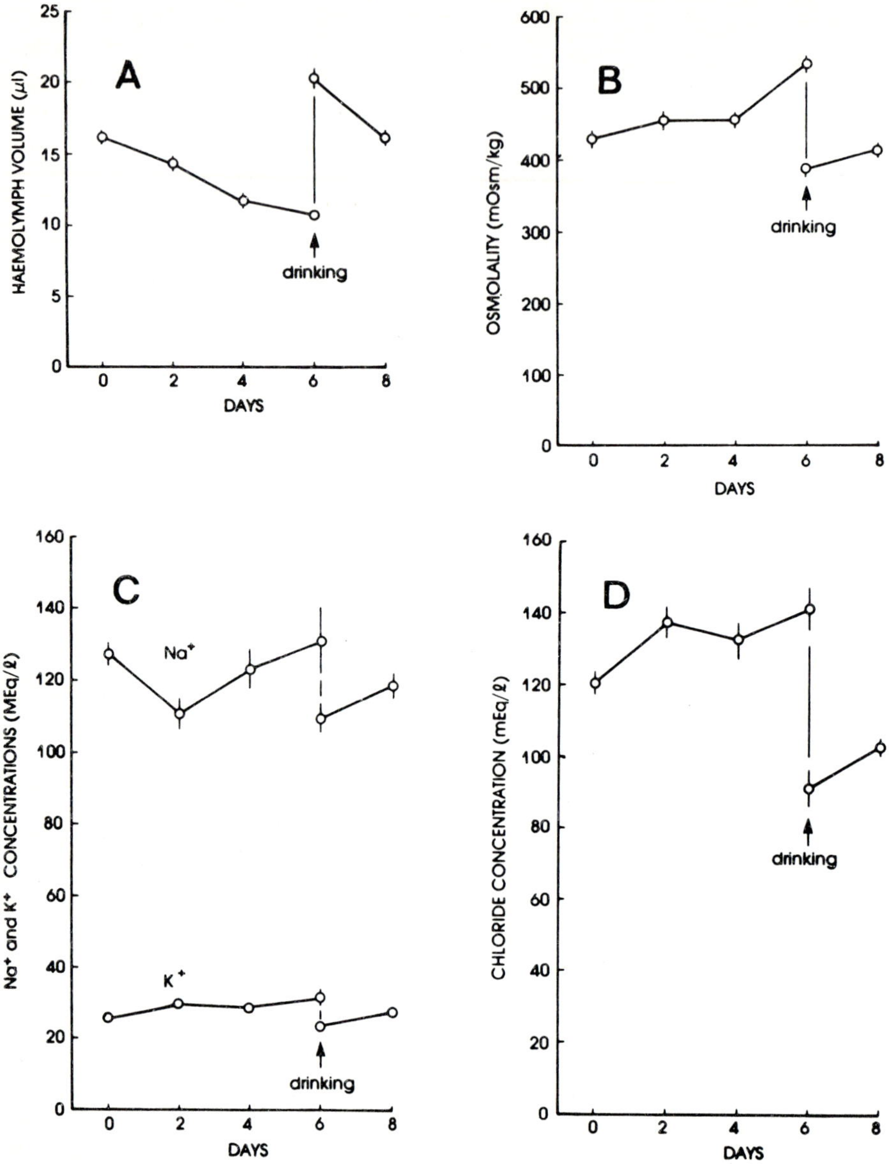

Fig. 4.7A-D. Effects of dehydration and rehydration on haemolymph in the tenebrionid beetle *Stips stali*. **A** Changes in volume (µl). The data apply to a standard beetle of initial weight 92.1 mg. **B** Changes in osmolality (mosmol kg^{-1}). **C** Changes in Na^+ and K^+ concentrations (mEq l^{-1}). **D** Changes in Cl^{-1} concentrations. *Arrows* indicate time of drinking. *Vertical lines* ± SE (n=10-12 for haemolymph volume; n=6 for the other parameters). (Naidu and Hattingh 1986)

Excess Na^+ may be stored in the tissues of dehydrated *Periplaneta americana* and mobilized by rehydration (Wall 1970). In *Onymacris plana*, regulation of Na^+ takes place during desiccation (Nicolson 1980), while Cl^- is partly regulated in *Trachyderma philistina* (Broza et al. 1976). Naidu and Hattingh (1986) also observed a regulation of Na^+, K^+ and Cl^- (Fig. 4.7CD) in *Stips stali.* Na^+ and K^+ returned to normal after drinking. The Cl^- concentration was only partly restored by rehydration, but returned to normal when the beetles fed.

In the East African tenebrionid *Rhytinota praelonga*, Na^+ makes up about 25 % of the total extracellular osmolality (Zachariassen and Einarson 1993). During dehydration the concentration of Na^+ in the haemolymph increases, but considerably less than expected from the reduction in haemolymph volume. This implies that Na^+ is one of the solutes that are removed from the haemolymph when the beetles become dehydrated. The cellular Na^+ content, however, remains constant during dehydration. In this way the total Na^+ content of the beetles is reduced, and it is suggested that the ions are excreted by the organism. This also implies that the beetles will have to replace Na^+ from a dietary source during rehydration (Zachariassen and Einarson 1993).

Free amino acids may also play a role in osmoregulation (Coutchié and Crowe 1979). In *Trachyderma philistina*, Broza et al. (1976) found active removal of amino acids from the haemolymph during dehydration and their quick replenishment after drinking. Simultaneous changes in haemolymph protein concentrations suggest a regulatory mechanism whereby amino acids are rapidly bound and liberated, but this was not confirmed in studies on *Stips stali* (Naidu and Hattingh 1986).

As pointed out by Hadley (1994), the maintenance of tissue water at the expense of haemolymph is of great physiological importance. In this way the functional integrity of the cells and their metabolism operate for as long as possible. In larvae of *Onymacris marginipenis*, Coutchié and Crow (1979) showed that the distribution of water reserves is controlled in such a way that more water is lost from the haemolymph than from the tissues. In *Rhytinota praelonga* there is a marked reduction, in both haemolymph volume and cell water content, during dehydration (Zachariassen and Einarson 1993). The reduction is faster in the haemolymph than in the intracellular water content, and when the haemolymph volume approached zero, the cell water content is reduced by only ca. 50 %. The beetles are able to tolerate almost complete loss of haemolymph, but apparently die when the extracellular fluid reservoir is exhausted. Further dehydration also results in the loss of intracellular water. The studies of Zachariassen and Einarson (1993) demonstrate that the ability to restrict the reduction of intracellular water, such as in *Rhytinota praelonga*, is an important physiological adaptation to an arid environment.

The removal of solutes from the haemolymph through the cryptonephridial system of tenebrionids was first demonstrated by Ramsay (1964) and Grimstone et al. (1968). During desiccation the perirectal fluid of *Tenebrio molitor* reaches concentrations of 3000 mosmol kg^{-1}. In this way a strong osmotic gradient is formed, and more water is removed from the rectum (Chap. 4.4.2). In spite of the high values of the perirectal fluid, the osmolality of the haemolymph did not exceed 800 mosmol kg^{-1}, illustrating a considerable capacity for osmoregulation.

4.5.3 Alpine Beetles

In tenebrionid beetles from the High Atlas Mountains, the cryptonephridial system plays a major role in the osmoregulation of the haemolymph (Gehrken and Sømme 1994). During desiccation, an exponential increase in solute concentrations was observed in the perirectal fluid of *Pimelia thomsoni*, *Pimelia* sp. and *Opatrum rhaticum*, resulting in concentrations of 2500-2750 mosmol kg^{-1} (Fig. 4.6). The exact mechanism by which water is returned to the haemolymph is not known, but it probably takes place through the Malpighian tubules against an osmotic gradient. As a result, the haemolymph osmolality of the three species remained much lower during desiccation, compared with the perirectal fluid. All three species tolerated relatively high levels of solutes before they succumbed. In *Pimelia thomsoni* this level exceeded 1000 mosmol kg^{-1}, and corresponding values for *Opatrum rhaticum* and *Pimelia* sp. were 850 and 750 mosmol kg^{-1}, respectively. A strong osmoregulation of the haemolymph was also observed in the chrysomelid beetles, *Timarcha* spp. (Fig 4.5). In contrast to the tenebrionids, the osmolality of the perirectal fluid showed only small increases, suggesting that a different mechanism is present in chrysomelids.

4.6 Conclusions

Several types of physiological mechanisms have evolved in terrestrial arthropods as adaptations to arid environments.

Evaporative water loss takes place through cuticular as well as respiratory transpiration, and water is also lost through the faeces. Different kinds of adaptation are present to reduce the losses through the three main routes of water loss.

Cuticular transpiration is important in most species but great differences in water permeability have been demonstrated. In different species of collembolans and mites there is a clear correlation between the dryness of their habitats and the cuticular water flux. In many insects, cuticular water loss is reduced due to more efficient surface wax layers. Desert scorpions have an almost impermeable cuticle.

Respiratory water loss accounts for a large proportion of the total evaporative water loss in several beetles. The formation of a subelytral cavity by the fusion of the elytra helps to reduce water loss from the abdominal spiracles in many species from hot arid regions.

Compensation for water losses is possible in several ways. Most insects have the ability to absorb water from the content of the rectum, and the capacity is increased in species from dry habitats. Through the development of the cryptonephridial system, the Malpighian tubules are brought into close contact with the rectum. Active uptake of ions from the haemolymph creates a strong osmotic gradient, and causes the water to flow from the content of the rectum to the Malpighian tubules. The cryptonephridial system is an important adaptation in desert beetles and is well developed in tenebrionids and other species inhabiting some of the most arid environments.

Water vapour can also be absorbed from the air with high efficiency through the cryptonephridial system, and a threshold as low as 81 % RH has been demonstrated in desert tenebrionids. Other mechanisms of water vapour absorption are known from desert as well as non-desert terrestrial arthropods. In the desert cockroach *Arenivaga investigata*, it takes place through a special organ on the hypopharynx. Apparently, water vapour absorption is also important in desert isopods.

The maintenance of a relatively constant osmotic pressure of the body fluids in terrestrial arthropods is achieved through osmoregulation. Insects, in particular, appear to have a considerable capacity of regulating the osmotic pressure of their haemolymph. Within certain limits, increase in osmolality is relatively low during desiccation, while haemolymph volume decreases as a function of decreasing water content. The ability to restrict the reduction of intracellular water at the expense of reduced haemolymph volume is an important physiological adaptation to an arid environment.

Osmoregulation is partly based on the removal of osmotically active substances from the haemolymph, and on their return during rehydration. Furthermore, the uptake of solutes from the haemolymph through the cryptonephridial system is important to maintain a low haemolymph osmolality.

In most species, an integration of several kinds of adaptation is found, and this makes it possible for them to survive the harsh conditions of arid environments. In addition to physiological mechanisms, behavioural (Chap. 6) and phenological adaptations (Chap. 7) are of major importance.

5 Anhydrobiosis in Terrestrial Invertebrates

5.1 Introduction

Many invertebrates adapt to physiological stress by reduced metabolism and enter a state of quiescence associated with retarded senescence (e.g. Evans and Perry 1976; Womersley 1981, 1987; Ushatinskaya 1990; Wright et al. 1992). Quiescence can be induced by different kinds of environmental stress, including extreme temperatures, dehydration, high salt concentrations and anoxia. During increasing stress, some organisms are able to shut down metabolism to an undetectable level.

This extreme kind of quiescence has been termed cryptobiosis, and is reversible at the onset of favourable conditions. Womersley (1981) pointed out that cryptobiosis should not be confused with diapause (Chap. 7.2). The latter is arrested development resulting from reduced hormone activities, which may persist even under favourable environmental conditions, and is temporarily irreversible.

Keilin (1959) divided cryptobiosis into four types: cryobiosis caused by freezing temperatures, osmobiosis by high solute concentrations, anoxybiosis by lack of oxygen and anhydrobiosis by desiccation. Cryobiosis, osmobiosis and anhydrobiosis may actually represent homologous adaptive states since all involve loss of water from the cells, resulting in a cessation of all metabolic functions. Anhydrobiosis is the most common form of cryptobiosis, and also the one that has been most investigated. Numerous organisms are able to survive almost complete dehydration; in many cases more than 99 % of the water content is lost (Crowe et al. 1992). As an example, anhydrobiotic specimens of the nematode *Aphelencus avenae* contain only 0.5 % water (Higa and Womersley 1993), while fully hydrated nematodes contain 75-80 %. According to Clegg (1979), metabolism in cystes of the brine shrimp *Artemia salina* is reduced stepwise with decreasing water content. Fully hydrated cysts contain 1.4 g H_2O/g dry weight, but normal metabolism is maintained at water contents as low as 0.6 g H_2O/g. Restricted metabolism is recorded during further dehydration to 0.3 g H_2O/g dry weight and below 0.1 g H_2O/g no metabolic events are observed.

Examples of anhydrobiosis can be found among numerous plants and animals (e.g. Womersley 1981, 1987; Crowe et al. 1992; Wright et al. 1992). Many organisms are capable of anhydrobiosis only as propagules or in specific developmental stages. Such organisms include plant seeds, yeast cells, bacterial and fungal spores, larvae of a few species of chironomid midges and cysts formed by the eggs of *Artemia salina*. Other organisms such as protozoans, rotifers, tardigrades, nematodes and springtails may enter anhydrobiosis in several life stages.

Due to their high ratio of surface area:volume, small animals like rotifers, tardigrades and nematodes are likely to lose water rapidly and are therefore strongly affected by changes in environmental moisture content. Apparently, these animals have not been able to evolve strategies of water conservation similar to those of terrestrial arthropods. The survival of rotifers, nematodes and tardigrades is based on their remarkable ability to undergo almost complete dehydration without being killed. All of them are essentially aquatic, inhabiting temporary pools and the interstitial water of soil and plants. During periods of drought these bodies of water are subjected to rapid evaporation and the interstitial fauna to anhydrobiosis which may be of prolonged duration. Rotifers, nematodes and tardigrades are typical inhabitants of alpine and polar areas, and are also found in hot lowland deserts. The environmental stress of these habitats makes it impossible for such animals to survive without highly specialized tolerance of desiccation.

Anhydrobiosis in collembolans has been studied in several species from different arid environments, but in this group other forms of resistance to drought have also evolved in this group (Chap. 3.1). At present, anhydrobiosis in insects is known from only a few chironomid species.

5.2 Nematodes

5.2.1 Tolerance to Desiccation

According to Womersley (1987), the most successful anhydrobiotic animals of which we at present are aware belong to the phylum Nematoda. Anhydrobiosis has been reported from a number of nematode species (e.g. Ellenby 1968,1969; Crowe and Madin 1975; Womersley 1987; Forge and MacGuidwin 1992a). Some species are able to survive for several years in anhydrobiosis (Aroian et al. 1993). Thus, the plant parasite *Anguina tritici* was reactivated from seed galls after 28 years (Fielding 1951) and *Tylenchus polyhypnus* survived in a herbarium for 39 years (Steiner and Albin 1946).

Nematodes are widely distributed in a diversity of habitats that are subjected to the effects of partial or extreme dehydration. As pointed out by Womersley (1987), it is important to assess the anhydrobiotic potential of an organism in relation to the dynamics of water availability in its natural environment. The use of severe desiccation regimes in the

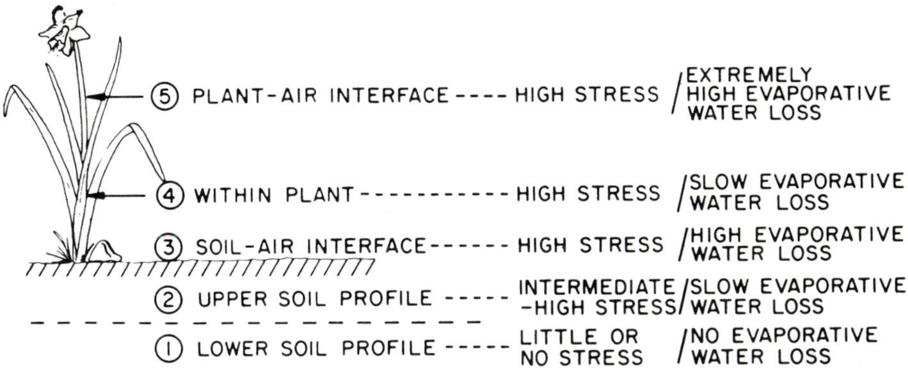

Fig. 5.1. Division of a generalized terrestrial habitat occupied by nematodes with respect to levels of dehydration stress imposed by rate of evaporative water loss. (Womersley 1987)

laboratory has led to some nematodes being classified as intolerant of desiccation, although they may have survived if allowed to dry slowly. Based on the rate of evaporative water loss and level of dehydration stress encountered, terrestrial habitats may roughly be divided in five regions (Fig. 5.1).

The lower soil profile represents a stable environment where essentially no evaporative water loss occurs. In the upper soil layer the water loss is slow, and the rate depends on the physical characteristics of the soil. Both plant parasitic and free-living species of this region may withstand desiccation if allowed to dry slowly. An example is larvae of the free-living nematode *Rotylenchulus reniformis*. In the laboratory, this species survived when dried slowly over 10-12 days at the natural rate of soil moisture loss (Womersley and Ching 1989). Survival was associated with coiling in the larvae, after which they were able to withstand severe dehydration at 80 and 40 % RH but not at 0 % RH. The results suggest that the tolerance to desiccation in *R. reniformis* is well in accordance with the conditions of its natural habitat.

Relatively little is known about nematodes inhabiting the soil-air interface (Fig. 5.1), but species occupying this region must be particularly well adapted to cope with frequent periods of rapid evaporative water loss. For nematodes parasitizing the aerial parts of the plants, anhydrobiosis is a routine part of their life cycle. Slow drying within the plant tissue, e.g. when the plants wither, may be associated with the formation of galls and the development of more resistant stages. In order to continue their life cycles, these nematodes must tolerate periods of desiccation at the plant-air and soil-air interfaces before infecting new plants and continuing development in their protected phase within the plant host itself. Another strategy of plant parasitic nematodes involves an exodus from senescent plants. Large aggregates of nematodes are formed at the plant-air interface, and slow drying is mainly accomplished by the behavioural adaptation of coiling and clumping.

5.2.2 Rates of Dehydration

Survival rates often increase with decreasing rates of dehydration, and a number of species possess means for impeding the rate of water loss. The most common strategies involve coiling and the formation of aggregations, although cuticular modifications may also be present. As pointed out by Womersley (1987), complex biochemical adaptations are involved at the cellular level, and the slow rates of drying may enable certain anhydrobiotes to prepare themselves metabolically for the onset of anhydrobiosis. On the other hand, some plant parasitic nematodes, such as *Anguina tritici*, are able to survive direct exposures to low RHs, and these species appear to be pre-adapted from a metabolical point of view.

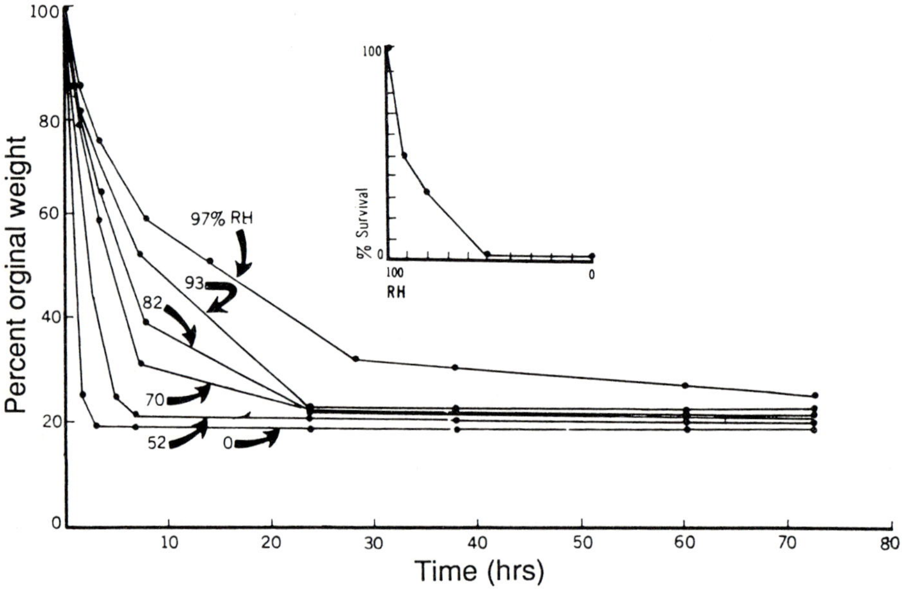

Fig. 5.2. Weight loss from pellets of *Aphelencus avenae* (ca. 100 mg wet weight) kept at the indicated relative humidities. *Inset* The relationship between survival at 0 % RH and the RH at which the animals were dried initially. (Crowe and Madin 1975)

The plant parasite *Ditylenchus dipsaci* forms aggregations during dehydration (Ellenby 1968). The worms on the periphery are killed while those in the centre survive, probably thanks to reduced rates of water loss. Many plant parasitic nematodes such as *Heterodera rostochiensis* form cysts with a protective proteinous shell containing a large number of eggs (Ellenby 1968).

The effect of different rates of dehydration has been clearly demonstrated in the free-living mycophagous nematode *Aphelenchus avenae* (Crowe and Madin 1975). During dehydration, *A. avenae* forms a tight coil spiral by which the external surface is reduced and the rate of water loss is slowed down. In the experiments of Crowe and Madin (1975), clusters of nematodes were initially dehydrated at six relative humidities (RH) from 0 to 97 %. Subsequently, they were kept at 0 % RH for 72 h before being rehydrated. The rate of water loss was rapid in dry air and an equilibrium was achieved after 3 h (Fig. 5.2). In air at 97 % RH the rate was much slower and equilibrium was not reached until 72 h. A direct relationship was found between the proportion of animals that survive dehydration and the RH at which they were kept before exposure to 0 % RH. More than 95 % of the nematodes initially kept at 97 % RH survived (Fig. 5.2), while less than 1 % survived when initially exposed to 0 % RH.

5.2.3 Nematodes from Hot Deserts

Most research on anhydrobiosis in nematodes has been carried out on plant parasitic species from temperate areas, but studies on nematode anhydrobiotes from desert soil suggest that they may be even more specialized than temperate forms (Freckman and Womersley 1982).

In the Chihuahuan Desert of New Mexico, nematodes spend most of their time in anhydrobiosis (Freckman and Womersley 1982; Freckman 1986). Their activity appear to be pulse-regulated by rainfall, after which the percentage of anhydrobiotic nematodes falls greatly. Using artificial irrigation, a general moisture threshold of 4-5 % soil water content appeared (Freckman and Womersley 1982; Freckman 1986). If the moisture content was lower than 4.7 %, more than 60 % of the nematodes were anhydrobiotic, while 40-100 % were active at higher water contents. The experiments indicate that desert nematodes are activated by rainfalls of ca. 25 mm. In their active state they are probably of importance in decomposition processes, but are otherwise without effects. Moisture alone may not result in increased population densities of soil biota and associated rates of decomposition, since lack of nitrogen may limit productivity in desert ecosystems.

Anhydrobiotic survival has also been reported in free-living nematode *Pangrolaimus* sp. from dry soils in Australia (Nicholas and Stewart 1985). Under experimental conditions this species successfully entered the anhydrobiotic state, provided the rate of dehydration was slow. Another *Panogrolaimus* sp. from Armenia was found to survive in the anhydrobiotic state for more than 8 years (Aroian et al. 1993).

5.2.4 Antarctic Nematodes

Few data are available on anhydrobiosis in nematodes from the Antarctic, although species inhabiting the fellfields and chalikosystems on the islands and continent are exposed to severe climatic conditions (Pickup and Rothery 1991; Wharton and Barclay 1993). In addition to freezing stress, nematodes and other terrestrial invertebrates have to survive exposure to arid conditions during the summer due to low precipitation, high winds and shallow soils. During the winter, vapour saturation deficit may be encountered due to lack of free water when the nematodes are surrounded by ice.

Panagrolaimus davidi, which inhabits mosses in the extreme continental environments of East Antarctica, can easily be cultured and used for laboratory experiments (Wharton and Barclay 1993). The nematodes did not survive direct exposure to 0 % RH in the laboratory, but slower rates of water loss at 99 and 76 % RH increased the percentage of coiling in the nematodes and subsequent survival at 0 % RH. Increasing the severity of desiccation resulted in an increased lag phase upon rehydration. Following exposure to 99 % RH, there was some recovery in the first hour, and maximum recovery was reached after 4 h. In samples exposed to 76 % RH, maximum recovery appeared after 24 h, and after 48 h in those exposed to both 76 % and 0 % RH. Although *P. davidi* depends on a slow rate of dehydration for survival, it is well adapted to the desiccation conditions of its natural moss habitat.

Pickup and Rothery (1991) found seasonal variations in the rates of water loss in the Antarctic nematode *Teratocephalus tilbrooki* from Signy Island in the maritime Antarctica. Although increased anhydrobiotic survival in other nematodes is often associated with increased resistance to water loss, this is not the case in *T. tilbrooki*. The species is found in cushions of mosses of the genus *Andreaea* in maritime antarctic fellfields. The degree of desiccation of their habitat may temporarily be extreme during the summer. In the laboratory, medium survival time of anhydrobiotic *T. tilbrooki* varied from 20 to 60 days at 5 °C and 60 % RH.

In comparison, *Ditylenchus* sp. inhabiting the thalli of *Usnea* lichens from the same fellfield had considerably lower rates of water loss. A median survival time of 99 days was recorded in this species at 5 °C and 60 % RH. When dehydrated and stored at 5 °C and 0 % RH, median survival time decreased to 31 days.

Although they differ in their tolerance to desiccation, Pickup and Rothery (1991) concluded that both species possess extremely well-developed anhydrobiotic capabilities. From continuous measurements of the water vapour deficit in their natural habitats, it appears that these species rarely encounter desiccation conditions as severe as those experienced in the laboratory experiments. An important conclusion from these studies is that the survival time in anhydrobiosis is limited and depends on the moisture conditions of the environment.

5.2.5 Accumulation of Trehalose

Most anhydrobioitc nematodes accumulate low molecular weight carbohydrates (Womersley 1987). In larvae of *Aphelenchus avenae*, Madin and Crowe (1975) found increased concentrations of glycerol and trehalose during desiccation. Levels of 55 μg mg^{-1} dry weight of glycerol and 110-130 μg mg^{-1} dry weight of trehalose were attained in anhydrobiotic larvae. According to Womersley (1981), preferential storage of trehalose is evident in the plant parasitic nematodes *Anguina tritici* and *Ditylenchus dipsaci*, reaching levels of 91 μg mg^{-1} and 46 μg mg^{-1} dry weight, respectively. Only small amounts of glycerol were found in these species, but *A. tritici* contained high levels of inositol. The trehalose content was reduced dramatically within 24 h in revived specimens of both *A. tritici* and *D. dipsaci*.

Although there is a strong relationship between levels of trehalose and anhydrobiotic survival in nematodes (Womersley 1981, 1987), available data are limited. The accumulation of trehalose is not unique to this taxa, but has been reported from other anhydrobiotic organisms as well. It has been suggested that the protective effect of trehalose includes the replacement of bound water around protein molecules and the inhibition of oxidative damage. As discussed in Chapter 5.6, trehalose has a stabilizing effect on cell membranes (Crowe et al. 1992). In some species of nematodes that survive rapid desiccation, however, little or no carbohydrate accumulation appears, suggesting that these species are already pre-adapted for dehydration at the cellular level. Apparently, variations in biochemical adaptations to anhydrobiosis exist in nematodes (Womersley 1987).

Recently, Higa and Womersley (1993) have demonstrated that accumulation of trehalose in itself is not sufficient for the anhydrobiotic survival of *Aphelenchus avenae*. Although levels of trehalose increased to 8-10 % of dry weight during preconditioning at 97 % RH, the nematodes were unable to survive direct transfer to either 75 % nor 40 % RH. In comparison to rapidly dehydrated nematodes, small clumps of *A. avenae* subjected to sequential dehydration showed high survival rates. In these experiments, the nematodes were also preconditioned at 97 % RH and were subsequently exposed to a regime of 93, 85, 75, 40 and 0 % RH for 24 h periods. Even at the lowest RHs, between 60 and 90 % of the nematodes survived, following rehydration. Glucose and glycerol levels remained low in all experiments, suggesting that these compounds are of little importance.

These results are apparently in disagreement with those of Crowe and Madin (1975), who found survival at 0 % RH following direct transfer from 97 % RH (Sect. 5.2.2). According to Higa and Womersley (1993), the difference may be explained by differences in clump size. Since Crowe and Madin (1975) used large clumps of approximately 100 mg wet weight, dehydration at 0 % RH may actually have been slower than anticipated in nematodes situated in the interior parts of the clumps. Higa and Womersley (1993) used clumps of ca. 10 mg fresh weight, which secure a more uniform dehydration regime.

In conclusion, Higa and Womersley (1993) pointed out that the synthesis of trehalose may signify preliminary preparation for survival, but its presence alone does not ensure survival in *Aphelenchus avenae*. Slow rates of drying are necessary for the nematodes to respond to different levels of hydration in such a way that adaptations can proceed in the required sequence. Presumably, this process is necessary for the ultimate stabilization of cellular structure in the anhydrobiotic state. Rapid dehydration apparently does not allow sufficient time at the different hydration levels for these adaptations to be fully implemented.

5.3 Tardigrades

The Tardigrada is a relatively small phylum of invertebrates, comprising approximately 600 species on a world basis (Pilato 1979). About 90 % of them live in fresh-water or terrestrial habitats, while 10 % of the species are marine. The terrestrial species are in principle also aquatic animals, living in the interstitial water of cryptogams and humic soils (Wright et al. 1992). Tardigrades are widely distributed and appear in enormous numbers in suitable habitats. According to Pilato (1979), 16 % of soil-inhabiting species have a cosmopolitan distribution, 54 % are endemic to smaller geographical areas while the remaining 30 % are found in one or more geographical regions.

5.3.1 Tun Formation

Cryptobiosis, including anhydrobiosis, in tardigrades has recently been reviewed by Ramløv (1989), Ushatinskaya (1990) and by Wright et al. (1992). It is interesting to note that the earliest observations of anhydrobiosis in tardigrades were described by Spallanzani (1776). Since they reminded him of turtles, he named the group "tardigrades", which means those that walk sluggishly. Other important early contributions to cryptobiosis in tardigrades were made by Doyère (1842) and Baumann (1922). Reviews of the older literature have been presented by May (1949) and Keilin (1959).

According to Wright et al. (1992), most of the recent advances in the cryptobiosis of tardigrades have been made in studies on anhydrobiosis, which is also the main topic of the present discussion. The following presentation is based mainly on the reviews of Ramløv (1989) and Wright et al. (1992), where more detailed references are to be found.

During dehydration, anhydrobiotic tardigrades contract into a structure resembling a small tun (Baumann 1922; Greven 1972; Crowe 1972,1975; Wright 1989a). Tun formation involves active longitudinal contraction and infolding of intersegmental cuticle which in this way is covered by the thicker dorsal segmental plates. The anterior segments, as well as the lobopodial limbs, are fully invaginated (Fig. 5.3). The tun formation in tardigrades is facilitated by several structural adapations of the cuticle. Concurrent with tun formation, wax extrusion covers the surface, and this may also help to reduce transpiration.

Tun formation is essential in the preparation for anhydrobiosis. It serves to protect internal organs, and more important, to reduce water losses through transpiration. The process of tun formation requires active metabolism. If tardigrades are dried under anoxia (Crowe 1972), they do not form a tun but shrink into a flat formless skin and do not survive. Similarly, if the animals are dried at low RHs, water loss is too fast and they are unable to form tuns.

In general, slow dehydration is necessary for tun formation in tardigrades. Initial rate of desiccation is essential, although great variations between species have been observed (Wright 1989a). As an example, the xerophilic species *Hypsibius oberhaeuseri* may accomplish tun formation at RHs down to 59 %, while the hygrophilic *Hypsibius dujardini* does not tolerate an initial RH below 78 %.

Transpiration declines gradually as the surface area is reduced during tun formation (Wright et al. 1992). A more dramatic reduction of water loss appears soon after completion of the tun. At this point, transpiration is reduced by a factor of 20-60 within a few minutes. In some species this slump in permeability may be reached in 20 min. Subsequently, transpiration declines slowly until scarcely measurable levels and, finally, an ametabolic state is reached. The tardigrade *Macrobiotus aerolatus* can tolerate water losses down to 2-3 % of its fresh weight (Crowe 1972).

The rapid change in rate of transpiration is followed by a fast decline in the permeability of the cuticle, associated with phase transitions in the cuticular lipids (Wright et al. 1992). This allows the animals to retain a fraction (5 to 15 %) of their initial body water. When dried at lower RHs, the change in transpiration is delayed, less body water remains, and the tardigrades cannot be reactivated (Wright 1989a). It has been suggested that a critical proportion of body water must be retained for sustained metabolism before the entry into the anhydrobiotic state is completed.

Water retention is improved in xerophilic species by more efficient tun formation. Tolerant species have thinner intersegmental cuticle, which facilitates rapid contraction, as well as thicker intracuticular plates and lipid layers (Greven 1972; Wright 1988, 1989b). These modifications apparently allow them to withstand more rapid initial transpiration at lower humidities.

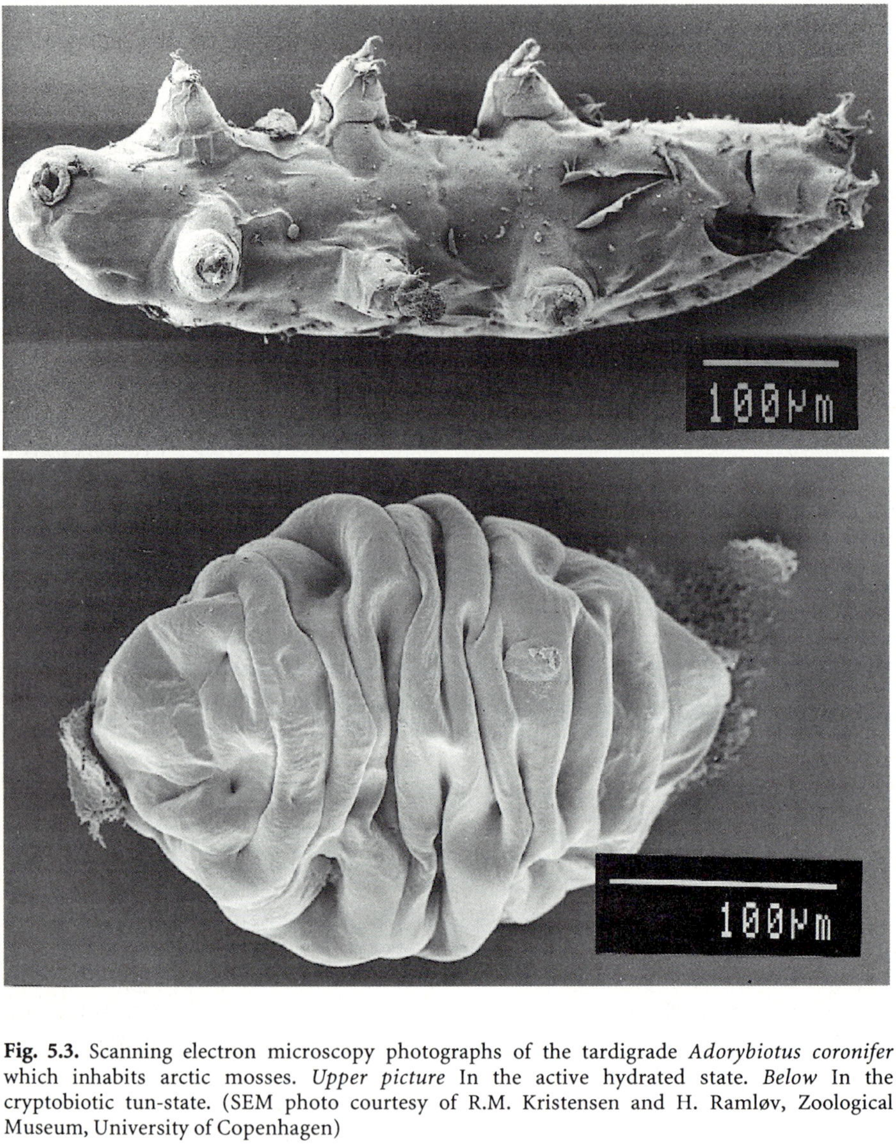

Fig. 5.3. Scanning electron microscopy photographs of the tardigrade *Adorybiotus coronifer* which inhabits arctic mosses. *Upper picture* In the active hydrated state. *Below* In the cryptobiotic tun-state. (SEM photo courtesy of R.M. Kristensen and H. Ramløv, Zoological Museum, University of Copenhagen)

5.3.2 Ecological Implications

Tardigrades are typically found in moist habitats, inhabiting mosses and humic soils. Their limited tolerance to the speed of initial dehydration probably restricts them to slowly desiccating habitats (Wright et al. 1992). Several factors in their environment may be of importance to their rates of water loss. As pointed out by Wright (1991), such factors include substrate heat capacity, drainage and exposure, insolation and habitat plant species. Desiccation regimes are slower in the soil than in mosses, and soil tardigrades may display lower tolerance.

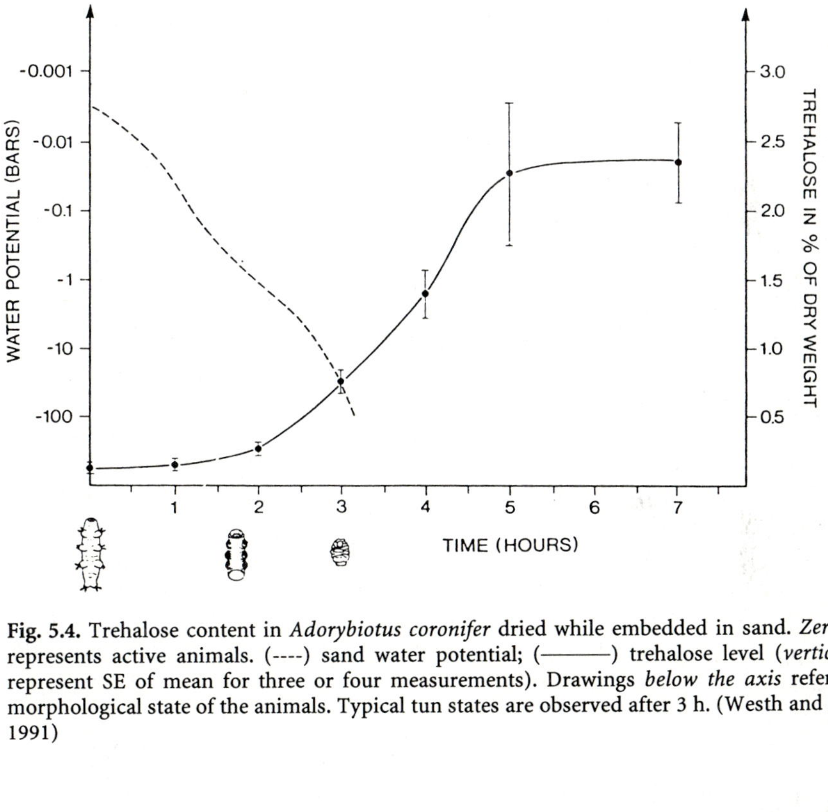

Fig. 5.4. Trehalose content in *Adorybiotus coronifer* dried while embedded in sand. *Zero hours* represents active animals. (----) sand water potential; (———) trehalose level (*vertical bars* represent SE of mean for three or four measurements). Drawings *below the axis* refer to the morphological state of the animals. Typical tun states are observed after 3 h. (Westh and Ramløv 1991)

5.3.3 Accumulation of Trehalose

Trehalose plays an important role in the survival of anhydrobiotic organisms (Crowe et al. 1992), and was initially reported from tardigrades by Crowe (1975). The first quantitative data of trehalose accumulation in tardigrades has been presented by Westh and Ramløv (1991), based on studies of *Adorybiotus coronifer*. In their experiments, the tardigrades were embedded in sand with an initial water content of approximately 30 %, which was subsequently dried in an airstream at 20 °C and 50 % RH. This resulted in a gradual decrease in water potential until equilibrium was reached after 3 h (Fig. 5.4). At the start of dehydration, the trehalose content of the tardigrades was 0.1 % of dry weight. During dehydration, a substantial increase took place, stabilizing at 2.3 % of dry weight after 5-7 h (Fig. 5.4). Under these experimental conditions, tun formation was completed in 3 h, but trehalose synthesis continued after this stage had been reached. The results confirm the assumption that water retention following the slump in permeability is of great importance for prolonged metabolic processes.

The levels of trehalose in this tardigrade species are lower than found in other anhydrobiotes (Westh and Ramløv 1991). On the other hand, the accumulation of trehalose is generally faster than observed in other organisms.

5.3.4 Recovery from Anhydrobiosis

When anhydrobiotic tardigrades are immersed in water they remain contracted as a tun for several minutes (Wright 1989a) before gradual extension of the body and the resumption of movement. Revival may be accomplished in less than 20 min, but more typically takes a few hours, in particular if the tardigrades have been subjected to severe desiccation. During rehydration, trehalose is rapidly lost, even when the animals are still in the tun stage (Westh and Ramløv 1991).

A striking characteristic of tardigrades is that they can remain in anhydrobiosis for several months or even years. Westh and Ramløv (1991) kept *Adorybiotus coronifer* in dry moss for 8 months, and Sømme and Meier (1995) found a high degree of revival in three dehydrated species of antarctic tardigrades after 8 years at -22 °C. Older literature includes numerous accounts of cryptobiosis in tardigrades, as reviewed by Baumann (1922), May (1949), Ramløv (1989) and others. The record is supposed to be 120 years in anhydrobiosis, after which the tardigrade resumed activity but died shortly afterwards.

5.4 Rotifers

Anhydrobiosis in adult Rotifera is known from species of the class Bdelloidea (Gilbert 1974; Ricci 1987). Bdelloid rotifers are remarkable in having obligatory parthenogenesis, and no males are found in the populations. Most species live in lakes, ponds and rivers, but several inhabit the interstitial water of mosses, lichens and soil, where they are periodically subject to desiccation. They are widely distributed in polar regions (Chap. 2.2.4), where they survive desiccation and freezing for extended periods. In contrast to bdelloid rotifers, species of the class Monogonata can reproduce sexually, and resting eggs are produced under certain environmental conditions (Gilbert 1974).

As in tardigrades and nematodes, slow drying is essential for the survival of bdelloid rotifers in anhydrobiosis. Some species have been revived after prolonged periods of desiccation. Thus, Rahm (1923) reported that *Macrotrachela musculosa* become active by wetting moss after 2-3 months and *Mniobia magna* and *Mniobia russeola* were revived after 2 1/2 years. Although investigations on anhydrobiosis has been limited to relatively few species, the ability appears to be widespread within the class (Gilbert 1974). Anhydrobiosis can be considered as a major adaptation of rotifers to inhabit environments with rapid and unpredictable changes between a frozen, wet and dry state. During anhydrobiosis, bdelloid rotifers are extremely tolerant of low temperatures.

Rotifers do not form cysts for entering anhydrobiosis, but some species secrete cases or houses into which they withdraw, and which may retard drying. As reviewed by Gilbert (1974), the drying processes of *Philodina roseola* include the withdrawal of head and foot, and contraction of the circular muscles at both ends of the trunk. The cuticle becomes wrinkled, and the body shrinks to approximately one-fourth of its original volume. The internal organs are closely packed in the resulting lemon-shaped structure, resembling the tuns of tardigrades. Observation on *Macrotrachela quadricornifera* showed that tissues and cells become condensed, maintaining their identity during drying (Ricci 1987). Loss of water in anhydrobiont rotifers is high, but some bound water is retained. It is not known how membranes of dried rotifers are stabilized in the absence of water.

5.5 Terrestrial Arthropods

There are few reports on anhydrobiosis in terrestrial arthropods, but examples are known from Collembola (Poinsot 1968; Poinsot-Balaguer 1976; Greenslade 1981) and Diptera (Hinton 1968; Miller 1969).

5.5.1 Collembola

In the mediterranean type of climate of southern Europe and certain parts of Africa and Australia, many species of Collembola survive the hot, dry summer as drought-resistant eggs (Wallace 1968; Greenslade 1981). In the laboratory, Wallace (1968) found that eggs of *Sminthurus viridis* tolerated 2 weeks at 41 % RH and 16 or 30 °C. The eggs of this and other species are activated by winter rain, and development is completed during the wet season. In Provence in France, most Collembola pass the dry summer in the egg stage, and most adults die because of the low water content of the soil (Poinsot-Balaguer 1976).

As reviewed by Greenslade (1981) and Greenslade and Greenslade (1984), some species of Collembola are able to enter anhydrobiosis in the juvenile and adult stages. Anhydrobiotic Collembola are distinguished by a shrivelled appearance and empty gut. The shrinking is similar to the structural changes that take place in rotifers and tardigrades at the beginning of the desiccation process. Anhydrobiosis is known from several Australian species of *Folsomides* (Greenslade 1981; Greenslade and Greenslade 1984). In dry weather, the adults are inactive in the surface soil layer and litter, while resumption of activity after rain leads to high catches in pitfall traps. *Proisotoma brisbanensis* (Fig. 5.5) from Koonamore in South Australia, with a mean annual rainfall of 195 mm, lives deeper in the soil than other Isotomidae, and requires heavier rainfall to be reactivated.

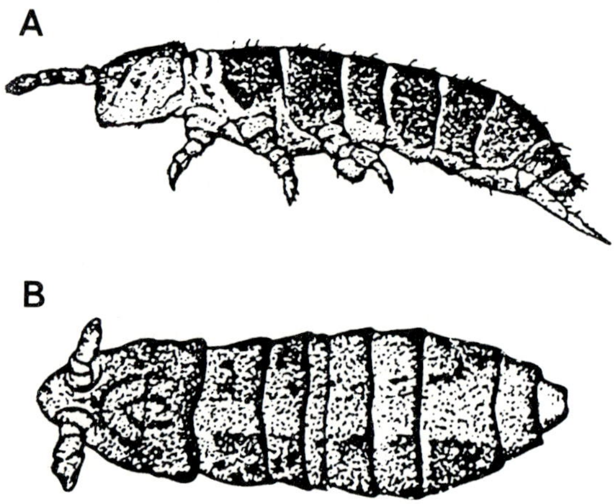

Fig. 5.5A, B. *Proisotoma brisbanensis*. **A** Active form **B** Anhydric form. (Greenslade 1981; drawing by J. Thurmer)

When the springtail *Folsomides variabilis* is subjected to slow, gradual desiccation, it forms small chambers in the soil, in which dehydration takes place (Poinsot 1968). Dehydration is accompanied by shortening and shrinking of the body. The first thoracic segment is hidden between the second segment and the head, and the first abdominal segment is covered by the third thoracic segment. The antennae are curved backwards on the head and thorax, and the legs stretched out along the central midline. The springtails may remain in the desiccated condition for several months, but are rapidly reactivated when water is added to the soil. Similar morphological modifications are also associated with anhydrobiosis in *Brachystomella parvula*, but the changes are reversed in both species as soon as the water content of the soil becomes more favourable (Poinsot--Balaguer 1976). During desiccation *F. variabilis* and *B. parvula* are able to resist a mean loss of 65 % of their total water content.

Anhydrobiosis is also accompanied by ultrastructural modifications at the cellular level in *Folsomides variabilis* (Barra and Poinsot-Balaguer 1977). In *Folsomides angularis* a hydrophile, labile substance is formed in connection with the transcuticular pores (Barra and Poinsot-Balaguer 1983), and this probably fascilitates the rapid absorption of water through the cuticle during rehydration.

Due to the fact that many Collembola are able to enter cryptobiosis, the number of species from desert areas may be underestimated (Greenslade 1981). Many investigators have used Tullgren funnel extraction and found that Collembola are restricted to the most humid sites of their studies. Since they will not respond to conventional extraction methods, anhydrobiotic species may not be recorded at all, and estimates of population densities may be far too low.

5.5.2 Chironomid Midges

The larvae of the chironomid midge *Polypedium vanderplanki* is the largest multicellular animal known to tolerate almost complete dehydration without ill effects (Hinton 1968). The larvae are found in shallow and exposed rock pools in Nigeria and Uganda. At the beginning of the rain season, the pools may be filled and dried out several times. During the dry season, they may also be occasionally filled with water for short periods. When the pools are dry, the larvae also dry out, and remain desiccated until the next rain. During the dry periods the upper 4-8 mm of the mud, where the larvae are found, may reach temperatures of 60 to 70 °C. Larvae of a related species, *Polypedium dewulfi*, live under similar conditions (Miller 1969), and appear to be identical with *P. vanderplanki* in their physiological mechanisms.

The degree of desiccation in larvae of *Polypedium vanderplanki* depends on the ambient humidity (Hinton 1968). At 60 % RH they have a water content of 8 %, but the content drops to 3 % at 1 % RH. When placed in water, the larvae are rapidly rehydrated until their normal moisture content of 85-90 % is achieved. Subsequently, they may start feeding within an hour.

According to Hinton (1968), anhydrobiotic larvae survived exposure to 106 °C for 3 h and immersion in liquid helium at -270 °C. In accordance with their natural environment, the larvae tolerated repeated desiccation and hydration in the laboratory. They were able to survive extended periods of anhydrobiosis. Larvae returned to water after 39 months of desiccation at room temperatures produced normal adults. Since the developmental stage, and accordingly the biological age of these larvae corresponded to only 3 weeks, a distinction has to be made between the biological age and calender age of cryptobiotic organisms (Hinton 1968).

5.6 Biochemical Aspects of Anhydrobiosis

Although anhydrobiosis is found in a variety of organisms, Crowe et al. (1992) have provided evidence to illustrate that the physical principles governing the stability of dry materials are universal. Their conclusion is based on the presence of high concentrations of trehalose in a number of organisms of different phylogenetic origin, such as bacteria, yeast cells, nematodes, brine shrimps and desert resurrection plants. The occurrence of trehalose in nematodes and tardigrades has been described above (Chaps. 5.2.5, 5.3.3). Less is known about biochemical adaptation to dehydration in the seeds of higher plants, but sucrose may be an analogue of trehalose.

Trehalose has been found to have a stabilizing effect on cell membranes and molecules. This has been discussed in a long series of publications by J.H. Crowe and his coworkers, and summarized in a recent review paper (Crowe et al. 1992). According to Crowe et al. (1992), morphological damage is inhibited in sarcoplasmatic reticulum isolated from muscles when dried in the presence of trehalose at concentrations near those found in anhydrobiotic organisms. Upon rehydration, the vesicles are able to transport Ca^{2+} at rates similar to those of fresh preparations. Similarly, when liposomes are dried with trehalose and rehydrated, vesicles are obtained that retain as much as 100 % of their original content. Those dried without trehalose leak all their content during drying and rehydration.

Trehalose may also stabilize phospholipids during dehydration, and Crowe et al. (1992) have developed a model to explain how this may be achieved (Fig. 5.6). Membrane phospholipids are hydrated and water may account for ca. 20 % of the weight of the hydrated bilayers. When water is removed, the packing of the head groups increases, and as a result the phase transition temperature (T_m) also increases greatly. For this reason, dry lipids may be expected to undergo a phase transition when placed in water at room temperature. Hydrated bilayers become temporarily leaky as they pass through a phase transition, and dry bilayers that are in the gel phase may be expected to leak during rehydration. The leakage is probably prevented by trehalose depressing the phase transition temperature of the dry lipids. This maintains them in the liquid crystalline phase in the absence of water.

Trehalose also preserves labile protein during drying (Crowe et al. 1992). Several molecules may stabilize proteins in solution, but only disaccharides are effective during extreme dehydration. As an example, proline stabilizes phosphofructo-kinase during partial dehydration. In contrast, if trehalose is added, the enzyme is completely stabilized, even after extreme dehydration.

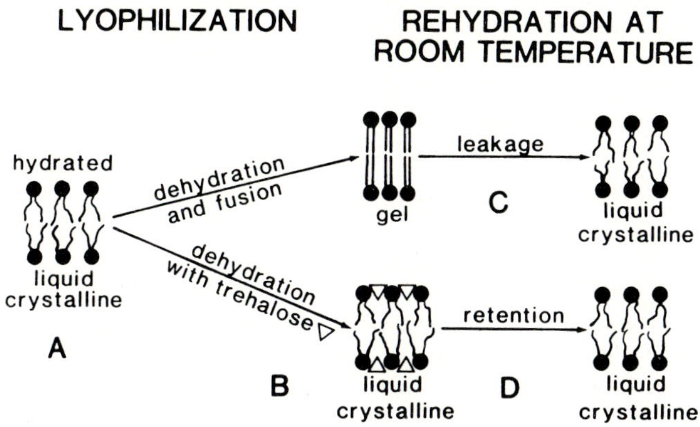

Fig. 5.6. Diagram illustrating the mechanisms by which trehalose is thought to stabilize dry phospholipid bilayers. It inhibits fusion during dehydration, but also reduces the gel to liquid crystalline transition temperature in dry phospholipids. As a result of the latter effect, passage through the phase transition during rehydration can be avoided. (Crowe et al. 1992)

It has been suggested that trehalose interacts by hydrogen bonding of -OH groups to polar residues in the protein, but it is not clear how this interaction leads to stabilization.

As pointed out by Crowe et al. (1990), the specific effect of trehalose during anhydrobiosis is in contrast with the low specificity of solute requirements for the stabilization of proteins during freezing. When cells or proteins in solution are frozen, a fraction of the water remains bound or unfrozen. The unfreezable water amounts to about 0.25 g H_2O/g dry weight. Many solutes are known to act as cryoprotectants (Chap. 9.4), and may inhibit destructive changes in solution properties during freezing and thawing. It is suggested that any solute that is preferentially excluded from the hydration shell of a protein is also a cryoprotectant. Thus, a number of polyols, sugars and amino acids, which are known to stabilize proteins in solution, also protect during freezing. Conversely, solutes such as urea, which preferentially binds to the protein, do not act as cryoprotectants.

In contrast with freezing, non-freezable water can be removed during dehydration. In this respect freezing and dehydration are fundamentally different stress factors (Crowe et at. 1990, 1992). For this reason, fundamentally different mechanisms are also required to prevent harmful effects. Stabilization of proteins during drying requires direct interaction with the protective substance. Several solutes used as cryoprotectants in biological cryopreservation have no stabilizing effect on proteins during extreme dehydration. The only solutes that do exhibit such properties are disaccharides, and among these trehalose is the most effective.

5.7 Conclusions

Many species of terrestrial invertebrates react to desiccating conditions by entering cryptobiosis, in which the metabolism is reduced to undetectable levels. Anhydrobiosis is the most common form of cryptobiosis, and is found in nematodes, rotifers, tardigrades, springtails and a few species of chironomid midges. The strategies of survival in these organisms are principally different from those of most drought-resistant terrestrial arthropods, in which methods of water conservation have evolved. Anhydrobiosis is an important adaptation to arid environments in small invertebrates, and makes survival possible under fluctuating moisture conditions. Anhydrobiosis is widely distributed among nematodes from a variety of habitats subjected to partial or extreme dehydration. It appears as an important way of survival in plant parasitic and free-living nematodes during periods of drought. Anhydrobiotic nematodes have been studied from hot desert areas as well as from cold habitats in the Antarctic. A slow rate of water loss is important for entering the dehydrated state. Means of impeding water loss include coiling and the formation of aggregations and cysts.

In tardigrades entering anhydrobiosis, a tun-like structure is formed by the contraction of the body. This is facilitated by structural adaptations of the cuticle. Slow dehydration is essential for tun formation in tardigrades, and wax extrusion on the cuticle surface may help to reduce transpiration. A fraction of 5-15 % of the initial body water is retained during anhydrobiosis. Although tardigrades are principally aquatic organisms, anhydrobiosis makes it possible for these animals to live in habitats with changing moisture conditions.

Collembola may spend dry periods as drought-resistant eggs, but several species enter anhydrobiosis in post-embryonic stages. During desiccation, shrinking in the springtails resembles the structural changes that take place in tardigrades. Anhydrobiotic Collembola are known from hot arid environments, but not from polar deserts. Finally, anhydrobisos is known from larvae of chironomid midges inhabiting temporary ponds in Africa. They are the largest invertebrates known to tolerate almost complete dehydration.

There are apparently large variations in the biochemical adaptations to anhydrobiosis in different invertebrates, but in many organisms a strong relationship to increased levels of trehalose are found. The protection given by this disaccharide is based on its stabilizing effect on cell membranes and protein molecules. This mechanism is fundamentally different from the cryoprotective effect of various solutes during freezing. Recent studies have shown that the presence of trehalose alone does not ensure anhydrobiotic survival in nematodes, but that slow rates of drying are required.

6 Behavioural Adaptations

6.1 Arthropods from Hot Deserts

6.1.1 Introduction

The importance of behavioural adaptations for the survival of desert invertebrates has been pointed out several times in the preceding chapters. Through their behaviour, many species are able to avoid the physical stress of hot and dry environments by seeking sheltered microhabitats. In addition to behaviour, physiological adaptations are required to avoid the risk of desiccation, while growth and development require modifications of the life cycles of many species. Behavioural adaptation alone is not sufficient for survival, and an array of survival strategies has been adopted by most desert species.

The microclimates in hot deserts may differ greatly within a few metres, and shelter is found underneath rocks and in the vegetation (Chap. 1.1.3). In spite of a dry surface, deeper layers of desert soils may contain considerable amounts of moisture. By digging burrows into the soil or under rocks, a number of species construct chambers in which the risk of desiccation is reduced. Some species remain in their burrows for long periods, while others come to the surface for short intervals during the day or night. To sustain growth and development, the utilization of resources is an essential requirement. At high temperature, energy requirements increase with increasing metabolism. To meet this demand, invertebrates have to exploit the energy resources that are available. The active life of many species is interrupted for long periods, such as those that occur during anhydrobiosis in nematodes, tardigrades and collembolans, or during diapause and aestivation in terrestrial arthropods. Reduced metabolic activity makes it possible for such species to avoid foraging and other activities during the hottest and driest seasons of the year.

Behavioural adaptations of desert invertebrates have been discussed by many authors. In the present section the more salient points of the behavioural adaptations of arthropods to hot desert are summarized, while the reader is referred to the comprehensive reviews by Crawford (1981) and Cloudsley- Thompson (1991) for more details.

6.1.2 Burrowing

Burrowing into the sand or soil is an effective way of avoiding extreme surface temperatures. In Arizona, the scorpion *Hadrurus arizonensis* retreats into burrows extending about 20 cm below the surface (Hadley 1970). In the summer, temperatures at this depth may fluctuate between 30 and 40 °C, while surface temperatures exceed 40 to 60 °C for most of the day. As pointed out by Cloudsley-Thompson (1991), the burrows of many desert arachnids and other terrestrial arthropods have a depth of 20 to 30 cm. At this depth, temperature fluctuates little on a daily basis, but is generally lower in winter than in summer. The moisture content of the soil increases with depth, which is of great importance to burrowing arthropods in enabling them to avoid desiccation. The air within the burrows of the desert woodlouse *Hemilepistus reaumuri* may remain continuously saturated at a depth of 30 cm (Coenen--Stass 1989). When the water content of the soil is low, the woodlice form aggregations that induce increased RH within their narrow chambers.

According to Cloudsley-Thompson (1991), Solifugae also depend on digging burrows for their survival, although they lack specific morphological modifications for this purpose. Removal of the substrate is performed by the joint action of the chelicerae and the legs. The burrows of Solifugae may reach a depth of 1 m, and are used not only for hiding during the daytime but also for ecdysis and deposition of the eggs.

Scorpions are often equipped with morphological adaptations that are suitable for digging (Polis 1990). The North American scorpion *Anuroctonus phaiodactylus*, for instance, uses its thick and heavy pedipalps for support while digging with the legs. In other scorpions, too, massive claws are used to form their deep burrows. Not all scorpions are able to dig: some inhabit cracks and crevices, or hide in plant litter on the surface. Some species also utilize burrows dug by other animals, such as lizards or small rodents.

Ants and termites are successful insects of desert regions, partly because of their ability to construct underground nests (Chap. 2.1.3). Due to the large number of individuals comprising a single colony, ants can excavate nests deep enough to provide high humidity and relatively constant temperature (Pierre 1958; Rissing and Pollock 1989). From their protected burrows, they forage over the surface, adjusting their behaviour in response to changes in the environment.

Termites may construct mounds above the surface in arid regions, but species living in the hottest areas tend to inhabit subterranean nests (Wallwork 1982). The mounds of *Amitermes* sp. in northern Australia are oriented in a north-south direction (Grigg 1973). In this way the largest faces of the mounds receive most heat from the sun in the morning and evening when the air temperature is low, while smaller areas are exposed to the sun at midday.

In addition to improved microclimatic conditions, the burrows of desert arthropods offer protection against predators and parasites. By hiding during daytime, many species may avoid heavy predation. Nocturnal activity, however, is not entirely without risks, since predators may be nocturnal as well.

As pointed out by Cloudsley-Thompson (1991), several desert arthropods are unable to dig burrows. Most tenebrionid beetles, for example, hide under rocks and stones, or dig into loose sand. They also enter the burrows of other animals, and some large species are known to shelter in the burrows of scorpions.

In conclusion, burrows are of great importance for the survival of many desert arthropods. They provide ameliorated microclimatic conditions and, to some extent, protect against predation. When surface conditions become more favourable, arthropods emerge from their burrows to resume foraging and other biological activities.

6.1.3 Circadian and Seasonal Rhythms

Emergence from their burrows and other hiding places by desert arthropods may be controlled both by responses to environmental factors and by biological clocks. Circadian rythms of locomotory activity appear to be particularly well developed in desert species (Cloudsley-Thompson 1991). Examples of circadian rhythms are found in many species of tenebrionids. Some day-active species are probably better adapted to tolerate exposure to high temperatures than nocturnal species. While most tenebrionids and carabids of inland hot deserts are nocturnal, more diurnal species may be found in cool coastal deserts.

Circadian rhythms are not constant throughout the year, but in many species change with the season. Day-active desert arthropods often show bimodal rhythms of activity during the hot summer months and a single midday peak in the cool season. As reviewed by Cloudsley-Thompson (1991), several tenebrionid beetles are known to seek shelter under stones and in the vegetation during the hottest months of the year, while surface activity can be observed at all times of the day in spring and autumn. According to Holm and Edney (1973), different species of desert insects in southwest Africa exhibit different seasonal and daily cycles. Seasonal climatic differences are so large that summer and winter faunas are almost entirely different.

The seasonal rhythms of *Hemilepistus reaumuri* have been studied in detail (Coenen-Stass 1989). In Algeria, no surface activity was observed in January when surface temperatures did not exceed 17 °C. In March, when temperature on the ground surface rose to 32 °C, the woodlice were observed in daylight and with a peak activity around noon. When daytime temperatures reached 70 °C in July, the woodlice emerged from their burrows only at dusk and shortly after dawn.

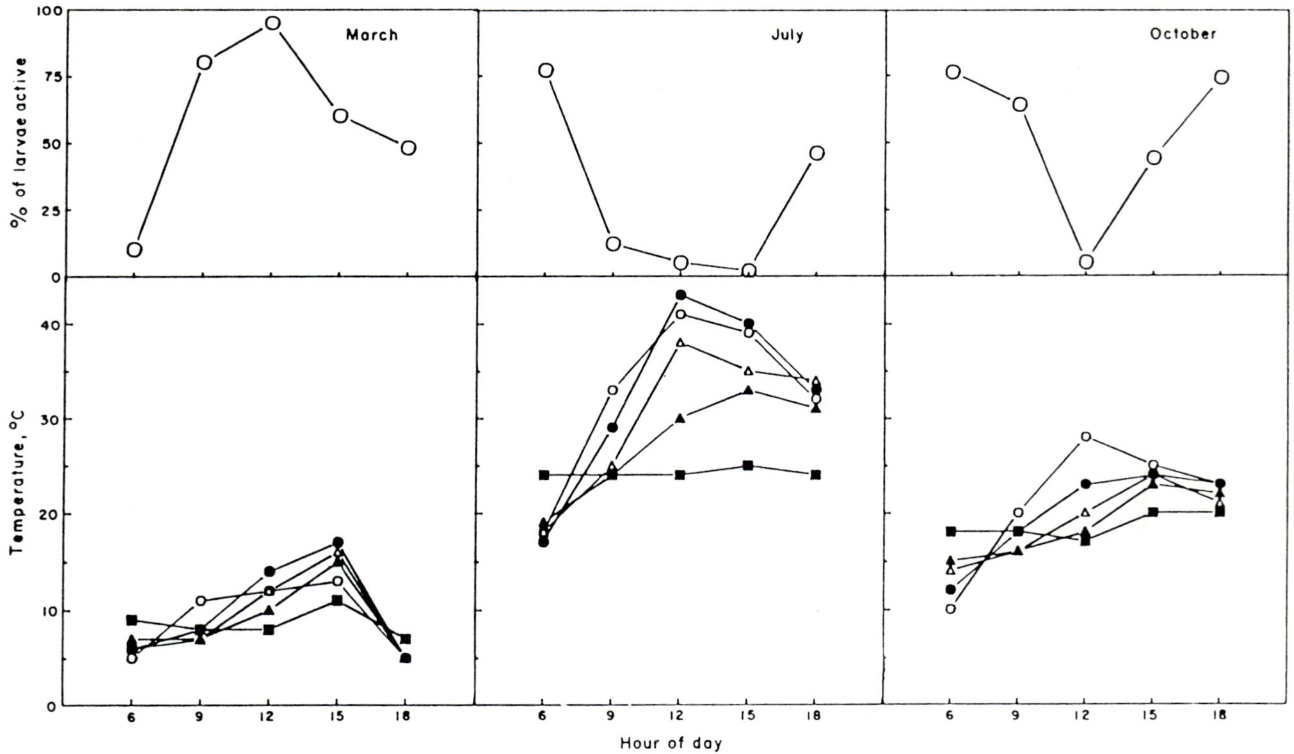

Fig. 6.1. Burrow temperatures and percent activity of tiger beetle larvae (*Cicindela marutha*) at 3-h intervals during the day in spring, summer and autumn. Data are based on 5-10 burrows, each on 2-5 dates per season. Temperature measurements: O 1 cm above ground; ● 3 cm below; Δ 6 cm below; ▲ 9 cm below; ■ 24 cm below ground. (Hadley et al. 1990)

Similarly, the third instar larva of the tiger beetle *Cicindela marutha*, which inhabits open areas with scattered vegetation in Arizona, shows great seasonal variations in activity (Hadley et al. 1990). The larvae dig burrows with an average depth of 30 cm, which provide a climatic gradient from the bottom to the surface. Water loss rates increase with increasing temperature and decreasing RH, and the larvae avoid desiccation by moving along the climatic gradient within their burrows. Maximum larval activity occurs at midday in March when surface temperatures average 13 to 17 °C (Fig. 6.1).

During the month of July, surface and upper burrow temperatures exceed 40 °C, and the larvae retreat to depths of which temperature and humidity are more moderate. Temperature measured at the bottom of the burrows remained approximately 24 °C throughout the day. In October, the larvae still exhibited a marked reduction in activity at midday, but activity was higher than in July. Temperatures in the burrows were intermediate in October, compared with those in March and July.

6.1.4 Thermoregulation

Behavioural thermoregulation is important in many terrestrial arthropods to maintain temperatures as close to optimal levels as possible under changing ambient conditions. When temperatures are low, e.g. in the morning or in the cooler season of the year, hot desert species utilize insolation to warm up. On the other hand, heat stress during periods of strong insolation is avoided by seeking shaded areas within their habitat. A primary method of thermoregulation in day-active desert arthropods is to shuttle between sunshine and shade. Selection of areas with suitable temperature conditions is often combined with altering the position of the body in relation to the sun's rays.

Many day-active species of tenebrionid beetles are known to shuttle between sun and shade (Cloudsley-Thompson 1991). Examples include *Onymacris* spp. and *Stenocara* spp. from the Namib Desert. *Onymacris rugatipennis* is active in sunny areas in the early morning when ambient temperatures are low, and is found primarily in the shade at higher temperatures in the middle of the day. *Stenocara phalangium* and *Onymacris* spp. have long stilt-like legs. When running, the body of *O. plana* is raised 1 to 1.5 cm above the hot ground, and this may represent a temperature difference of 15 °C (Holm and Edney 1973). This beetle is one of the swiftest-running arthropods ever recorded, but it is not known if it runs as a means of cooling its body.

According to Heatwole and Harrington (1989), several tenebrionid beetles from the Tunisian steppe are nocturnal during the summer and day-active in the winter months. The dark beetles use solar radiation to warm up and obtain body temperatures higher than air temperatures by 5 °C or more.

Thus, mean body temperature of *Pimelia obsoleta* increased to 28 °C, of *Blaps gigas* to 31.0 °C and of *Prionotheca coronata* to 33.9 °C.

Posture and orientation also play an important part in the thermoregulation of desert arthropods (Cloudsley-Thompson 1991; Heinrich 1993). Locusts orient their bodies at right angles to the sun's rays at low ambient temperatures in the morning (Uvarov 1966). In the middle of the day, they face the sun, thereby reducing the area of body surface exposed to the sun. At low temperatures, the nymphs of *Schistocerca gregaria* form aggregations - this probably reduces convective heat loss - and orient their bodies perpendicular to the sun's rays. At temperatures higher than 40 °C, they climb onto the vegetation, thereby enhancing heat loss.

According to Whitman and Orsak (1985), the grasshopper *Taenipoda eques* of the Chihuahuan Desert adopts a number of thermoregulatory postures. By "flanking" it exposes the broad side of its body to the sun so that it absorbs heat. "Stilting" is used to raise the body above the ground to avoid excessive heat, while "crouching" is used to absorb heat by pressing the ventral side of the body against a warm rock or some other substrate. Orientation along the shady side of plant stems is also employed to avoid insolation.

6.2 Polar and Alpine Terrestrial Arthropods

6.2.1 Introduction

Behavioural adaptations are presumably of great importance to terrestrial polar arthropods, although information on this topic is relatively limited. A striking feature of arctic areas is the disappearance of flying insects under unfavourable weather conditions. On calm and sunny days, however, swarms of midges and flies can be seen, butterflies are active in their selected habitats, and a number of insects visit flowers for nectar. Their behaviour allows the insects to take advantage of warm and calm periods at any time of the day (Downes 1965; Danks 1990). With the rapidly changing climatic conditions of polar areas, the opportunistic behaviour of arctic insects is of fundamental importance for development and reproduction.

6.2.2 Mating and Oviposition

Although there are exceptions, the formation of swarms in arctic nematocerous Diptera is restricted by the prevailing harsh conditions (Danks 1981). Thus, swarming is reduced and mating takes place on the ground in the high arctic empidid *Rhamphomyia ursinella* at Lake Hazen, Ellesmere Island (81°, 47' N). In contrast, *Rhamphomyia nigrita* from the same area has a mating sequence normal for empidids. Swarms of chironomids are frequently observed in many species (Oliver 1968), but sometimes close to the ground (Danks 1981). In mosquitoes, mating may take place without swarming, and the high arctic species *Aedes nigripes* and *Aedes impiger* at Lake Hazen are exceptional in that adult activity persists for several weeks. Both species are able to complete ovarian development without taking a blood meal.

In several species the adult life span is greatly reduced (Downes 1965; MacLean 1975; Danks 1981), and in this way the risks involved in mating behaviour are omitted. Two of the three most abundant craneflies at Barrow, Alaska have brachypterous females, but the males are winged and locate the females for mating. In females of *Pedicia hannai* not only the wings, but the legs, eyes, antennae and mouth parts are reduced as well, while the sexual organs are fully developed (Tjeder 1963). Alaskan tipulids emerge from the pupae with fully developed ova, copulate within hours and lay their eggs within 1 or 2 days. In the Norwegian alpine blackfly *Prosimilium ursinum* most individuals fail to emerge from the pupae, and the eggs are shed directly into the stream (Carlsson 1962).

Females of the flightless antarctic midge *Belgica antarctica* also mature their eggs in the pupal stage (Edwards and Baust 1981). Males aggregate at the open surfaces where eclosion takes place, and attempt to copulate soon after emergence of the females. Following insemination, the female discharge all her eggs at one site within 1 or 2 days. In surface aggregations of adult *B. antarctica*, males predominate throughout the emergence period with an overall sex ratio of 6:1. Although females live for up to 7 days and males for 9 days, this cannot account for the observed surface ratio, and the sex ratio before eclosion is close to 1:1. Edwards and Baust 1981) concluded that the male swarming behaviour, characteristic of winged chironomids, apparently is compressed onto the ground surface in this flightless antarctic species.

6.2.3 Microarthropods

Through their behaviour, microarthropods select the most favourable microhabitats, and humidity is a factor of great importance to most species. The antarctic mite *Alaskozetes antarcticus* forms clusters of adults and juveniles under rocks (Cannon 1986). In this way they are protected from desiccation and feed on microorganisms that need moisture for their growth.

In the chalikosystems around the Syowa Station in east continental Antarctica, Ohyama (1977) found that prostigmatid mites prefer sand with medium water content, and are not present when the ground is either too wet or too dry.

According to Booth and Usher (1985), several physical and chemical factors can influence the location of soil arthropod aggregations. In the dense moss turves of Signy Island in the maritime Antarctic, the vertical distribution of the two main species of Collembola was different. *Cryptopygus antarcticus* occurred predominantly in the upper 1.5 cm, and showed a preference for certain water contents. *Friesia woycienchowskii* was absent from the surface layer, but was distributed down to 9 cm or deeper. Percentage water content of the moss had little influence on the distribution of this species. Other abiotic factors may affect the distribution of both species, and predation could also influence their aggregations.

6.2.4 Behavioural Thermoregulation

The annual heat budget of the polar regions is small, and in terrestrial arthropods an increase from the average conditions may be of great ecological and physiological importance (MacLean 1975). Thermoregulation by endothermy, sun basking and choice of locally warm sites are methods by which insects are able to raise their body temperature.

6.2.4.1 Endothermal Thermoregulation

Many insects are able to raise their body temperature above the ambient temperature by internal heat production (Heinrich 1993). Endothermic heat regulation is found in bumblebees and a variety of moths, including sphingids, geometrids and noctuids. The heat production results largely from the thoracic flight muscles. During preflight warmup, antagonistic flight muscles are activated simultaneously, producing heat but little wing movement. During flight, heat is produced by the rapid contraction of flight muscles, and large moths achieve thoracic temperatures of 10 to 20 °C above the ambient during flight (Kingsolver 1985; Heinrich 1993).

Few arctic insects are known to have the ability of endothermic thermoregulation. The arctic bumblebee *Bombus polaris*, however, is of large size, has long dense hairs and dark colours (Richards 1973), and presumably depends on endothermal heat production, as do other bumblebees. At Lake Hazen, Ellesmere Island, *B. polaris* is able to fly at low temperatures, although peak flight activity corresponds closely to maximum diel air temperature. In the subarctic area of Abisco, Sweden, queens of several *Bombus* spp. forage for nectar from *Salix* at temperatures as low as -1 and -2 °C, while the workers become active at 2 to 6 °C (Lundberg 1980).

6.2.4.2 Thermoregulation in Butterflies

In butterflies, the principle way of thermoregulation is by behavioural orientation, taking up a posture relative to the position of the sun (Kingsolver 1985). Basking in the sun is a common form of adaptation in arctic butterflies (Downes 1965; Kevan and Shorthouse 1970; Danks 1981, 1990), as well as in alpine species (e.g. Kingsolver 1983, 1985; Shapiro 1985; Sømme 1989). It is a striking feature of many arctic areas that butterflies appear when the sun comes out, as in north-east Greenland, where several species are adapted to high arctic conditions (Böcher 1992).

Thermoregulation by basking in butterflies is closely associated with melanism. Dark pigments may be located at the base of the underside of the wings, or on the dorsal body and wing base surfaces (Fig. 6.2).

Fig. 6.2. Upper side of a fritillary butterfly showing melanistic colours of body and base of wings. (Photo L. Sømme)

Several kinds of basking positions are recognized in butterflies (Clench 1966; Danks 1981; Kingsolver 1985). In lateral basking, the wings are folded over the back. The butterflies lean to one side so that the sun shines directly on the melanistic underside of the wings. Arctic fritillaries (*Boleria* spp.) are dorsal baskers, in which the wings are fully opened and the dorsal surfaces of the thorax and wings are positioned perpendicular to the sun. Body basking is similar to dorsal basking, except that the forewings at least are opened only at a small angle. Finally, the basking posture in the white *Pieris* spp. butterflies has been described by Kingsolver (1985) as reflectance basking. In this case the dorsal thorax is oriented towards the sun and the white wings function as solar reflecting plates. The region of wing that contributes to the reflection of radiation to the body is smaller at large wing angles than when the wings are held at a small angle.

Colias butterflies are lateral baskers and the role of wing colour and orientation in their thermoregulation has been reviewed by Kingsolver (1985). According to Hovanitz (1950), parallelism of phenotypes is found in the dark and sombre colours of all *Colias* spp. from the Arctic and from alpine areas wherever the weather is predominantly cold. In the high Arctic, in *Colias hecla*, which is found at Lake Hazen (Kevan and Shorthouse 1970), the ventral surfaces of the wings, particularly the hind wings, are darker than the bright yellow dorsal surfaces. In basking, the butterflies make use of one ventral hind wing surface at a time.

The increased melanism of *Colias* spp. in colder habitats makes it possible for them to achieve body temperature excesses above air temperatures of 15 to 80 % greater than lighter forms. This is reflected in increased flight space, and the range of meteorological variations at which flight can occur, in *Colias meadii* at 3600 m in the Colorado Rocky Mountains as compared with *Colias philodice eriphyle* at 1700 m (Kingsolver 1983, 1985). Alpine *Colias* butterflies have not become adapted by lowering the temperature limits of their range of voluntary flight activity, and species from the Colorado Rockies are unable to fly at body temperatures below 30 to 40 °C. By placing *Colias* spp. males in a fixed basking position, Kingsolver (1983) found that their body temperature increased rapidly with solar radiation, and dropped in shade. The temperature fluctuations of fixed specimens corresponded closely to the flight activity of free-living butterflies.

6.2.4.3 Contact Warming

In arctic butterflies, basking is often combined with ground contact, securing heat from a warm substrate. At Lake Hazen, *Boloria* spp. and *Plebius aquilo* use bare ground for contact warming (Kevan and Shorthouse 1970). Observations on the activity of high arctic butterflies show that they prefer warm and relatively windless sheltered places. On sunny days, they fly in the warmest air close to the ground, such as in gullies and creeks and in small depressions in hillsides.

Several other terrestrial arthropods actively select sites with favourable temperature conditions. The corollas of arctic plants, such as *Dryas integrifolia* and *Papaver radicatum*, act as solar reflectors in sunshine (Kevan 1975). Heat is concentrated on the sporophylls of the flowers, and temperatures within the corollas may be 5 to 8 °C higher than ambient temperatures. Basking insects are frequently to be observed in the flowers and, since they remain there for a considerable time, their increased heat budget may speed development of their oocytes. Among insects visiting the corollas of *Dryas*, Kevan (1975) measured considerable temperature increases in *Aedes* mosquitoes, the empidid *Rhamphomyia nigrita*, the syrphid *Carposcalis carinata* and the blowfly *Boreellus atriceps*. In general, the capacity for warming is greater in dark and hairy insects. The arctic lygaeid bugs *Nysius groenlandicus* show a similar form of behaviour in which the females achieve extra heat from the insolated seed heads of their host plants (Böcher 1975, 1976). Arctic plants, such as *Saxifraga oppositifolia*, which form characteristic dense cushions in arctic areas, have temperatures well above ambient when the sun is shining, and provide heat and shelter for invertebrates.

At Lake Hazen, *Aedes* mosquitoes oviposit around midday at sites that are directly insolated (Downes 1965; Danks 1981,1990). In this way the eggs are placed on south-facing banks where they will experience earlier snowmelt and higher temperatures the following spring, when the eggs hatch. Early egg hatching increases the chances of completing the life cycle during the short arctic summer. Mosquito larvae aggregate in the warmest parts of pools, and move clockwise around the pool during the day as the warmest point changes according to the changing position of the sun.

6.2.4.4 *Gynaephora groenlandica*

The woollybear caterpillar *Gynaephora groenlandica* is endemic to Greenland and the northernmost Canadian Arctic Islands. At Alexandra Fiord, Ellesmere Island, Kukal and Kevan (1987) found that this lymantriid moth has a life cycle with a duration of up to 14 years (Chap. 7.1.2). The larvae cease foraging prior to mid-summer, and this apparently contributes to the prolonged life cycle. During the summer two-thirds of the larvae are killed by parasitism, and it has been suggested that feeding is abrupted because the larvae have to seek shelter from the parasites.

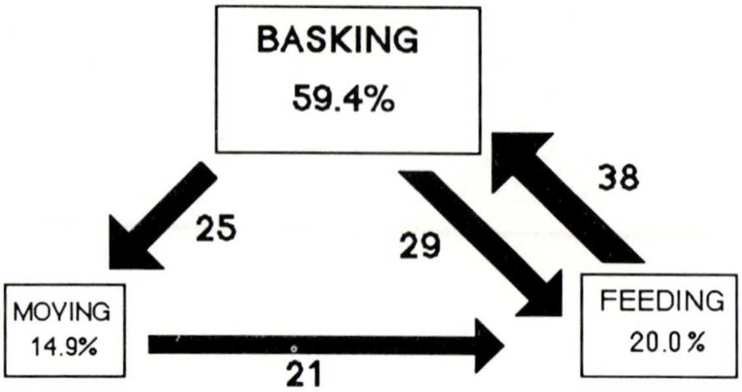

Fig. 6.3. Frequency and sequential organization of major behaviours in larvae of *Gynaephora groenlandica*. Based on observations of eight larvae. *Relative sizes of squares and arrows* indicate the percentage of time spent in a particular behaviour *(within squares)* and also the number of times one behaviour followed another *(arrows)*. (Kukal et al. 1988)

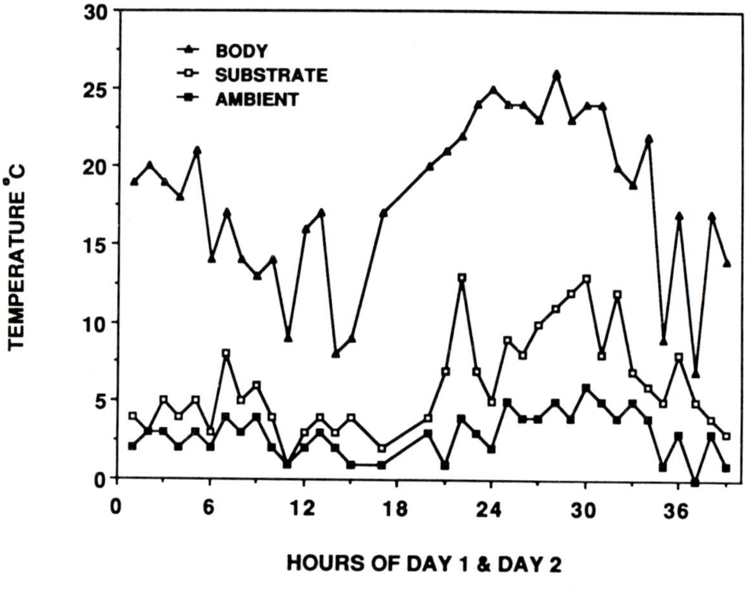

Fig. 6.4. Diel changes in mean body temperature of eight larvae of *Gynaephora groenlandica* compared with ambient and substrate temperatures recorded on an overcast day followed by a sunny day on the open tundra at Alexandra Fiord, Ellesmere Island. (Kukal et al. 1988)

During periods when the larvae are feeding actively, their body temperature is greatly increased by basking in the sun. Kukal et al. (1988) showed that approximately 60 % of their time is spent in basking, while only 20 % is used for feeding (Fig. 6.3). The remaining time is occupied by moving from one place to another, and by short periods of immobility in the shade.

During basking, the larvae orient their bodies towards the sun. The body temperature is raised considerably above the ambient temperature and the temperature of the substrate on which the larvae sit (Fig. 6.4). Changes to moving and feeding cause considerable drops in body temperature. The effect of basking is an increased larval growth rate and efficiency of assimilation although assimilation efficiency decreases with increased maintenance metabolism and ingestion rates at high temperatures. Kukal et al. (1988) concluded that the larvae maximize their assimilation efficiency by selective and seasonal feeding, and also by raising their rate of digestion while basking. By cessation of feeding early in the season, the larvae may avoid an energy deficit resulting from high maintenance metabolism when the quality of their food plant, *Salix arctica*, later declines.

6.2.4.5 Thermoregulation in Alpine Grasshoppers

There are few arctic grasshoppers, but thermoregulation has been studied in several species from cold alpine areas. Chappel (1983) compared *Melanoplus sanguinipes* from alpine meadows at 3800 m in White Mountains with *Trimerotropis pallidipennis* from Coachella Valley in California. Both species exhibited postures typical of thermoregulation in grasshoppers; crouching against a warm substrate to gain heat, or stilting on their legs when temperatures are too high. In the field, *T. pallidipennis* usually stilted to avoid the extreme temperature of the desert surface. At high altitudes, *M. sanguinipes* would sit quietly with its body axis perpendicular to the sun's rays, and was frequently found in a crouched position. Despite considerable differences in environmental temperatures, both species maintained relatively constant body temperatures throughout the day. While the environmental temperatures differed by at least 20 °C, the body temperatures of *T. pallidipennis* averaged only 7.7 °C higher than those of *M. sanguinipes*. This is reflected in the relationship between body and air temperatures at the site of capture (Fig. 6.5). The temperature excess was much higher in *M. sanguinipes* than in *T. pallidipennis*. While *M.sanguinipes* remains in the open areas in sunshine and gains heat by crouching, *T. pallidipennis* avoids excessive insolation and reduces body temperature by stilting.

In the Colorado Rockies the two acridid grasshoppers *Arphia conspersa* and *Trimerotropis suffusa* are sympatric at about 3100 m (Gillis and Possai 1983). Differences in their ability to thermoregulate and their different reactions to ambient temperatures make it possible for them to utilize the same resources at different times of day. The thermal niche breadth was found to be broader in *A.conspersa* than in *T. suffusa*, but due to its larger size and more sophisticated thermoregulatory repertoire, *T. suffusa* attains higher body temperatures. To maximize the input of heat, both species orient the long axis of their bodies perpendicular to the sun's rays. The males use noisy display flights as part of their pre-mating activities. Sexually displaying males of both species thermoregulate better than non-displaying males. Increased flight activity as a result of thermoregulation will increase reproductive fitness in intraspecific competition for females. Thermoregulation has also been observed in the grasshopper *Aeropedellus clavatus* at 3690 m in the Colorado Rockies (Hadley and Massion 1985).

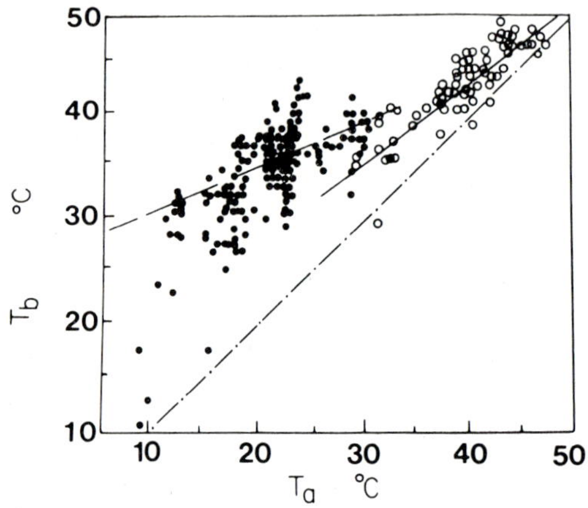

Fig. 6.5. Relationship between body temperature (T_b) and air temperature (T_a) at the site of capture of *Trimerotropis pallidipennis (open circles)* at 250 m in Deep Canyon and *Melanoplus sanguinipes (closed circles)* at 3800 m at Barcroft. *Dashed and dotted line* represents $T_b = T_a$. (Chappell 1983)

6.2.5 Avoidance of Heat and Cold by Alpine Tropical Arthropods

High variability of daily changes, contrasting with the relatively low variability of yearly cycles, is a major feature of tropical mountain climates. At high altitudes in equatorial mountains, high daytime temperatures alternate with freezing night temperatures (e.g. Azocar and Monasterio 1980; Sarmiento 1986). These diel cycles are more pronounced during the dry season, and minimum temperatures decrease with increasing altitude. Due to the thin and clear air at high altitudes, solar insolation is strong during daytime. Similarly, radiation during the night is faster than in the lowlands, resulting in ground surface temperatures several degrees centigrade below freezing.

The amount of precipitation in tropical mountains varies greatly, contrasting from alpine deserts to areas with daily rainfall. In some areas, such as the paramos of the northern Andes Mountain Range, low air humidities are encountered during the dry season when evaporation exceeds the amount of precipitation (Azocar and Monasterio 1980).

To insects and other terrestrial arthropods the fluctuating climatic conditions at high altitude in tropical mountains offer special problems of adaptation. To survive in exposed habitats, many species have to both tolerate freezing temperatures during the night, and be able to avoid desiccation during daytime. Unlike temperate species, tropical alpine insects have no time for preparation before the onset of low temperature (Sømme 1989). Actively feeding animals are particularly vulnerable to freezing due to the presence of ice-nucleating agents (Chap. 9.3.5) in the gut. High ground surface temperatures and low RH of the air during daytime represent other critical factors for many species.

As pointed out by Sømme (1989), most tropical alpine arthropods do not possess a sufficient degree of cold hardiness to survive exposure to the ambient night temperatures. In many species, the degree of resistance to desiccation is also restricted. Another strategy or manner of adaptation found in such species is to escape from the temperature extremes by hiding under rocks or seeking sheltered microhabitats within the leaves of plants. Several species occupy such protected habitats permanently. In plants with dense layers of leaves, ambient temperature fluctuations are greatly buffered (Fig 6.6). Temperatures within the leaves remain constantly low, restricting the activity and rate of development of insects and other terrestrial arthropods. Other species are active in the open during favourable periods, and seek protected microhabitats only to avoid extreme cold and drought.

6.2.5.1 African Mountains

Few studies have been carried out on the adaptation of terrestrial arthropods to the diel temperature fluctuations of tropical mountains, but Sømme and Zachariassen (1981) found that several beetles and other insects at 4200 to 4600 m on Mount Kenya showed no special adaptations to low temperature. Apparently, these insects spend the nights in protected habitats beneath stones and fallen trunks of plants. On this mountain, the senescent leaves of rosette plants of *Senecio* spp. and the drying inflorescence of *Lobelia teleki* also offer protection. Chironomid larvae, however, living in rain water collected in the *Senecio* rosettes, survived freezing during the night (Table 9.1), while the supercooling ability of some collembolans and heteropterous bugs was sufficiently high for them to avoid freezing. At 4985 m, where vegetation is scarce and no protection against cold is offered, the aphid *Branchycaudus helichrysi* supercooled to -25 °C.

In three species of tenebrionid beetles collected at 2000 to 2500 m on Pico del Teide, Tenerife, Ottesen and Sømme (1987) found very low rates of water loss (Fig. 3.3), while the carabid beetle *Calathus ascendens* relies on sheltered habitats in plant litter to avoid desiccation. The beetles are sufficiently cold-hardy to survive low temperatures during the night at these altitudes, but may seek protected microhabitats during extended periods of cold weather during the winter.

6.2.5.2 Andean Mountain Range

The paramos in the northern part of the Andean Mountain Range are high altitude-belts between the forest at about 3000 m and the perpetual snow fields above 4500 to 4800 m (Monasterio 1980). In these areas the giant *Espeletia* spp., rosette plants, offer protection against extreme climatic fluctuations for a large number of terrestrial arthropods (Fig. 6.6).

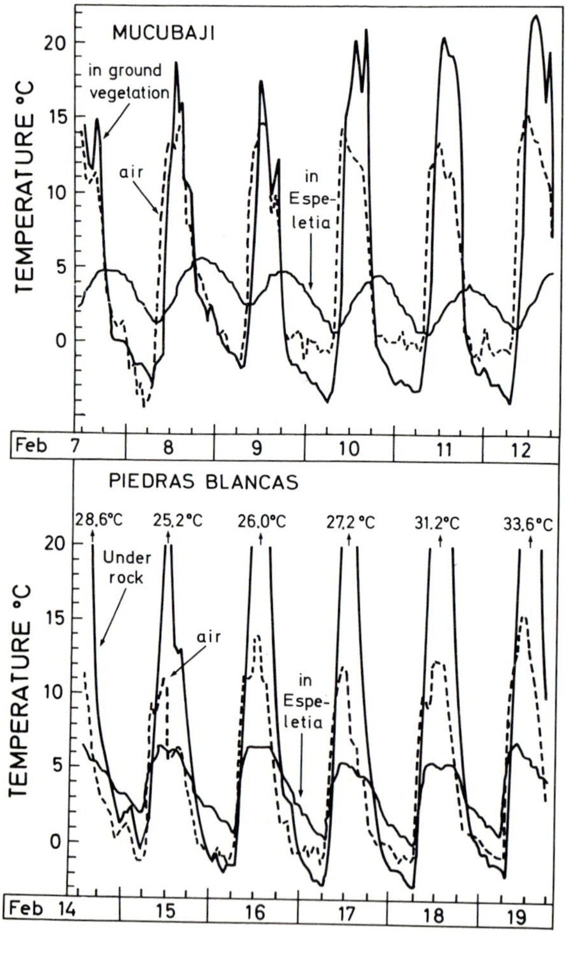

Fig. 6.6. Daily temperature fluctuations in senescent leaves of *Espeletia* spp. at ground level and in the air 15 cm above the ground. Data are hourly recordings from the paramos of Mucubaji and Piedras Blancas, Venezuela in February 1984. (Sømme 1986c)

According to Sturm (1978), the fauna of layers of senescent leaves surrounding the stems of these plants consists of about 90 % Acari, 3 % Collembola and 5-7 % of other arthropod species. The number of arthropods in the leaves of one plant with a stem 1.2 m high was estimated to be 130 000 (Garay 1981).

At the paramos of Mucubaji (3,600 m) and Piedras Blancas (4200 m) in Venezuela, the temperatures fluctuate between 20 to 25 °C during the day and minimum temperatures in the range of -5 to -12 °C have been recorded (Azocar and Monasterio 1980). Within the moist senescent leaves of *Espeletia* spp. the temperature never drops below 0 °C (Smith 1979). Recordings at Piedras Blancas carried out by Sømme (1986c) clearly illustrate how the diel temperature fluctuations are buffered within the senescent leaves (Fig. 6.6).

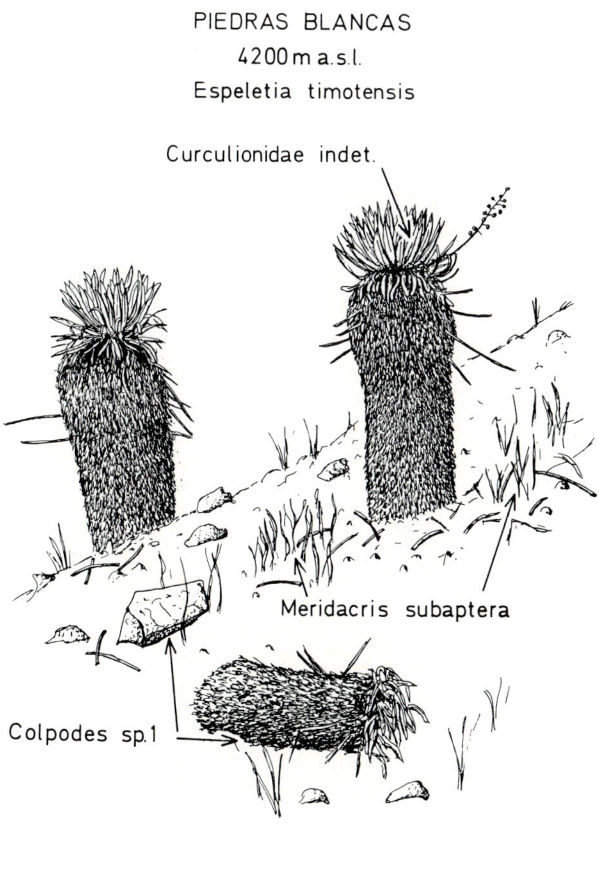

Fig. 6.7 Microhabitats of terrestrial arthropods in the Andean paramos. Some species find sheltered habitats in the senescent leaves of *Espeletia* spp. or under fallen trunks and rocks. Others, such as the grasshopper *Meridacris subaptera* survive in short alpine grass. (Sømme 1986c)

Different insects from this area react differently to low temperatures and desiccation (Sømme 1986c). Two carabids, *Colpodes* spp., sheltering under fallen stems of *Espeletia* (Fig. 6.7), were unable to tolerate freezing temperatures in the open ground. Their high rates of water loss (Fig. 3.4) also suggest that they would succumb to desiccation on warm days. Contrary to the carabid beetles, the acridid grasshopper *Meridacris subaptera*, which inhabits the short grass between the giant rosette plants (Fig. 6.7), shows degrees of freezing tolerance and resistance to dehydration sufficient to survive the ambient conditions.

At Pico Espejo (4750 m) in the Venezuelean Andes, Edwards (1986) observed that several species of arthropods seek shelter in the scree and thereby avoid extreme temperature fluctuations. Some of the carabid beetles were seen foraging after dusk when the surface temperature of the rocks had decreased, and continued to do so until the surface temperature approached 0 °C.

Fig. 6.8. Dry habitat of carabid beetles at 4500 m at Mount Chimborazo, Ecuador, characterized by scattered vegetation and dry soil. (Photo L. Sømme)

At elevations of 4250 to 4800 m on Mt. Chimborazo at 1°S in Ecuador, carabid beetles also appear to be active after sunset, but probably before nightly minimum temperatures are reached (Sømme, unpubl.). Hourly recordings at 0.5 cm below the surface in soil and sand gave a minimum of -4 °C during the night and a maximum of 50 °C during daytime. At a depth of 2 to 5 cm, fluctuations were greatly ameliorated. These recordings were conducted for only 1 week in March 1994, and lower temperatures may be expected on clear nights at other times of the year.

Three species of carabid beetles from Mt. Chimborazo were not particularly well adapated to low temperatures. Their mean supercooling points (Chap. 9.2.1) were in the range of -4 to -6 °C, and they were killed by freezing at these temperatures. When exposed to 5 % RH and 18 °C in the laboratory, their rates of water loss were relatively high (Fig. 3.4). Apparently, the beetles would not be able to survive the extreme heat and drought of the soil surface during daytime. In accordance with these findings, the carabid beetles were never encountered in the open during the day. All species were found in moist sand under rocks and at the base of cushion plants. During the night, all three species were caught in pitfall traps placed in the open spaces between scattered plants (Fig. 6.8). The carabid beetles of Mt. Chimborazo are apparently active during the early part of the night, before the temperature becomes too low. Presumably, their nocturnal activity is important for hunting as well as for other activities, such as mating and oviposition.

7 Phenological Adaptations

7.1 Life Cycles

7.1.1 Hot Desert Invertebrates

Growth and development among the terrestrial invertebrates of hot desert areas are limited by extreme temperature, low precipitation and associated reduced food availability. Breeding and the initial production of offspring are particularly vulnerable to the desert climate and in many invertebrates fecundity is reduced during drought years or the dry season (Riechert 1979).

In desert invertebrates, breeding is typically associated with rain. The adult stage of many insect species is less active during the hot season and the life cycle is continued by less active stages living in more protected habitats. As pointed out by Cloudsley-Thompson (1991), a vernal or rain fauna appears at the time of inflorescence when the desert is transformed by an abundance of plant and animal life. The flowers are visited by species of Lepidoptera, Coleoptera and Diptera, and predators feed on an abundance of food. The density of the scorpion *Paruroctonus mesaensis* in California is highly correlated with insect abundance (Polis 1980).

The activities of many ground-living terrestrial arthropods are increased by rainfall. Tenebrionids of the Namib and Kalahari Deserts have higher densities 2-3 months after the rainy season (Louw 1983). The leaf-cutter ant *Agromyrmex versicolor* of the Chihuahuan Desert embarks on reproductive flights (Riechert 1979), and Namib Desert ants show an outburst of foraging and nest construction after rainfall (Marsh 1988). In the Sahara Desert the giant velvet mite *Dinothrombium tinctorium* appears after rain (Cloudsley-Thompson 1962). The mites are day-active while the sand is moist, but retreat to burrows under dry conditions. In the spider *Agelenopis aperta* females of populations from New Mexico lay their eggs just prior to the rains (Riechert 1979). Most of the adults are killed by flooding during the summer, but the eggs survive. The spiderlings emerge concurrently with an increase in the abundance of prey following rainfall.

The life cycles of desert arthropods are regulated by seasonal changes in photoperiod and temperature as well as by rainfall. Dormancy with decreased activity and metabolism is common in the life cycles of many species (Sect. 7.2).

As pointed out by Crawford (1981), desert species have in one way or another solved the general problem of survival, although there are great variations in their life cycle strategies. Deserts are risky places in which to reproduce, and well-adapted life histories take advantage of chance events. It may be useful to spread the risk of reproduction over time and combine reduced breeding with an increase in longevity and reproductive life span. Although many species have life cycles in accordance with these principles, many others are short-lived, and have evolved different strategies.

According to Walwork (1982), the ideas embodied in the concept of r and K selection can be applied to certain groups of desert soil fauna. It appears that the unpredictable desert environment produces both r and K strategists. The desert wood-louse *Hemilepistus reaumuri* and certain species of oribatid mites represent examples of r strategists, while scorpions are good examples of K strategists. In other groups of arthropods, this generalization cannot be applied. Among millipeds, wood lice, collembolans and tenebrionid beetles, some species are probably r strategists and others are K strategists. Some species appear to occupy positions in between. Finally, there are taxa that do not seem to fit easily into either group. In the harsh desert environment several strategies may be required to favour the adaptations required for survival.

In reviewing life cycle strategies in desert invertebrates, Crawford (1981) distinguished between species with multivoltine, univoltine and those with prolonged life cycles. This approach is also adopted in the present discussion.

7.1.1.1 Multivoltine Life Cycles

An array of terrestrial desert arthropods may have multivoltine life cycles (Crawford 1981), but most species have not been studied in detail. Taxa with the potential of more than one generation per year include those that have high turnover rates in moderate climates. Mirid bugs, aphids, thrips and many small Lepidoptera belong to this group. Among beetles, coprophagous and necrophagous species as well as small carnivorous species, such as staphylinids, may be multivoltine. Among other taxonomic groups, prostigmatid mites and springtails may also have multivoltine life cycles.

Well-known examples of multivoltine desert insects include species with the capability of rapid maturation and development. Flies and other insects with larvae that feed on carrion and dung belong to this category. The Australian bushfly *Musca vetustissima* is known for its ability to colonize grazing pastures in the spring, and to complete development from egg to maturity in 5 days (Hughes et al. 1972).

Dispersal is important for multivoltine species to escape deteriorating food resources and find new ones (Crawford 1981). Adults of *Drosophila nigrospiracula*, whose larvae feed on bacteria and yeast in necrotic saguaro cacti, have exceptionally high rates of dispersal (Johnston and Heed 1976).

In conclusion, multivolitine species may have the capacity to synchronize a brief life span with ephemeral resources, combined with rapid breeding, dispersal or dormancy under unfavourable conditions and rapid assumption of activity when appropriate conditions return (Crawford 1981).

7.1.1.2 Univoltine Life Cycles

Many desert invertebrates, including isopods, solifugids and several species of spiders, have annual generations (Crawford 1981). Among the insects, grasshoppers are often univoltine, as are also a number of beetles, hemipterans and neuropterans. A reduction in larval instars appears as an important adaptation to arid conditions in the carabid *Thermophilum sexmaculatum* from the Sahara Desert (Paarmann 1979). Moisture is available for only short periods and the larvae pupate in the first or second instar. Only the adults are able to survive the long dry season of the desert.

The timing of reproduction is important for development and survival in univoltine species. Both migration and diapause are strategies that coordinate reproduction with the presence of food resources and availability of oviposition sites. Brooding of the young may increase reproductive success, and parental care is found in univoltine desert spiders and isopods (Crawford 1981). Female solifugids deposit their eggs in burrows and the newly hatched young are nearly immobile and incapable of feeding until they moult. Females of African solifugids probably remain with their offspring for some time (Cloudsley-Thompson 1977).

Species of *Hemilepistus* (Isopoda) are univoltine. They have a complicated pattern of behaviour and social life. Survival is associated with the ability to dig burrows in which they spend the winter and escape excessive heat during summer (Chap. 6.1.2). Their burrows, brooding and other forms of interactive behaviour apparently reduce the risk incurred by inhabiting an unpredictable desert environment (Crawford 1981).

7.1.1.3 Prolonged Life Cycles

According to Crawford (1981), prolonged life cycles are found among herbivores such as desert snails, millipedes, cryptostigmatid mites and insects, and among carnivorous species such as spiders, centipedes and scorpions.

Some of the long-lived insects are iteroparous, e.g. thysanurans and several tenebrionid beetles that can live as adults for many years. On the other hand, beetles such as cerambycids, bupestids and scarabaeids may have short adult lives, and spend most of their time in the larval stage feeding on roots and wood.

Iteroparous breeding is also common among other desert invertebrates (Crawford 1981). Pulmonate snails have long life spans and lay small clutches of eggs spread out over several years. *Sphincterochila zonata* and *Sphincterochila prophetarum* from the Negev Desert live for more than 10 years, and the North American *Sonorella* spp. for 15-20 years. Similarly, large spirostreptid millipeds from arid regions probably breed for a number of summers, while *Orthoporus ornatus* from New Mexico may live for 10 years.

Dormancy is common in herbivorous and detritivorous desert invertebrates with prolonged life cycles. As an example, the Moroccan pulmonate snail *Otala lactea* can remain dormant for 2-4 years (Machin 1967). At high temperatures, lengthy dormancy may result in the depletion of energy reserves. In addition to reduced metabolic rates during dormancy, a switch from carbohydrate to protein and lipid metabolism has been recorded (Crawford 1981). Tissue catabolism during aestivation is found in the millipede *Orthoporus ornatus* and the Texan snail *Bulimulus dealbatus*.

Among predaceous arthropods with prolonged life cycles, several species have evolved some degree of brooding (Crawford 1981). Maternal care may account for the small egg clutch sizes found in scorpions, centipedes and wolf spiders, many of which are iteroparous.

Large predaceous arthropods often require a long time to reach maturity. An example is afforded by the scorpion *Paruroctonus mesaensis* which needs 19-24 months before it is ready for reproduction (Polis 1980). Having reached the adult stage, the females can live for at least 5 years more. Equally long life spans have been reported in scolopendromorph centipedes (Lewis 1972), and the North American wolf spider *Lycosa carolinensis* (Shook 1978).

As pointed out by Crawford (1981), availability of food resources for long-lived desert carnivores is irregular. Indirectly, chances of capture are associated with the rainfall that results in the appearance of potential prey. The adaptations of carnivores to life in the desert include the ability to live for long periods without food. Dormancy, from which arousal is difficult, would be a disadvantage for these animals. When prey is available, carnivores must respond quickly in order to utilize the opportunities for growth and reproduction.

7.1.2 Polar Species

In polar areas the season for growth and development in terrestrial arthropods is limited to a few summer months. Long and severe winters put an almost complete stop to the activity of all forms of life, perhaps with the exception of fresh-water invertebrates below the ice of lakes and rivers. Even during the short summers, low temperatures reduce the speed of development. In arid polar and alpine areas this is further enhanced by lack of moisture. Indirectly, cold weather and lack of moisture also inhibit growth by restricting the availability of food resources. In some cases, parasitism may greatly limit the active feeding of arctic insects (Kukal and Kevan 1987).

To comply with conditions like these, selection for modified or specialized life cycles has occurred in most, if not in all, terrestrial arthropods. While most temperate and tropical species have one or more generations per year, polar species often require 2 or more years to complete their life cycles. In some species, univoltine life cycles are maintained as a result of special adaptations, but multivoltine life cycles are rare, or more probably non-existent. The speed of growth and development in many species is increased by local microclimatic conditions, e.g. by the warming of their habitats during periods of insolation. In the case of several fresh-water and soil-inhabiting arthropods, life cycles are flexible, being shorter under less severe than under more extreme climatic conditions.

The literature on the life cycles of polar and alpine terrestrial arthropods has been reviewed by several authors (e.g. Downes 1965; Mani 1968; MacLean 1975; Danks 1981; Block 1984a; Sømme and Block 1991). Although the life cycle strategies of many species have been studied, full details are missing in most cases. Danks (1981) lists a number of species in which the life cycles have been estimated to have a duration of several years.

7.1.2.1 Prolonged Life Cycles

A prolongation of life cycles to cope with the harsh environment is commonly found in arctic species (Downes 1965; MacLean 1975; Danks 1981) and in several microarthropods from the Antarctic (Block 1980, 1984a; Burn 1981, 1984). Similarly, many alpine species have prolonged life cycles (Mani 1968; Danks 1981; Sømme 1989). In the Arctic, the proportion of species that takes more than 1 year to development increases with latitude.

In polar regions, the life cycles of Collembola and Acari may be exceptionally long. At Truelove Lowland, Devon Island, Addison (1977) found that the collembolan *Hypogastrura tullbergi* has a 3-5-year life cycle, compared with 2-12 months in temperate regions. The species is able to overwinter in any developmental stage, and its long-term breeding success does not depend on the environmental conditions of any particular year. Similar observations have been made by Burn (1981) for the widespread antarctic collembolan *Cryptopygus antarcticus*, which overwinters in all post-embryonic stages at Signy Island in the Maritime Antarctic. The species undergoes up to four moults during summer and may reach maturity in the third year. Individual longevity was estimated to be between 3 and 7 years. In contrast, another antarctic species, *Parisotoma octooculata*, which has synchronous oviposition, overwinters in the egg stage and in one of the last instars (Burn 1984). Maturity is achieved in its second year of life. It is suggested that *Cryptopygus antarcticus* has evolved a more flexible life strategy suitable for unpredictable environments, whereas *Parisotoma octooculata* exhibits characteristics that are more common among temperate springtails. In *Cryptopygus sverdrupi*, which is known only from cold and arid inland nunataks in Dronning Maud Land, the life cycle is thought to last for up to 12 years (Sømme 1986a).

A number of antarctic terrestrial mites have prolonged life cycles. Examples include the prostigmatid mites *Ereynetes macquariensis*, *Nanorchestes berryi* and *Eupodes* spp., all of which have 2-3-year life cycles in the Maritime Antarctic (Booth and Usher 1986).

All post-embryonic stages of the antarctic oribatid mite *Alaskozets antarcticus* overwinter, and a mixed population of larvae and nymphal instars is found in the field (Block 1980). Individual mites usually spend 1 year between moults, which are relatively synchronized and take place early in the summer (Convey 1994). The adults do not oviposit until their second summer, and may live for more than 2 years. Iteroparous breeding has been observed in adults with an extended life span (Block 1980). Based on these studies, Convey (1993) concluded that *Alaskozetes antarcticus* has a flexible life cycle that probably takes at least 5 years.

Iteroparous breeding has also been demonstrated in the oribatid mite *Edwardzetes elongatus* from South Georgia (West 1982). Relatively few eggs are deposited during each breeding season. An alternative reproductive strategy has been detected in the oribatids *Erobrachythonius oudemansi* and *Plathynotrus skottsbergi*, which develop to semelparous adults under favourable environmental conditions, but each of them has two resting stages.

In the Arctic, the proportion of species that takes more than 1 year to complete development increases with latitude. Prolonged life cycles have mainly been attributed to the effect of low temperature, but may also result from reduced food supply (Danks 1981). A number of arctic fresh--water insects have prolonged life cycles (MacLean 1975; Danks 1981; Butler 1982), but since they fall outside the scope of this book, they will not be included in the present discussion.

Among terrestrial insects, Kaufmann (1969, 1971) showed that the two Alaskan beetles *Upis ceramboides* and *Pterostichus brevicornis* can hibernate in any developmental stage beyond the egg. According to Ring and Tesar (1981), the beetle *Pytho americanus* has 2-3-year life cycle in the

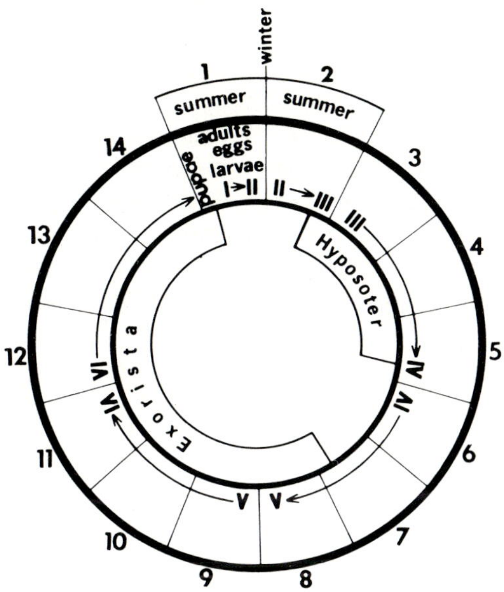

Fig. 7.1. Life cycle of *Gynaephora groenlandica* and its two parasitoids. The *outer circular segments* represent 14 summers of development interrupted by winters (*crossbars*). Different stages in the life cycle of *G. groenlandica* are indicated within the sectors, and the *Roman numbers* refer to larval instars. The cumulative number of years required for the development is indicated on the *outside of the circle. Two inner circular segments* represent the development and emergence of two parasitoids restricted to a particular life phase of *G. groenlandica*. (Kukal and Kevan 1987)

Arctic. In Greenland most of the beetles are univoltine (Böcher 1988), but at least four species are biennial with larval and adult hibernation.

The lymantriid moth *Gynaephora groenlandica* offers an unusual example of a prolonged life cycle in an arctic species. The moth is endemic to high arctic areas of Canada and Greenland (Kukal and Kevan 1987; Kukal 1991). The species survives in an environment with wide climatic fluctuations, low temperatures and xeric conditions. At Ellesmere Island (78°N), growth and development are confined to an abbreviated period during each summer, and the life cycle may last for up to 14 years (Fig. 7.1). The adult, egg and early larval instars comprise a small part of the entire life cycle, and are dependent on energy stored by the older larval instars. During the summer, more than two-thirds of the larvae are killed by parasitoids. In response to this, active feeding of the larvae is restricted to the early part of the summer when parasitoid populations are low. The *G. groenlandica* larvae stop feeding in early July and hide in the ground, probably to prevent parasitism, although most of the favourable summer season is lost in this way.

7.1.2.2 Univoltine Species

In spite of the harsh environment, some species of terrestrial arctic arthropods are able to complete their development in 1 year. Most of these univoltine species belong to groups that can only pass the winter in a particular stage, e.g. as egg, pupa or adult (Danks 1981). Univoltine species are relatively rare, and it appears that each of them has developed specialized adaptations, partly through behaviour and better utilization of favourable microhabitats. In the Arctic, the best-known examples of univoltine species include mosquitoes that develop in shallow pools (Downes 1975; Danks 1981).

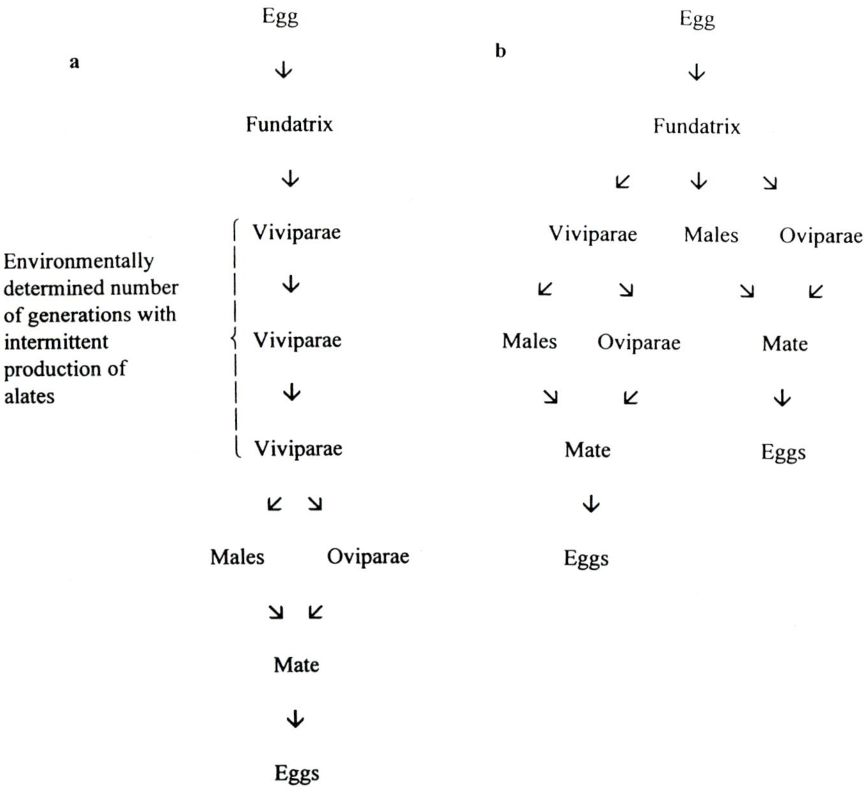

Fig. 7.2a,b. Comparison of the life cycle of the arctic aphid *Acyrthociphon svalbardicum* from Spitsbergen (*right* **b** with that of a generalized holocyclic temperate species *left* **a**). (Strathdee et al. 1993)

A univoltine life cycle is also known from the bug *Nysius groenlandicus* (Böcher 1976), which overwinters in the diapausing egg stage. This species prefers dry habitats in Greenland, and is distributed from the farthest south to Thule and Peary Land in the north. During severe summers, Böcher (1976) observed that the bugs did not reach maturity, and that local populations might be temporarily exterminated.

The lives of univoltine species of psyllids on arctic willow are closely synchronized to the development of the catkins (MacLean 1983). Life cycles of arctic aphids are somewhat abbreviated and simplified compared to those of temperate species (Danks 1981), but there is at least one generation per year. An example is *Acyrthosiphon svalbardicum*, which has a life cycle that is well adapted to the thermal environment at Svalbard (Strathdee et al. 1993). This species has four morphs; fundatrix, viviparae, males and oviparae. The fundatrix gives rise directly to the sexual morphs, a process that is essentially prevented in temperate species (Fig. 7.2).

The oviparae and males of *A. svalbardicum* mate and lay eggs in the middle of the summer, at a time when chances for survival are high. Under favourable thermal conditions, the aphid is able to include a second generation of sexual morphs in its life cycle (Fig. 7.2). Provided that this third generation is able to reproduce before conditions became adverse, the overwintering egg population will be increased.

Arctic bumblebees exhibit several adaptations for survival in a high arctic environment (Richards 1973). In *Bombus polaris* only a single brood of workers is developed prior to the production of sexual forms. *Bombus hyperboreus* is a nest parasite of the former species, and the shortening of its life cycle is achieved by elimination of the worker caste.

7.1.3 Alpine Species

7.1.3.1 Prolonged Life Cycles

A number of alpine species have life cycles with a duration of 2 years or more (Mani 1968; Danks 1981). As reviewed by Sømme (1989), several carabid beetles have been studied in detail. According to DeZordo (1979a), a majority of the carabids from Obergurgl, Tyrol, at 2000-3000 m need at least 2 years for their development (Fig. 7.3). Among these are the snow-bed species Nebria germari and Nebria castanea, as well as Bembidion bipunctatum from 2650 m and Pterostichus jurinea, Calathus melanocephalus and Patrobus assimilis from 2000 m.

At 2650 m, *Amara quenseli* has a more flexible life cycle, and overwinters in all larval instars as well as in the adult stage (DeZordo 1979b). The life cycle is completed in 2 or 3 years (Fig. 7.3). The eggs are usually deposited late in summer, and the first winter is spent in the first or second instar. Some individuals reach the adult stage the next summer, while others probably overwinter as third instar larvae. The latter complete their life cycle the following summer and overwinter for the third time as adults.

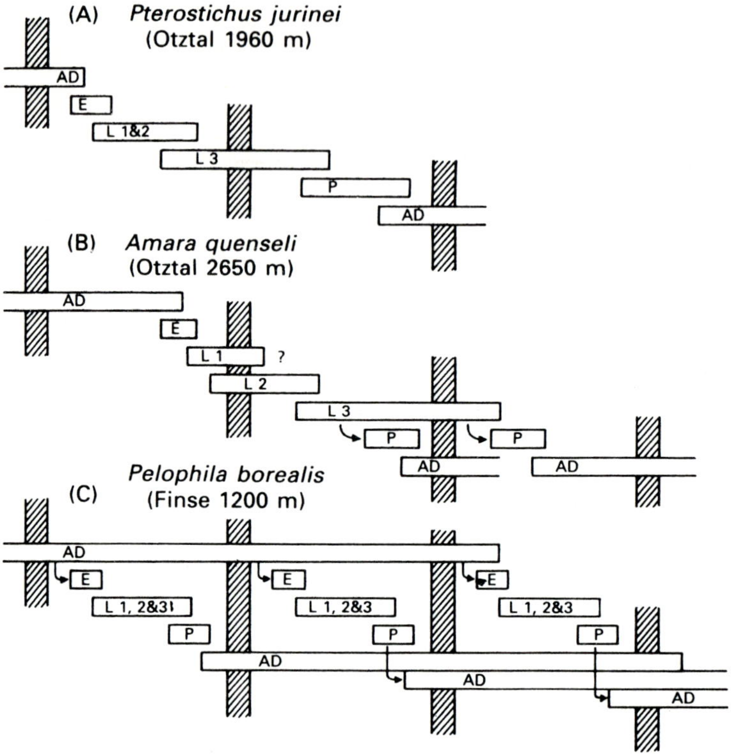

Fig. 7.3A-C. Life cycles in three species of alpine carabid beetles. **A** *Pterostichus jurinei*. **B** *Amara quenseli*, both with prolonged life cycles (DeZordo 1979a). **C** *Pelophila borealis* with univoltine life cycle and prolonged longevity as adult. AD Adults; E eggs; L larval instars; P pupae. Vertical hatched columns symbolize winters. (Sømme 1989)

Wolf spiders of the genus *Pardosa* are widely distributed in alpine regions, and some species are found in the Arctic. The *Pardosa* spp. spiders exhibit different types of life cycles (Fig. 7.4). In the Rocky Mountains, *Pardosa ourayensis* which lives close to the timberline, completes its life cycle in 1 year, whereas *Pardosa concinna* and *Pardosa tristis* from higher elevations require 2 years (Schmoller 1970).

A two-year life cycle is also found in *Pardosa nigra* from 1500-2000 m in the Tatra Mountains, the Slovak Republic (Lomnicki 1963), whereas *Pardosa palustris* from Hardangervidda, Norway, takes 3 years to complete development (Steigen 1975). Life cycles of 3 years are also found in *Pardosa cincta* and *Pardosa saltuaria* from Hohen Tauern, Austria (Czermak 1981), and *Pardosa mixta* has a 2-year life cycle at 1900 m and a 3-year life cycle at 2400 m.

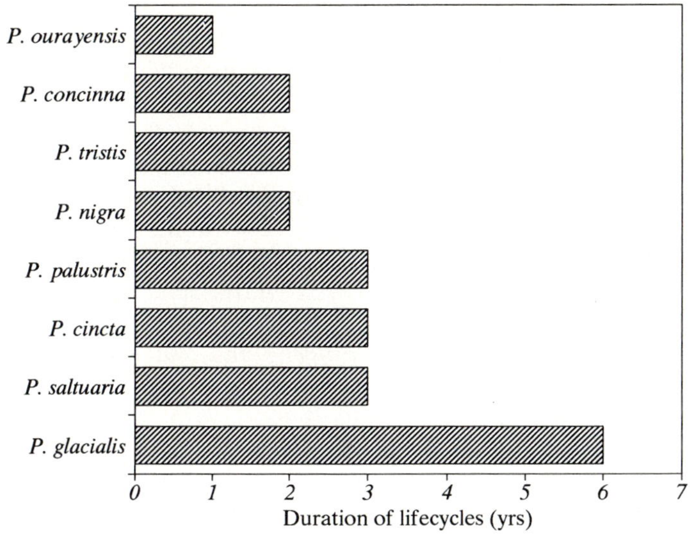

Fig. 7.4. The duration of life cycles in different species of alpine and polar *Pardosa* wolf spiders. See text for further explanation

Among *Pardosa* spp. with different life cycles, different stages may overwinter. This suggests that selection for a prolonged life cycle has taken place independently in different mountain areas (Sømme and Block 1991).Pre-adaptation of a common ancestral form may have increased the probability for this selection to occur in related species. The large number of *Pardosa* spp. found in alpine regions also reflects the ability of these spiders to adapt to extreme climates. In the Arctic, the life cycle of *Pardosa glacialis* at Lake Hazen, Ellesmere Island, has a duration of 6-7 years (Leech 1966) (Fig. 7.4).

The soil fauna is an important part of the alpine ecosystem, and many species find more sheltered conditions in the soil and upper layer of litter. Some of them are dependent on high ambient moisture content, while surface-living species may be exposed to severe desiccation. As reviewed by Sømme (1989), some alpine *Tipula* spp. from North America and Europe have 2-year life cycles. In the Alps, Meyer (1983) found that the psychidid moth *Epichnopterix ardua* overwinters in its 2nd and 5th instars in alpine meadows at 2000 m. Several species of Collembola from alpine habitats are also reported to have 2-year life cycles, although univoltine species are found above the timberline (Sømme 1989). Solhøy (1975) suggested that some of the alpine oribatid mites from Hardangervidda, Norway, may have prolonged life cycles when compared with lowland species. According to Schatz (1985), the oribatid mite *Oromurcia sudetica* from the Tyrolean Alps takes at least 2 years to develop from egg to adult, and all instars are present in the field at all times.

7.1.3.2 Univoltine Life Cycles

Several species of alpine terrestrial arthropods complete their life cycles in a single year, and univoltism appears to be more common than in the Arctic (Sømme 1989; Sømme and Block 1991). Under alpine conditions, large temperature gradients are found within short distances, and microhabitats with favourable conditions are frequently encountered - in particular in mountains of middle and lower latitudes. Thus, the univoltine wolf spider *Pardosa ourayensis* (Fig. 7.4) inhabits warmer and more productive microhabitats in the Rocky Mountains than do semi-voltine species from the same altitude (Schmoller 1970).

Among beetles, the carabid *Pterostichus diligens* and the staphylinid *Philolonthus aerosus* from the Tyrolean Alps have univoltine life cycles (DeZordo 1979a,b). According to Sømme and Block (1991), the carabid *Pelophila borealis* from Finse, Norway, completes development from egg to adult during the summer, and overwintering takes place in the adult stage (Fig 7.3). If summer conditions are exceptionally cold and harsh, the whole population may run the risk of extinction; but the adults probably live for 2 or 3 years and are able to deposit new eggs each spring. This compensates for the eventual loss of immature stages during a bad summer. In Alaska, adults of the carabid *Pterostichus brevicornis* also overwinter twice (Kaufmann 1971).

As a compensation for the short summers at 4000 m in the Rocky Mountains, the grasshopper *Aeropedellus clavatus* has one nymphal stage less than normal, and adults emerge after the fourth stage (Alexander and Hilliard 1964). The species undergoes egg diapause which may last for 2 or 3 years, combined with early hatching after termination. The early appearance of the first nymphal stage is of great survival value, extending the summer season at high altitudes as much as possible. Univoltine life cycles are also known from grasshoppers in the Snowy Mountains, Australia (Dearn 1977). In *Praxibulus* sp., *Kosciuscola cognatus* and *Kosciuscola usitatus*, the number of eggs per female decreases with latitude, while the number of eggs per egg batch increases. According to Dearn (1977), the increase in eggs per batch may be related to the general physiological strain of egg production on the females, and the increased risk of mortality following egg laying. If the probability of future reproductive success is low, then natural selection will favour females that increase their immediate reproductive efforts.

7.1.4 Conclusions

Deserts, alpine and polar areas are risky places in which to live and reproduce, and adjustments of life cycles are important adaptations in invertebrates to cope with the unfavourable conditions. Large variations in life cycle strategies are found in terrestrial arthropods from such environments. While climatic conditions in hot deserts are both unstable and unpredictable, the changing seasons of alpine and polar areas are more regular. In spite of these differences, prolonged life cycles are a form of adaptation in terrestrial arthropods from all three biomes. In desert species, prolonged life cycles are sometimes combined with iteroparous breeding, spreading the risk of reproduction over time. Long-lived species possess the possibility to utilize occasional and irregular resources of food over extended periods. In alpine and polar regions, prolonged life cycles are an adaptation to short and cold summers so that most species are unable to complete their entire development in one season.

In hot deserts, as well as in polar and alpine areas, univoltine life cycles depend on special adaptations. Migration, dormancy, exact timing of reproduction and brooding of the young are factors of importance for the completion of development during one season. Univoltine life cycles are known from a number of alpine species, but become more rare with increasingly harsh polar conditions. Overwintering in diapause, utilization of warm microclimatic niches, and a reduction in the number of developmental stages are characteristic of univoltine species from cold areas.

Multivoltine life cycles are probably common in desert terrestrial arthropods that are able to explore ephemeral resources, but are extremely rare or almost non-existent in polar species.

7.2 Diapause and Quiescence

7.2.1 Introduction

Diapause in insects and other terrestrial arthropods has been extensively studied and the voluminous literature reviewed by several authors (e.g. Beck 1980; Behrens 1984; Denlinger 1986,1991; Tauber et al. 1986; Danks 1987; Zaslavski 1988; Leather et al. 1993). Diapause in terrestrial arthropods is a physiological suppression of growth and development. The arrest occurs at a specific stage in the life cycle, and for most species at only one stage. Depending on the species, this may be the egg, larva, pupa or adult.

Diapause is an important adaptation to unfavourable environmental conditions, and is found in species from temperate and arctic areas, as well as in species from tropical areas, including hot deserts. As pointed out by Tauber et al. (1986), the diapause stage is central in synchronizing the life cycle with seasonal changes in the environment. In this way diapause regulates the timing of growth, development and reproduction both before and after dormancy. Induction of diapause takes place in advance of the expected adverse conditions, and ensures that development is halted in the stage that is best adapted to survive environmental stress, such as cold or drought. In this respect diapause differs from quiescence that is an immediate response to adverse conditions (Chap. 5).

The state of diapause in insects and other terrestrial arthropods offers some of the best examples of how closely many organisms are tuned to their environment. As pointed out by Callaghan et al. (1993), future climatic changes may have an impact on diapausing insects, since a relatively small divergence from the normal photoperiod and temperature regime may alter the induction and termination of diapause.

Although there are important similarities with regard to the induction, maintenance and termination of diapause in terrestrial arthropods (Tauber et al. 1986), there is also a variety of complicated patterns. The variation reflects the strong selective value of diapause in protecting an organism against adverse conditions. Presumably, diapause has evolved independently in a large number of species. On the other hand, many non-diapausing species are able to cope with environmental stress based on other life cycle strategies. Without diapause, there is greater overlap of generations, and several stages may have to endure harsh environmental conditions during overwintering or aestivation.

The documentation of diapause in a given species requires extensive field observations and confirmation by experimental work. For this reason, only a limited number of species have been studied. Judging from the increasing number of publications in the field, Masaki (1983) predicted that by the end of the century the photoperiod response in insects will have been studied in more than 2000 species. Although this is a high figure, it also illustrates that, for most species, we do not known if the response to unfavourable conditions is true diapause or merely quiescence. Diapause has mainly been studied in terrestrial arthropods from temperate zones, where many species enter diapause in preparation for the winter. Summer diapause is found in species from hot temperate areas (Masaki 1980) and tropical insects may enter diapause in preparation for unfavourable environmental conditions, such as periods of drought (Denlinger 1986).

Where a distinction between diapause and quiescence cannot be made, the term dormancy has been used to include both possibilities (Denlinger 1986; Danks 1987). According to Tauber et al. (1986), dormancy is a general term referring to a seasonally recurring period in the life cycle of a plant or animal during which growth, development and reproduction are suppressed. When dormancy occurs during the summer or winter, these periods are termed aestivation and hibernation, respectively.

A short description of the principles of diapause is presented in the following section, but the reader is referred to the more comprehensive reviews that have already been cited for more detailed accounts. Examples of diapausing species are discussed below in relation to the adaptation of terrestrial arthropods to hot and cold arid environments.

7.2.2 Principle Features of Diapause

Diapause includes several stages of development as outlined by Tauber et al. (1986) and Danks (1987). The induction of diapause takes place through environmental cues acting on the sensitive stage in the life cycle. This may be the same stage that enters diapause, but it is usually one of the previous stages. The duration of the sensitive period varies greatly among different species. The reaction to diapause-inducing stimuli includes changes in the neuroendocrine system, metabolism, behaviour and morphology of the insect. Tauber et al. (1986) recognized three essential features of the pre--diapause period: (1) the regulation of growth and development so that the diapausing stage is reached in advance of adverse conditions, (2) the accumulation of metabolic reserves required for survival during and after dormancy and (3) the development of behavioural changes resulting in movement to the dormancy site and the provision of protection during dormancy.

Decreasing photoperiod is the most reliable sign of approaching winter conditions and the main cue for diapause induction in temperate and arctic species. For induction of summer diapause in species from warmer regions, the situation is reversed and based on long days (Masaki 1980). In addition to photoperiod, temperature and decreasing nutritional quality are important factors for the induction of diapause in many species.

The reaction to changes in the duration of photoperiod may be extremely sensitive (Zaslavski 1988). A difference of less than 1 h may sometimes determine whether an insect will enter diapause or not. The classical studies of Danilevski (1961) serve to illustrate the close interaction between photoperiod and temperature in the induction of diapause. According to Danilevski (1961), cited from Zaslavski (1988), the percentage of the population of the noctuid moth *Acronycta rumicis* that enters diapause under different photoperiods increases with decreasing temperatures (Fig. 7.5). In addition, the percentage of diapause is genetically determined in populations from different latitudes. The population from Abkhasia at 43°N will not enter diapause at photoperiods above 15 h, while diapause in the population at St. Petersburg at 60°N is induced by photoperiods up to 20 h or longer, when compared at the same temperature (23 °C). In this way, the northern population, which has only one generation per year, enters diapause early in the season in preparation for the early onset of winter. In the southern population, two or three generations may be completed before diapause is induced by shorter daylength.

Following induction, environmental factors also affect the intensification of diapause (Tauber et al. 1986). In most cases there is a relatively rapid decline in metabolic and behavioural activity as diapause intensifies.

Individuals that have entered diapause will not respond to conditions that are normally favourable for growth and development. In this way the

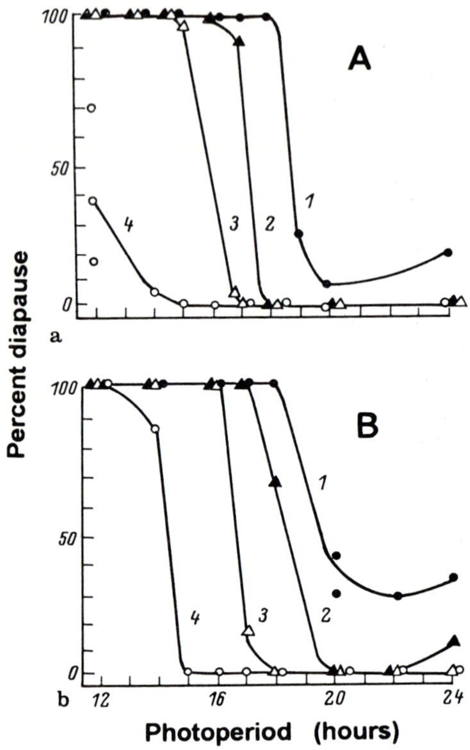

Fig. 7.5. Effect of temperature and geographical location on the induction of pupal diapause in *Acronycta rumicis*. **A** Larvae of a population from Belgograd at constant temperatures; *1* 15 °C; *2* 20 °C; *3* 25 °C; *4* 30 °C. **B** Larvae from different regions developed at 23 °C; *1* Leningrad (60°N); *2* Vitebsk (55°N); *3* Belgograd (50°N); *4* Abkasia (43°N). (Zaslavski 1988)

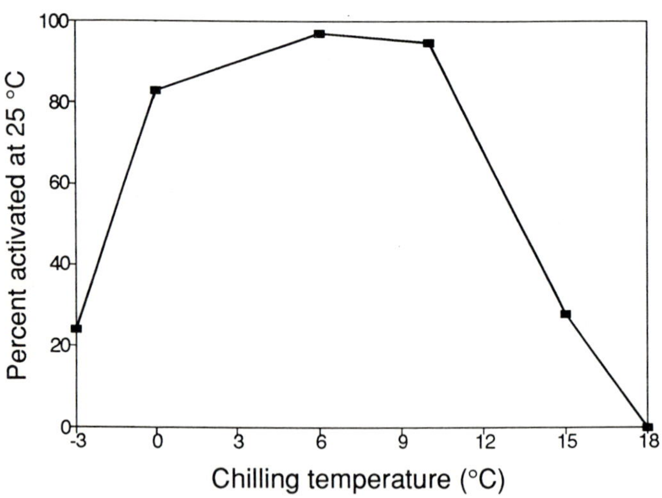

Fig. 7.6. The termination of prepupal diapause in the codling moth *Cydia pomonella* at 25 °C, following chilling at different constant temperatures for 4.5 months. (After Zaslavski 1988)

exposure of stages more sensible to adverse conditions is avoided. In diapausing insects, physiological changes take place that eventually lead to the completion of diapause. These changes are commonly termed diapause development (Tauber et al. 1986).

The relation between diapause duration and environmental conditions is complex. In some insects, diapause is maintained by short photoperiods, but terminates when the photoperiods are increased (Behrens 1984). In other species the photoperiodic response changes or decreases during diapause. Diapause is maintained at high temperatures in a number of terrestrial arthropods. Based on laboratory experiments, it appears that exposure to cold conditions terminates diapause, although Tauber et al. (1986) warn against the conclusion that chilling also acts to terminate diapause in nature. In their opinion, diapause-terminating stimuli are difficult to identify in natural populations. Some species of insects cease to respond to diapause--maintaining factors at a certain time during the winter, and diapause ends gradually or spontaneously during late fall or early winter. In other species, the occurrence of a specific diapause-terminating stimulus may be required. The effect of chilling on the termination of diapause under experimental conditions varies with both time and temperature in different species (Zaslavski 1988). In many species, such as the codling moth *Cydia pomonella*, a bell-shaped curve reflects that the effect of chilling first increases, and then declines as the temperature is lowered (Fig. 7.6).

Transition from the diapause state to non-diapause levels of metabolism, activity and development is a gradual process, which can be characterized as post-diapause development (Tauber et al. 1986).

7.2.3 Dormancy in Tropical Insects

Diapause is known from a number of a number of tropical insects (Denlinger 1986), but relatively few studies have examined experimentally the occurrence of diapause in insects from hot deserts (Tauber et al. 1986). Many desert-inhabiting species enter a state of dormancy, but the significance of diapause versus non-diapause seasonal adaptations is largely unknown. Prolonged dormancy, with a duration of one or more years, is an important adaptation of some desert-inhabiting insects to cope with highly irregular seasonal patterns.

In cases where diapause has been elucidated, the physiological basis appears to be similar for temperate and tropical species (Denlinger 1986), but different environmental regulators may be involved. Seasonal variations in environmental factors are usually less distinct in tropical than in temperate areas. In the desert, both the occurrence and persistence of abiotic factors of importance for insect growth are variable and unpredictable. Favourable conditions, such as moisture supplies after rainfall, are often of short duration.

While photoperiod is a key factor in inducing diapause in temperate regions, variations in daylength are much smaller in tropical areas (Denlinger 1986). At 20° N or S the shortest and longest daylength differ by 2 h 25 min, while the difference is only a few minutes at 1° N or S. Light intensity, however, may change on a seasonal basis in relation to cloudiness and rainy seasons. Annual variations in temperature are also much smaller in the tropics than in temperate areas, but daily variations may show great seasonal differences. In this way, temperatures during part of the day can be useful indicators of seasonal changes.

As pointed out by Denlinger (1986), rainfall pattern is the most conspicuous aspect of tropical seasons. Rainy seasons are regulated by the prevailing winds, which change in direction according to the position of the earth on its orbit around the sun. In many tropical areas there are distinct patterns of dry and wet seasons, although rainfall pattern may be extremely variable from year to year.

In addition to abiotic factors such as light, temperature and rain, insect dormancy is influenced by changes in biological factors. The wet season results in a periodic flush of vegetation of great consequence for the entire ecosystem. For many insect species, rainfall following a period of dryness represents the time for termination of dormancy. In this way insect growth and reproduction are intimately timed to the occurrence of rainfall (Tauber et al. 1986).

Although the factors responsible for diapause induction and termination are poorly known for most tropical terrestrial arthropods, many species probably react to a combination of seasonal stimuli. The abiotic factors of the habitat, as well as availability of food and oviposition sites may be influenced profoundly by seasonal rhythms (Denlinger 1986).

7.2.3.1 Diapause in Grasshoppers and Locusts

Migration and diapause are two alternative strategies of grasshoppers and locusts to adjust their life cycles in accordance with inhospitable periods (Farrow 1990; Joern and Gaines 1990). Migration may be an effective response to rapidly changing but aperiodic conditions, whereas diapause and other forms of suspended development are more likely to occur if the variability of the environment is seasonal and predictable.

Most species of grasshoppers and locusts experience periods when the environment is inhospitable, e.g. cold winters or hot periods of drought. Changes in the impact of natural enemies and the patchiness of their habitats also contribute to seasonal variations in environmental favourability (Farrow 1990). Dormancy appears to be widespread among tropical locusts and grasshoppers (Farrow 1990). Among 70 species of acridids of the African Sahel, with both univoltine and multivolitine life cycles, reproductive quiescence has been found in about 55 % during the dry season

(Fishpool and Popov 1984). Of the remaining species, some spend the dry season as dormant eggs, while others are able to reproduce continuously.

Large variations are found in the life cycles of different grasshopper species, and in some cases within the same species, to cope with their particular environments. A generalized life cycle is presented in Fig 7.7 (Joern and Gaines 1990), which includes opportunities for diapause and migration. Diapause has been documented in all stages (Tauber et al. 1986).

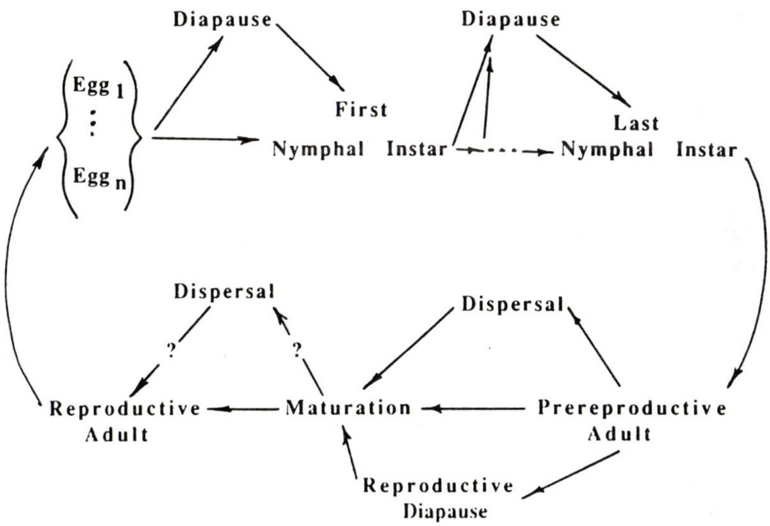

Fig. 7.7. Generalized grasshopper life cycle. Diapause may occur at a variety of stages, although typically only once for a particular species. Egg diapause may take several pathways within a generation, and even among eggs from a single egg pod. (Joern and Gaines 1990)

Most species of locusts are able to breed continuously, but some examples of adult diapause have been reported. The migratory locust *Locusta migratoria migratoroides* requires suitable moist refuges for continuous breeding, but in northern populations, photoperiodic reaction results in imaginal diapause (Cassier 1968; Zaslavski 1988). In the red locust *Nomadacris septemfasciata* of tropical Africa, breeding is confined to the rainy season (Norris 1959). Both north and south of the equator new adults emerge at a time when day length is diminishing, and undergo prolonged diapause during the dry season. According to Popov et al. (1984), the desert locust *Schistocerca gregaria* is able to develop and reproduce continuously without diapause during any of the stages of its life. Sexual maturation, however, may be delayed. Carlisle et al. (1965) found that contact with terpenoids from the desert shrub *Commiphora myrrhae* hastens the attainment of reproductive activity. In the desert, bud burst of aromatic shrubs may provide the environmental cue by which these locusts gear their breeding season to the rains.

Egg diapause is recorded in several species of grasshoppers and locusts from arid and semi-arid environments (Uvarov 1966), and is known to occur in different embryonic stages. Eggs of diapausing grasshoppers are often equipped with dense wax layers that reduce water loss (Slifer and King 1961). Thus, eggs of the Australian grasshopper *Austroicetes cruciata* are highly resistant to both desiccation and water loss (Andrewartha and Birch 1954) and remain in diapause for 9 months. Similarly, the African grasshopper *Zonocerus variegatus* spend 6-7 months in egg diapause during the dry season (Page 1980).

The relation between non-diapausing and diapausing eggs in grasshoppers is often complicated. Some species, such as the Australian plague locust *Chortoicetes terminifera*, may even deposit eggs of both categories within the same batches (Wardhaug 1986). Chances of survival for some of the offspring of this species are increased by spreading the dangerous process of hatching over an extended period. Migration and egg diapause are combined in this species to withstand periods of drought. Diapause induction is directed by complex interactions of environmental cues, acting on both the parents and on the eggs (Hunter and Gregg 1984; Wardhaug 1986). Decreasing photoperiod favours entry into embryonic diapause, modified by temperature and moisture conditions.

The time required for egg development is also highly variable in the Senegalese grasshopper *Oedaleus senegalensis*. It depends on season, rainfall and whether the eggs are in diapause or not (Cheke 1990). Some of the eggs in a pod do not hatch with the first rain of the season, and under identical laboratory conditions the time from when an egg pod is moistened till hatching of the first nymph varied from 11 to 901 days. Within the same egg pod, hatching was protracted, with more than 3 years between the hatching of the first and last egg.

According to Matthée (1951), *Locustana pardaline* lays both diapause and non-diapause eggs, some of each in the same egg pod. The eggs are surrounded by a keratin-like membrane secreted by the serosal cells. The membrane permits uptake of water under moist conditions, but prevents water loss when it is dry. Non-diapausing eggs hatch in 9 days in moist soil, but become quiescent under dry conditions. Diapausing eggs, on the other hand, require at least 45 days of dry conditions before they respond to moisture and eventually hatch, but they usually remain in diapause for extended periods.

The examples show that dormancy is common in eggs of many grasshoppers and locusts. Apparently, a variety of complicated strategies have evolved which delay and spread hatching over extended periods.

Strategies like these are clearly excellent adaptations to the unpredictable environmental conditions of hot deserts, and make it possible for many species to survive even years of drought. Following rainfall, opportunistic development takes place, but part of the population will remain in dormancy as a guarantee against extinction should the moist period be of short duration.

7.2.3.2 Diapause in Other Desert Insects

As in acridids, prolonged diapause has been reported from other desert insects. During periods of lack of rain, the Californian ethmiid moths *Ethmia scylla* and *Ethmia semilugens* may remain in diapause for up to 4 years on yucca (Powell 1974). Prolonged pupal diapause accounted for the survival of populations of the butterfly *Pieris napi microstriata* during a Californian drought in the 1970s (Shapiro 1979). In another butterfly, *Papilio alexanor*, from arid areas in southern Europe and the Middle East, the larvae feed exclusively on the flowers of the perennial herb *Ferula* sp. (Nakamura and Ae 1977). In years when flowering does not occur, the pupae of this species remain in diapause.

The neotropical pierid butterfly *Eurema daira* appears as both dry season and wet season morphs in the lowlands of Costa Rica (Young 1982). The dry season morph enters a reproductive diapause, while part of the population of the wet season morph migrates to higher and wetter environments and may return to the lowlands at the end of the dry season. The mixed strategy of diapause and migration insures against complete elimination of the species. Migration is also important in tropical milkweed bugs, *Oncopeltus* spp. Dingle (1978) found no diapause in these species, although the temperate *Oncopeltus fasciatus* undergoes photoperiodically induced diapause. In Central and South America, the bugs leave seasonally dry environments and migrate to wetter habitats where milkweed plants are available.

In California, the lacewing *Chrysopa mohave* is well adapted to the local dry summer and sparse food supply (Tauber and Tauber 1973). The autumnal phase of diapause is maintained by day length, while the winter-spring phase is maintained by the absence of prey, even though abiotic factors may be favourable for breeding.

Egg diapause is a common way of overwintering in temperate species of insects, but appears to be more unusual in tropical species. As pointed out by Masaki and Nagase (1992), however, egg diapause in the cricket *Dianemobius fascipes* species complex may have a tropical origin. In temperate strains of the cricket, egg diapause is programmed by the parental photoperiod. Delayed hatching is much rarer in tropical strains, but may be induced by low temperatures. This results in greater variability of hatching time and may serve as a risk-spreading tactic to cope with the unpredictable fluctuating tropical environment. The presence of egg diapause has also been demonstrated in the collembolan *Sminthurus viridis* (Wallace 1968). This species has been introduced to Western Australia, part of which has a Mediterranean climate with dry summer seasons and wet winters. Egg batches deposited in the spring are covered with a smooth coating of soil, probably giving increased protection against desiccation. The termination of diapause requires 2-3 months under dry field conditions before the eggs are ready to hatch as a response to moisture.

7.2.4 Dormancy in Polar Species

Although the literature on dormancy among terrestrial arthropods is very extensive, little work has been carried out on arctic and antarctic species (Danks 1981; Tauber et al. 1986). Many polar insects and other terrestrial arthropods have life cycles that last for several years, but it is not known whether one or several of the overwintering stages enters diapause or not. Conditions for the induction of diapause may be poorer in polar areas than in temperate regions, since photoperiodic cues are weaker or lacking due to continuous daylight (Downes 1965; Tauber et al. 1986). Temperature, however, may replace photoperiod as the primary cue for both initiation and termination of dormancy in arctic species. Nevertheless, light conditions cannot be excluded entirely. Although the photoperiod in itself has a constant duration of 24 h, light intensities change during the day, and the spectral composition of the light varies with the season. Patterns of daily light intensity are important for photoperiodic rhythms in temperate insects (Remmert 1980), and increase in red light during the late summer is a cue for arctic plants to prepare for the winter.

In some species of arctic insects, lack of diapause has been clearly documented. The life cycle of the high arctic moth *Gynaephora groenlandica* may last up to 14 years, but Kukal (1991) stated that there is no evidence of diapause in any of the stages. *Chironomus* spp. midges from ponds at Barrow, Alaska normally have a 7-year life cycle (Butler 1982), but larval development is greatly accelerated at 15 °C. Growth corresponding to 4 years in the field was accomplished in the laboratory without developmental arrests in 40 days, indicating that at least there is no obligate diapause in the early instars. Some polar springtails and mites with prolonged life cycles apparently also overwinter without diapause (Addison 1977; Burn 1981; Convey 1994) (Chap. 7.1.2).

Some arctic species enter dormancy when they reach a certain stage, and do not continue development until the next spring. Well-documented examples of diapause are rare, but in Greenland the arctic bug *Nysius groenlandicus* undergoes obligatory diapause in the egg stage. This is probably induced by the day-length experienced by the adult female (Böcher 1975). There is also good evidence of diapause in fully grown larvae of chironomids from high arctic ponds (Danks and Oliver 1972). At Lake

Hazen, *Aedes* spp. overwinter in the egg stage, and *Culiseta alaskensis* as adults (Danks 1981). Development is arrested in the autumn in both cases, which ensures emergence of the right stage in the following spring.

In the blowfly *Protophormia terranovae*, which has a distribution including arctic areas, the winter is passed by adults in diapause (Wood and Nordin 1976). High levels of glycerol are accumulated in this species as protection against low temperatures.

7.2.5 Conclusions

Diapause has not been extensively studied in tropical and polar terrestrial arthropods. Some examples of it are known, but more information is needed. Compared with temperate areas, environmental cues for the induction of diapause may be weaker than elsewhere in polar regions and in areas close to the equator. In both polar and tropical areas, other strategies of dormancy may be more common. Egg diapause is found in many tropical grasshoppers and combined with quiescence results in delayed and temporally dispersed hatching to cope with unpredictable environmental conditions. Polar and alpine species species with prolonged life cycles may be non-diapausing, while univoltine species probably spend the winter in diapause.

8 Water Balance During Overwintering

8.1 Introduction

As will be discussed in the present chapter, overwintering in insects and other terrestrial arthropods involves several problems. To cope with winter conditions, an array of survival strategies have evolved. The selection of overwintering sites is of primary importance in order to ensure optimal environmental conditions throughout the winter (Danks 1991). Survival during extended exposures in temperate and polar areas requires adaptations at the biochemical level, and lower extreme temperatures must be tolerated (Chap. 9).

During overwintering, many species are at a risk of freezing due to the penetration of ice through the cuticle. In species that cannot survive ice formation in their tissues, this inoculation will have fatal consequences. Most species overwinter in an unfrozen state, and remain supercooled when ambient temperature falls below the melting points of their body fluids. Due to differences in water vapour pressure, water in supercooled insects may be lost to the surroundings, resulting in different degrees of dehydration.

Apparently, overwintering strategies involve not only adaptations to cold but adaptations to arid conditions as well. Ring and Danks (1994) have recently argued that cryoprotection and resistance to desiccation are, in fact, overlapping adaptations. While adaptations such as solute production, regulation of water content and selection of overwintering sites have been linked to cold hardiness, the same factors may also protect against desiccation or enhance resistance to dehydration. As pointed out by Ring and Danks (1994), many overwintering insects are exposed to dry cold air for a long time, and for this reason resistance to desiccation is required. The accumulation of polyols and other low molecular weight substances, considered as cryoprotectants, also reduces water loss in insects exposed to arid conditions. The amount of water that is available for loss contributes to the potential tolerance of desiccation. In selecting protected overwintering places, many insects avoid extreme low temperature conditions, but the maintenance of their water balance during the winter is equally important.

8.2 Choice of Overwintering Sites

8.2.1 Environmental Factors

As emphasized above, the choice of overwintering sites is of major importance for the survival of insects and other terrestrial arthropods of temperate and polar areas. A number of factors are involved, as reviewed by Danks (1978, 1991) and Leather et al. (1993). Most studies have been concerned with temperature conditions during overwintering, and less attention has been paid to the moisture conditions. Temperate and polar species spend very long time intervals in their overwintering sites, and it is of vital importance that the moisture remains sufficiently high to avoid dehydration.

Several species of terrestrial arthropods overwinter above the snow cover. Eggs of many insects, such as Lepidoptera and Homoptera, are found on twigs and branches of trees. Mites and springtails, as well as larvae of different insects, overwinter under the bark on stems of trees. To survive in these habitats, the animals must be adapted to both low and fluctuating temperatures, and to low air humidities during periods of frost. Most terrestrial arthropods, however, seek more protected overwintering sites in soil and litter under the snow, which is an excellent insulating material. A snow layer of 20-50 cm or more usually stabilizes soil surface temperatures around 0 °C, but temperatures as low as -10 °C may occur under the most severe climatic conditions (Danks 1991). As an example, carabid beetles and other alpine insects overwintering under 50-100 cm of snow cover at Finse, Norway, experience temperatures in the range of 0 to -2 °C, although air temperatures above the snow may fall to -35 °C (Østbye and Sømme 1972). The insulation effect of snow depends not only on the thickness of the cover, but also on its density. Heavy snow that has been subjected to remelting has a thermal conductivity four times higher than that of light fluffy snow (Danks 1978).

Temperature is also influenced by other physical factors, such as the moisture conditions and depth of soil. The first exposures to frost in the autumn are delayed or eliminated in deeper sites (Danks 1991). Freezing is further delayed by high water content due to the release of heat by the formation of ice. Warming in superficial sites below the snow precedes warming of the air in the spring. In this way, heat stored in the earth ameliorates conditions beneath an insulating layer of snow. The litter layer on top of the soil buffers the temperature, partly due to the insulating effect of trapped air. Hiding in rock piles, in stumps, under fallen logs and in mammal burrows also offers protection from fluctuating temperatures and the extreme lower values.

Springtails and mites are predominant species of alpine and polar soil ecosystems (Cannon and Block 1988; Sømme and Block 1991). They occur in large numbers in the soil, and also inhabit moss and lichens in exposed places. In habitats with little soil, such as in extreme antarctic conditions, microarthropods aggregate close to streams of melt water and under rocks. In preparation for overwintering, some species have been observed to select the most humid sites. Several alpine species overwinter on windswept mountain ridges, where they become enclosed by ice but miss the protection of a deep snow cover against low ambient temperatures.

8.2.2 Cues for Overwintering

The ability of terrestrial arthropods to find their overwintering sites depends on how they respond to environmental cues. Although there are many observations on insects preparing for overwintering, there is a lack of experimental evidence in this respect (Danks 1991). Several environmental factors, including light, temperature, moisture and gravity may influence the selection of the overwintering site. As pointed out by Leather et al. (1993), it is important for the insect to have an appropriate suit of responses to available cues, which will enable it to select the best site.

During preparation for overwintering, many species move from one place to another within their habitat and thus find the most suitable microhabitat conditions. In flying insects large-scale features in the surroundings may serve as guides. Many species exhibit negative phototaxis, which helps them to find hollows, caves and crevices, or openings in the leaf litter covering the ground (Danks 1991; Leather et al. 1993). Insects entering their overwintering sites may orient themselves to dark or shaded areas that contrast with the background.

Apparently, only a few species use temperatures directly in choosing overwintering sites, but high arctic *Aedes* mosquitoes oviposit on the warmest available south-facing slopes (Corbet and Danks 1975). In this way, the overwintering eggs hatch as early as possible in the following summer (Chap. 6.2.4). Other species may respond to lower temperatures to find the most protected places. An example is adults of the carabid beetle *Pterostichus brevicornis* that live in stumps in Alaskan forests (Baust and Miller 1970). In the autumn the beetles dig further into the colder central parts of the stumps that are better buffered against the most extreme winter temperatures.

Studies on bumblebees suggest that hygrotaxis is important in choosing overwintering sites (Leather et al. 1993). A number of other insects may also seek habitats with high moisture content and thereby reduce the risk of desiccation. On the other hand, too wet sites may not be suitable, and many species prefer well-drained microhabitats. According to Pienkowski (1976), the alfalfa weevil *Hypera postica* has very sensitive hygroreceptors which help it to aggregate in places with high humidity.

In entering their overwintering sites in crevices, under bark, etc., many insects respond to touch. Thigmotaxis is often combined with negative phototaxis. Response to gravity, or geotaxis, also favours the entry of insects into sites in litter and soil (Danks 1991). Positive geotaxis is often associated with a digging reflex, as observed in the Colorado beetle *Leptinotarsa decemlineata* by de Wilde (1969). Most insects penetrate less than 15 cm into the soil, apparently because conditions for spring emergence are delayed in deeper sites compared with more exposed sites that thaw earlier (Danks 1978).

8.2.3 Anaerobiosis

Although not directly related to the problems of water balance, some terrestrial arthropods are exposed to conditions of low oxygen tension during overwintering. It is well known that many insects intermittently can survive long periods of hypoxia or anoxia (Wigglesworth 1972), and this is the case in several alpine and polar species which are enclosed by ice during overwintering (Sømme 1989). According to Conradi-Larsen and Sømme (1973), the winter habitat of the alpine carabid beetle *Pelophila borealis* from Finse, Norway, is flooded by adjoining rivers, and the beetles subsequently become enclosed in ice formed in the soil. Presumably, the beetles remain without access to oxygen for several months. The temperature of the ice rarely drops below -1 to -2 °C due to a protective cover of snow. Under laboratory conditions, all specimens of *P. borealis* survived 127 days' exposure in an atmosphere of nitrogen at 0 °C, while 30 % died after 156 days. Several other beetles overwinter under similar conditions (Sømme 1974a). *Byrrhus pilulae* (family Byrrhidae) survived 120 days and the two curculionids *Lepyrus arcticus* and *Otiorrhynchus dubius* survived 60 and 120 days respectively.

As a result of anaerobic metabolism during anoxia, lactate was accumulated in the haemolymph of *Pelophila borealis*, reaching levels of 30-35 mM. The concentration of alanine increased three fold, while levels of pyruvate and alpha-glycerophosphate did not change. In the other species, concentration of lactate varied from 13 to 47 mM. High levels of anaerobiosis were also found in the chrysomelid beetle *Melasoma collaris* from 1250 m at Finse (Meidell 1983). Sixty percent of the beetles survived anoxia for 120 days, and the concentration of lactate increased to 50 mM.

Following anaerobiosis, the rate of oxygen consumption in *Pelophila borealis* increased to twice that of the normal rate, lasting for several days (Conradi-Larsen and Sømme 1973). The presence of anaerobic conditions in the natural habitat of the beetles was confirmed by the presence of lactate and an oxygen debt in beetles removed from their overwintering sites in the middle of the winter.

The overwintering habitats of terrestrial arthropods on windswept mountain ridges are also covered by ice, resulting in an oxygen deficiency in the minute air spaces surrounding the animals. As reviewed by Sømme (1989), specimens of the springtails *Tetracanthella wahlgreni*, *Xenylla maritima* and *Anurophorus laricis* from Finse survived up to 90 days of anoxia at 0 °C, and *Tetracanthella afurcata* and *Onychiurus vontoerni* from the Tyrolean Alps up to 60 and 24 days, respectively. Similarly, the oribatid mites *Carabodes labyrinthicus* from Finse and *Fuscozetes intermedius* from the Alps survived 120 and 90 days, respectively. Anaerobic metabolism with lactate as an end product was demonstrated in *Tetracanthella wahlgreni*.

The ability to withstand anaerobic conditions is also evident in antarctic microarthropods. Sømme and Block (1982) showed that *Cryptopygus antarcticus* survived anoxia (30 % survived for 28 days in nitrogen at 0 °C) better than *Parisotoma octooculata*, which were all killed within 8 days under the same conditions.

Anaerobiosis is apparently an important adaptation to winter survival in several alpine and polar terrestrial arthropods, and may be more widespread than is presently known. Anaerobic conditions occur in different microhabitats following saturation of the soil by rain in the autumn and water from melting snow on sunny days in the winter, resulting in the formation of a solid layer of ice at freezing temperatures. Many species apparently survive the winter in litter and porous soil, encapsulated in the ice. More studies on anaerobiosis in terrestrial arthropods would be of great interest, and should particularly be seen in relation to other strategies of survival, such as the maintenance of water balance.

8.3 Moisture Content of Overwintering Sites

8.3.1 Lack of Fluid Water

The low precipitation in many polar areas affects the biology of terrestrial arthropods in several ways. Lack of water during the summer retards growth and development, and subjects organisms to desiccation stress. During the winter, water is present in the form of ice and snow, which cannot directly affect the water balance of insects and other terrestrial arthropods. Warm-blooded animals may replenish their body water by eating snow, but poikilotherms lack the heat capacity to melt snow, even if they were able to consume it. In this way, terrestrial arthropods are surrounded by frozen water but are unable to replenish the water supply of their bodies. For an overwintering insect the amount of precipitation in the form of snow is of no importance, since no water is available in the liquid phase. Consequently, a number of temperate and polar insects may experience desert-like conditions in their overwintering sites. Their lack of free water is comparable to the situation of terrestrial arthropods in hot arid deserts. It may be stretching the term too far to claim that overwintering insects live in a desert, but in their case the microclimatic moisture conditions are more important than the general precipitation of the region itself.

Although free water is not available, high relative humidities are found in the air within the soil and litter below the snow cover. In general, rates of water loss in insects are greatly reduced at subzero temperatures (Edney 1977). Still, to survive winter conditions that may last for 6 to 9 months, many terrestrial arthropods depend on high relative humidities of the air within their microhabitats. Species overwintering on twigs and branches of trees are exposed to increasingly dry air with decreasing temperatures below freezing. Although little information is available, it is reasonable to assume that these highly exposed species are particularly resistant to desiccation.

8.3.2 Water Vapour Pressure in Frozen Habitats

8.3.2.1 The Risk of Freezing

Frozen soil and litter have a porous structure, partly due to the nature of the substrates themselves. During freezing, the formation of ice crystals causes expansion, adding to the porosity of the soil and litter. In some habitats, such as windswept mountain ridges, a firm layer of ice is formed on the surface, partly due to the temporary melting of ice in periods of strong solar insolation.

A number of invertebrate species are enclosed in habitats like these during the winter. Similar conditions may be found in other closed hibernacula, e.g. in stumps or under the bark of trees. In the soil, most invertebrates probably occupy small hollows or pores completely surrounded by frozen substrate. Some contact with ice cannot be avoided and for a number of species there is apparently a risk of inoculative freezing (Chap. 8.4). For most soil animals the right moisture content before freezing is of great importance. Forge and McGuidwin (1992a) found higher survival rates in juveniles of the nematode *Meloidogyne hapla* exposed to prefreezing soil water potentials of -2050 to -1100 kPa than in those exposed to lower potentials when the soil was frozen at temperatures from -1 to -4 °C (Fig. 11.2). An explanation for this difference could be a greater risk of inoculative freezing or that the nematodes are crushed by the expansion of ice in moist soils. Another possibility is that physiological changes during acclimation to low water potentials makes nematodes better adapted to freezing conditions.

8.3.2.2 Water Vapour Saturation Deficit

In addition to the risk of freezing, many invertebrates also run the risk of dehydration. In the air-filled porous structure of soil and litter, the water vapour pressure is in equilibrium with ice. For animals that overwinter in a supercooled state, e.g. springtails, mites and freeze avoiding insects, this represents a problem. The water vapour pressure in equilibrium with supercooled water is higher than that in equilibrium with ice, and a gradual water loss by evaporation will take place in unfrozen invertebrates.

The equilibrium of the water vapour pressure with an unfrozen organism also depends on the concentration of solutes in the body fluids given by the equation:

$$f = \frac{55.556}{(55.556 + osm)},$$

where f is the molal fraction of water in the body fluid and osm is the osmolal concentration of solutes. The molal concentration of 1 kg of pure water is 1000/18 = 55.556. As an example, at a concentration of 1 osmol, the fraction (f) will be 0.982, which corresponds to a relative humidity (RH) of 98.2 %.

The water vapour pressure deficit (ΔP), which is the difference between the vapour pressure of the body fluid and that of the air, can be calculated as follows, according to Lundheim and Zachariassen (1993):

$$\Delta P = \left(\frac{55.556}{(55.556 + osm) - RH/100} \right) \times Pw$$

where P_w is the vapour pressure of pure water.

As a consequence of evaporation, the osmolality of the haemolymph increases, resulting in a lower vapour pressure deficit. At higher osmolalities, water losses will be gradually reduced, and an equilibrium will finally be established, as long as the ambient RH does not change. To supercooled insects and other terrestrial arthropods overwintering in the soil, it is essential that the water losses do not exceed their tolerance of desiccation. Many species accumulate glycerol and other low molecular weight substances which increase their cold hardiness during overwintering (Chap. 9.2). Since these substances increase the haemolymph osmolality, they also reduce the equilibrium vapour pressure, and protect against dehydration as well as against cold. According to Ring and Danks (1994), molar concentrations of polyols and sugars in the range of 0.5 to 5 are normal in overwintering insects, and a number of examples are presented. At such concentrations, the solutes will reduce the vapour pressure of supercooled solutions below that of ice at the same temperature.

On the other hand, if an overwintering organism is frozen, no loss of water will take place (Zachariassen 1991a; Lundheim and Zachariassen 1993). In this situation the body fluid is in vapour pressure equilibrium with the ice within and outside the body. Regardless of temperature, the organism will not lose water in the frozen state (see below).

8.3.2.3 Dehydration in Earthworm Cocoons

According to Holmstrup (1992) the overwintering cocoons of the lumbricid earthworm *Dendrobaena octaedra*, containing the embryos, are killed by freezing, but are able to survive in a supercooled state. As explained below, they tolerate substantial dehydration which results in decreased melting points in equilibrium with the ambient temperatures.

Cocoons placed at 0 °C and 100 % RH lost less than 5 % of their fresh weight in 13 days (Fig. 8.1). At 3 °C and 97.5 % RH, however, rapid water losses were recorded until a loss of ca. 60 % of their fresh weight was reached. At this level no further losses took place. When the cocoons were kept suspended over ice at -3 °C, at which temperature they remained unfrozen, dehydration took place at a slower rate. After 5 to 6 days the water loss was stabilized at 55-60 % of fresh weight. The experiment at -3 °C resembles the conditions in frozen soil, and clearly illustrates that unfrozen invertebrates may suffer substantial water losses under natural conditions. All cocoons of *Dendrobaena octaedra* survived this degree of dehydration.

Fully hydrated cocoons have a total water content of ca. 3 g g dw^{-1} (= g water per g dry weight), but not all water is available for dehydration. Holmstrup and Westh (1994) showed that approximately 0.5 g g dw^{-1} is osmotically inactive, which indicates that it is bound to other molecules, e.g. proteins. A small proportion of the bound water becomes active during dehydration. The melting point of fully hydrated cocoons is -0.2 °C, and decreases with the surrounding temperature. Under experimental conditions at -4.4 and -6.6 °C, equilibrium was established within 15 days. Two-thirds of the cocoons survived a loss of 85 % of the total water content, which corresponds to 97 % of osmotically active water, and a remaining water content of 0.45 g g dw^{-1}. Injuries observed at still lower temperatures may be related to the loss of all osmotically active water. The results show that freezing of the cocoons is prevented by equilibrating of the body fluid melting points with the surrounding temperature (Chap. 11.4).

The results with *Dendrobaena octaedra* were later confirmed in cocoons of five species of lumbricid earthworms (Holmstrup 1994). In an atmosphere above ice at -2 to -3 °C, all cocoons lost water until the vapour pressure of their fluids was in equilibrium with the ice. The mean equilibrium water loss was between 55 and 60 % of fresh weight for the five taxa.

Holmstrup (1992) also placed some cocoons of *Dendrobaena octaedra* suspended above ice at -21 °C. In this situation the water losses of batch 1 and 2 (Fig. 8.1) were stabilized at less than 15 % of fresh weight. The results indicate that the cocoons froze after a short time and did not lose any more water in the frozen state. In batch 3 some water losses continued to a level of 25 % of fresh weight, and this difference is not fully understood.

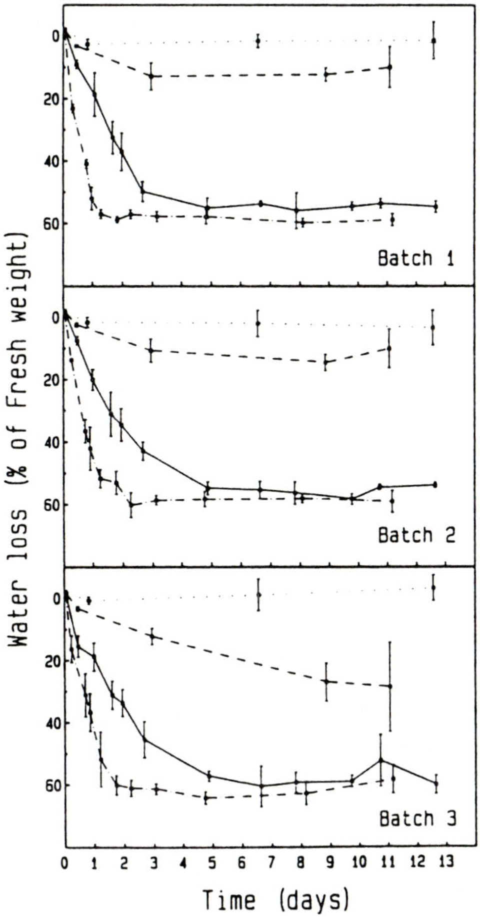

Fig. 8.1. Water loss expressed as percent of fresh weight of *Dendrobaena octaedra* cocoons in relation to time at 0 °C and 100 % RH (*dotted line*), -21 °C and presence of ice (*dashed line*), -3 °C and presence of ice (*solid line*) and 3 °C and 97.5 % RH (*dotted-dashed line*). Each *point* represents the mean (± SD) of six cocoons. *Batch 1* cocoons with early stage undifferentiated embryos; *batch 2* cocoons with early stage differentiated embryos; *batch 3* cocoons with almost fully developed embryos. (Holmstrup 1992)

8.3.2.4 Dehydration in Frozen Insects

The effect of freezing on the rate of water loss is also known from insects. In unfrozen larvae of *Pytho deplanatus* (Coleoptera, Pythidae), the water content was reduced from 69 to 30 % during 5 weeks at -15 °C (Ring 1982). Both this species and larvae of the related *Pytho americanus* are found under the bark of fallen spruce trees in the Rocky Mountains. *P. americanus* overwinters in the frozen state, and the water content (ca. 65 % of fresh weight) of these larvae did not change during exposure to -15 °C. Similarly, Ring (1982) showed that frozen larvae of *Xylophagus* sp. (Diptera, Xylophagidae) are particularly resistant to water loss, and Baust and Nishino (1991) pointed out that larvae of *Eurosta solidaginis* (Diptera, Tephritidae) do not lose water in the frozen state.

The effect of freezing on water losses has recently also been clearly demonstrated by Lundheim and Zachariassen (1993) in two species of beetles. In their experiments, freeze tolerant larvae of *Pytho depressus* (Pythidae) and freeze avoiding adults of *Upis ceramboides* (Tenebrionidae) were divided into two groups, one of which was kept frozen and the other supercooled at -5 °C. The larvae of *Pytho depressus* were kept at ca. 5 % RH above silica gel. Under these conditions, the water vapour pressure deficit (ΔP) between ambient air and the haemolymph of supercooled beetles was calculated as 3.044 mmHg and as 3.013 mmHg in frozen specimens. The corresponding rates of water loss were 1.40 and 0.998 % of body weight day^{-1}. Although overwintering beetles will not encounter dry conditions like this in nature, the experiment demonstrates the fundamental difference between a frozen and an unfrozen organism. This was even more clearly illustrated in adults of *Upis ceramboides* which were kept at water vapour pressure in equilibrium with ice. ΔP was calculated as 0.089 mmHg for supercooled specimens and was virtually zero for the frozen ones. Under these conditions, the supercooled beetles lost 0.054 % of their body water day^{-1}, while no detectable water losses were recorded in the frozen beetles.

Based on several species of beetles, Lundheim and Zachariassen (1993) also concluded that rates of water loss through transpiration are lower in beetles that remain supercooled compared with freeze tolerant species. Thus, the freeze avoiding species have developed more efficient water-saving systems to reduce the risk of desiccation in their hibernacula.

In conclusion, insects that tolerate freezing are better adapted to avoid desiccation during overwintering in hibernacula surrounded by ice. The unfrozen part of their haemolymph is in water vapour pressure equilibrium with the surrounding air, and the equilibrium is not affected by temperature. In unfrozen organisms, a slow dehydration takes place until equilibrium is reached with air, as in the case with earthworm cocoons at -3 °C (Holmstrup 1992). The ability to tolerate the loss of a proportion of their body water may also explain the survival of certain nematodes at high ΔP (Forge and McGuidwin 1992a). For supercooled insects and other invertebrates, the ambient temperature is of great importance, since the water vapour pressure of ice decreases with decreasing temperatures. Thus, the risk of dehydration will increase with lower temperatures in the hibernacula. Survival rates of overwintering invertebrates that are unable to survive freezing of their body fluids are in this way affected by a combination of temperature and desiccation tolerance.

8.3.3 Longevity and Saturation Deficit

In general, cryptostigmatid mites survive much longer at low temperatures than at high temperatures (Cannon and Block 1988). In accordance with this, the antarctic mite *Alaskozetes antarcticus* survived longer periods at 0 °C than at 4 °C. Madge (1964) found that longevity in mites is directly proportional to water losses, which is again related to the saturation deficit of the surrounding air. In winter-collected *Alaskozetes antarcticus*, the weight loss at 4 °C increased with decreasing humidities from 100 to 26 % RH (Fig. 8.2), associated with reduced water content (Cannon 1986). At 100 % RH all the mites survived for 49 days, and the experiment clearly demonstrated the importance of a low saturation deficit during overwintering.

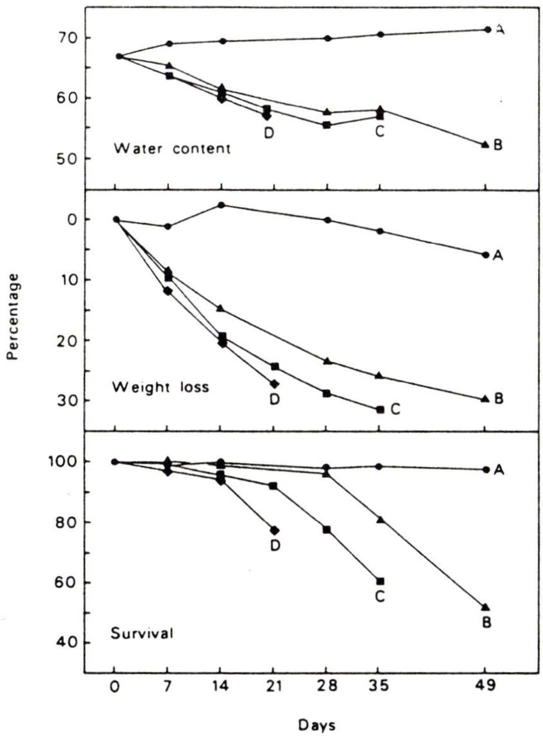

Fig. 8.2. Percentage body water contents, weight loss and survival for adult *Alaskozetes antarcticus* at 4 °C and four different RH. *A* 100 % RH (*filled circles*); *B* 55 % RH (*filled triangles*); *C* 42 % RH (*filled squares*); *D* 26 % RH (*filled diamonds*). (Cannon 1986)

In many terrestrial arthropods cryoprotective substances are lost at high ambient temperatures (Chap. 9.2). Accordingly, *Alaskozetes antarcticus* maintained at 4 °C and 100 % RH lost most of its glycerol content within 49 days (Fig. 8.3).

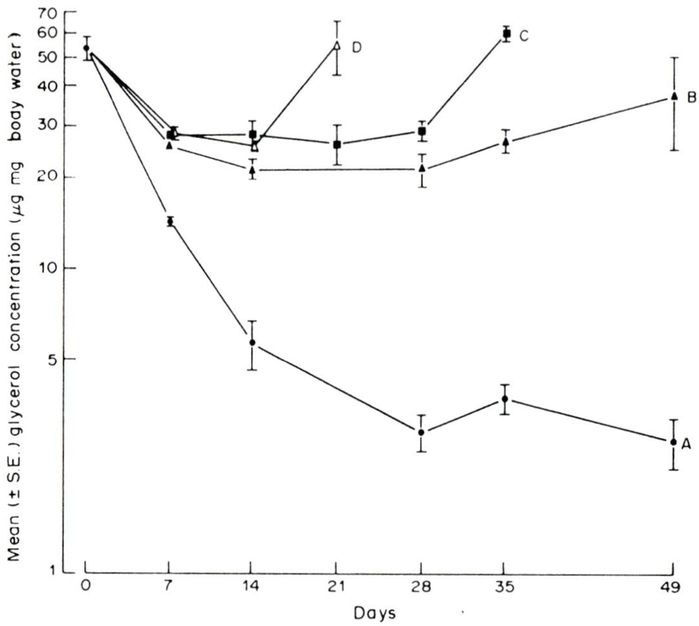

Fig. 8.3. Mean (± SE) glycerol contents (μg mg^{-1} body water) of *Alaskozetes antarcticus* at 4 °C and four different RH. A 100% RH (*filled circles);* B 55 % RH (*filled triangles*); C 42 % RH (*filled squares);* D 26 % RH (*filled diamonds).* (Cannon (1986)

The loss of glycerol also resulted in a decrease in supercooling capacity. In contrast with those kept at 100 % RH, mites kept at lower RHs lost glycerol initially, but following a certain level of dehydration, the content increased. Apparently, a metabolic change was triggered during the dehydration process. Although the data of Cannon (1986) do not show this explicitly, the increased levels of glycerol probably delay the rate of water loss in the mites, giving better protection against dehydration. With higher glycerol levels, mites from the drier treatments also maintained their ability to supercool. The fact that glycerol production is triggered by reduced water contents strongly supports the argumentation that such substances also protect against desiccation.

8.4 Inoculative Freezing

8.4.1 Contact with Ice and Inoculation

Species overwintering in the soil, under bark and in other moist habitats, are more or less surrounded by ice. Those that depend on supercooling to survive (Chap. 9.3) are therefore faced with another problem. During direct contact between ice and the body surface of a supercooled insect, ice may gradually invade the fine pore channels of the cuticle, resulting in an 'inoculative freezing' of the body fluids (Salt 1961; Sømme 1982; Lee 1991).

Many insects survive a certain degree of freezing, following inoculation of ice (Sømme 1974; Tanno 1977), although at lower temperatures the same species depend on supercooling. The tenebrionid beetle *Bolitophagus reticulatus*, which hibernates in the tinder fungus *Polyporus fomentarius*, offers an example (Gehrken et al. 1991). The fungus frequently has a high moisture content, resulting in inoculative freezing of adult beetles at subzero temperatures. In mid-winter frozen beetles survive temperatures down to - 5 °C, while specimens in dry tinder-fungus tolerate -30 °C or lower in the supercooled state. In other insects, inoculative freezing at high subzero temperatures is a requirement for survival (Chap. 9.4.5).

In alpine and polar habitats, the ground may be frozen into a more or less compact layer, partly by cooling from above and partly by the raised level of permafrost during the winter. Insects, collembolans and mites are literally enclosed by ice (Sømme 1982) in the upper few centimetres of soil and litter. The porous structure of frozen soil and litter contains numerous microscopic cavities offering sanctuary for these animals. Although little information is available on the conditions within such microhabitats, it is difficult to understand how contact with ice is avoided.

In general, mites and springtails possess a hydrophobic cuticula, and it is possible that species enclosed by ice in their overwintering sites are better protected against the inoculative penetration of ice. In fact, Young and Block (1980) showed that specimens of *Alaskozetes antarcticus* may become embedded in ice during the winter and yet resume activity upon rewarming. This was later confirmed by Cannon (1987), who found that these mites had a survival rate of 78 % after 21 days in ice at -15 °C. Shimada et al. (1992), however, showed that inoculation takes place at lower temperatures, in both adults and tritonymphs. At -20 and -24 °C a much higher percentage of tritonymphs was killed when cooled on wet filter paper compared with those on dry filter paper. Survival on dry filter paper was possibly due to the high supercooling capacity of some of the mites (Chap 9.3.2).

Other reports clearly demonstrate that contact with ice may be fatal. When supercooled specimens of the springtail *Gomphiocephalus hodgsoni* were touched by ice at -11 °C, they froze and died (Fitzsimons 1971). Pan and Shimada (1991) found different responses to inoculative freezing in mites from the South Shetland Islands in the Maritime Antarctic. In their experiments, freshly collected mites were placed on dry or moist filter paper in small bottles, which were then cooled to temperatures in the range -3 to -15 °C. In the prostigmatid mites *Rhagidia gerlachi* and *Steretydeus villosus*, mortality increased greatly among specimens on moist filter paper, suggesting that inoculative freezing took place. Moisture did not affect the survival of the mesostigmatid mite *Gammasellus racovitzai*, and this may be due to a less permeable cuticle.

8.4.2 Protection Against Inoculative Freezing

Overwintering in the soil often takes place in hollows that are prepared in advance. Some insects spin special cocoons, while others dig out small chambers. Cocoons are common in a variety of moths and parasitoid Hymenoptera, while silken hibernacula are prepared by different butterflies (Danks 1991). As an example, larvae of the bollworm *Heliothis zea* and the tobacco budworm *Heliothis virescens* construct subterranean chambers where they overwinter as pupae (Eger et al. 1982). The carabid beetle *Pelophila borealis* hardens the walls of its earthen chambers (Østbye and Sømme 1972) while other carabid beetles spend the winter in small hollows or cells dug out under rocks.

The function of some of these structures is not known, but they may serve several purposes. To some extent they give mechanical protection and allow for better air circulation. More important, they probably protect against excessive moisture. In this way direct contact with ice will be less when the water freezes, and consequently the risk of inoculative freezing will be reduced.

In summarizing their own and previous results, Hoshikawa et al. (1988) related differences in overwintering strategies to the depth of overwintering sites of 11 insect species from Hokkaido, Japan (Fig. 8.4). In all species, contact with moisture under experimental conditions caused inoculative freezing at temperatures higher than the supercooling points. While species that survived freezing were able to overwinter in the upper soil layers, most of the species that depend on the avoidance of freezing seek sites from 2 to 12 cm below the surface. Many of these species overwinter in cocoons or earthen cells, which prevent contact with moist soil, and consequently inoculative freezing. Species that lack this protection are cold avoiding, and dig to depths of 5 to 30 cm. An interesting example is the elaterid beetle *Agriotes fuscicollis* which overwinters in both the larval and adult stages. The adults remain supercooled in their pupal cells at 10 cm, while the larvae avoid inoculative freezing by digging to depths of 15 to 30 cm. Without snow cover at Hokkaido, very low winter temperatures are recorded in the upper soil layers, but temperatures increase with depth. As pointed out by Hoshikawa et al. (1988), by descending 10 cm underground from the snowless mid-winter ground surface, insects will experience a 10 °C increase in temperature, which is comparable to a horizontal migration of more than 100 km southwards.

Red wood ants offer another example of insects that avoid severe winter cold. *Formica* spp. usually overwinter deep below their mounds, where temperatures rarely fall below freezing. In their experiments Erpenbeck and Kirchner (1983) showed that workers of *Formica polyctena* and *Formica rufa* depend on supercooling to survive freezing temperatures. Increasing mortalities occurred with increasing time at temperatures below zero. High moisture content of the nest material greatly increased mortality, in particular below -5 °C.

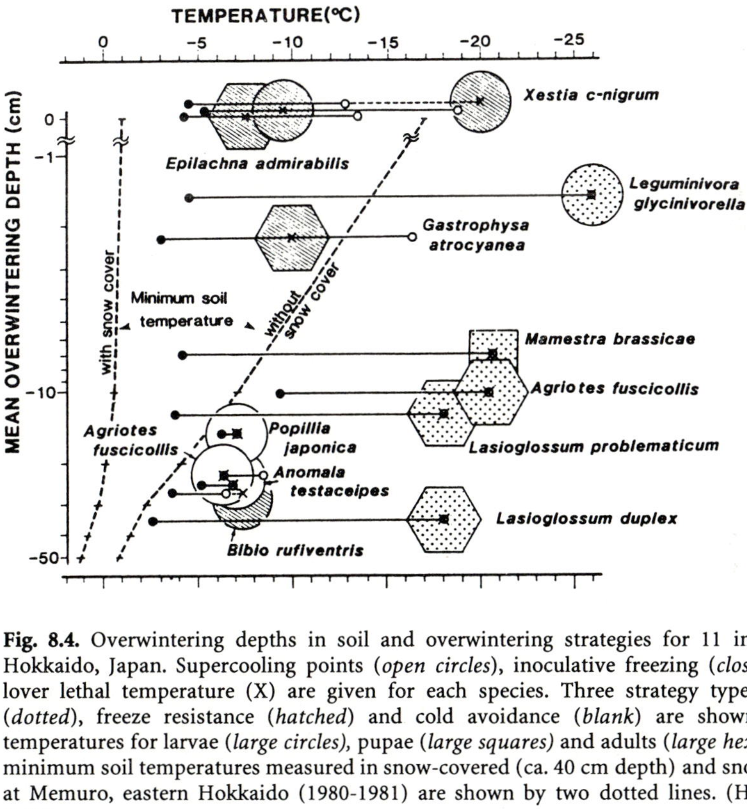

Fig. 8.4. Overwintering depths in soil and overwintering strategies for 11 insect species in Hokkaido, Japan. Supercooling points (*open circles*), inoculative freezing (*closed circles*), and lover lethal temperature (X) are given for each species. Three strategy types, supercooling (*dotted*), freeze resistance (*hatched*) and cold avoidance (*blank*) are shown at the lethal temperatures for larvae (*large circles*), pupae (*large squares*) and adults (*large hexagons*). Annual minimum soil temperatures measured in snow-covered (ca. 40 cm depth) and snow-free grounds at Memuro, eastern Hokkaido (1980-1981) are shown by two dotted lines. (Hoshikawa et al. 1988)

Gall insects often overwinter within their galls, where the plant tissue offers a protective shield. Leaves with galls frequently fall to the ground, where the gall insects overwinter in the plant litter. In Alaska, the gall wasp *Diplolepis bicolor* overwinters in multi-compartmental galls, most of which fall to the ground in the autumn (Miller 1982).

Some galls, such as those of the goldenrod fly *Eurosta solidaginis* (Baust and Nishino 1991) remain on the plants during the winter, protruding above the snow cover of the ground. A delicate balance exists between the larva and its gall. During autumn the gall affords a thermal buffer to extreme temperatures. At the onset of subzero temperatures, the moist gall tissue (with a water content of 65 %) freezes slowly, and the larva is maintained at 0 °C for several hours or days. This is the optimal temperature for sorbitol production in the haemolymph, and sorbitol is later required for the larva to tolerate freezing of their body tissues. Ice growth in the gall tissue initially begins in the inner layers and proceeds outward, causing a saturation deficit in the air within the larval chamber. In this way, ice crystals are not formed on the surface of the larvae, and inoculative freezing is prevented at a time when the larvae have not yet become freeze tolerant. According to Baust and Nishino (1991), larval frass in the chamber acts as an efficient ice-nucleating agent, causing inoculation through the cuticle. As the autumn progresses, the gall tissue dehydrates, but water losses are not increased since larvae are frozen at this time.

Inoculative freezing may be prevented by the presence of thermal hystersis proteins (antifreeze proteins) (Chap. 9.3), as shown by Gehrken (1992) in the bark beetle *Ips acuminatus* and the cerambycid beetle *Rhagium inquisitor*. Both species risk contact with ice under the bark of live or dead coniferous trees, respectively. In the laboratory, both species were invariably killed by freezing when inoculated by contact with external ice, but freezing was completely prevented in *I. acuminatus* by the presence of thermal hysteresis proteins. In *R. inquisitor* the capacity of antifreeze proteins to inhibit freezing was much smaller, but freezing occurred at lower temperatures in specimens with antifreeze proteins than in those without.

8.5 The Effect of Dehydration on Cold Hardiness

As pointed out previously (Chap. 8.3), the water content of freeze avoiding insects is usually reduced in the overwintering stage. In many species, a reduction in water content is associated with the initiation of diapause (Danks 1987). Any increase in the concentration of solutes will lower the haemolymph melting point of an insect (Zachariassen 1985). Since its supercooling capacity is increased with lowered melting points (Chap. 9.3) the cold hardiness of the insect is improved.

Different views have been expressed regarding the effect of dehydration on cold hardiness. Payne (1926) found a considerable increase in the supercooling capacity of *Synchroa punctala* larvae dried over $CaCl_2$, while other authors thought that water losses must be extreme before the increase in super-cooling ability is of any ecological importance (Salt 1961; Sømme 1982).

In *Cryptopygus antarcticus* and *Alaskozetes antarcticus* there is little evidence of seasonal patterns in supercooling capacity corresponding to relative water contents (Cannon and Block 1988). *C. antarcticus* has a large amplitude in water content fluctuations and a water content of less than 40 % of fresh weight is tolerated.

As pointed out by Zachariassen (1985), the solute concentration is inversely proportional to the water content, and an increased supercooling capacity can be calculated when the effect of changes in osmolality on the supercooling point (Chap. 9.3.) is known. A normal solute concentration of insect body fluid is ca. 500 mosmol kg^{-1}, and if half of the water is lost the concentration will increase to 1000 mosmol kg^{-1}. Since the supercooling points of insects are lowered twice as much as the haemolymph melting points (Fig. 9.8), the decrease will be -1.86 °C. In several insects accumulation of low molecular weight cryoprotectant substances takes place in preparation for overwintering. If the body fluid concentration of an insect has a value of 2000 mosmol kg^{-1}, the same reduction in water content will raise the concentration to 4000 mosmol kg^{-1} and lower the supercooling points by -7.4 °C.

In the larvae of *Pytho deplanatus*, the water content decreases from 69 % of fresh weight in non-acclimated specimens to 30 % in fully cold-acclimated ones (Ring 1982). This dehydration is accompanied by an increase in glycerol concentration to 13.2 % and of sugar concentrations to 5.5 % of body fresh weight.

In other species, decreasing water content may trigger the production of glycerol and other cryoprotectant substances, as seen in *Alaskozetes antarcticus* (Cannon 1986). The freeze avoiding bark beetle *Ips acuminatus* did not accumulate ethylene glycol under experimental conditions until the water content had fallen below 52 % of fresh weight (Gerkhen 1984). In an outdoor population of this species the water content fell to 49 % of fresh weight during the winter.

From these examples it appears that the relation between the water balance and cold hardiness is of great importance to many insects and other terrestrial arthropods. Some degree of dehydration is to their advantage, resulting in increased osmolality of the body fluid, which, again, increases the supercooling capacity. In some species, reduced water content may trigger the production of low molecular weight cryoprotective substances.

8.6 Conclusions

The moisture conditions affect overwintering of terrestrial invertebrates in several ways. Humidity is one of several factors that determine the selection of overwintering sites with suitable microclimatic conditions. Many species construct hibernacula in the soil or build cocoons which give protection against mechanical stress and excessive moisture. Microarthropods overwinter in the porous structure of soil and litter, and due to the expansion of ice, the porosity increases when soil moisture freezes.

When insects and other terrestrial arthropods are in contact with ice, freezing may take place through the cuticle. Different kinds of galls and hibernacula may protect against such inoculative freezing. It is not completely understood how springtails and mites avoid inoculation in frozen soil and litter. Apparently, the cuticle of many species is not easily penetrable by ice, and direct contact with ice crystals may be limited within their porous environment.

In hibernacula surrounded by ice, the water vapour pressure of invertebrate body fluids is higher than in the air, as long as the animal is not frozen. Overwintering invertebrates experience lack of liquid water comparable to that of hot desert species. Due to the water vapour pressure deficit, a gradual dehydration takes place, resulting in an increased osmolality of the body fluids. Since the vapour pressure of solutions is lower than that of pure water, an equilibrium with the surrounding air is eventually obtained. Many overwintering arthropods have high concentrations of low molecular weight organic substances in their haemolymph. Substances like sugars and glycerol are known to protect against cold, and will protect against desiccation as well since the water vapour pressure is lowered. When the ambient temperature decreases, the water vapour pressure of ice also decreases, and this may result in excessive dehydration in supercooled animals.

The dehydration of lumbricid earthworm cocoons is exceptional, since the melting point of the body fluids is in equilibrium with the ambient temperature. The cocoons can tolerate the loss of 85 % of their total water, which is higher than is known from any other invertebrates, except those that enter anhydrobiosis.

In frozen insects the situation is different, since the water vapour pressure of the unfrozen fraction of their haemolymph is in equilibrium with the ambient water vapour pressure of ice, regardless of temperature. Insects that survive freezing of their tissues are also protected against water loss during overwintering.

9 Cold Hardiness in Terrestrial Arthropods

9.1 Introduction

During the last few decades a vast number of articles have been published within the field of cold hardiness in insects and other terrestrial arthropods. The intriguing question as to how poikilotherm animals are able to survive temperatures far below the melting points of their haemolymph has puzzled many biologists. Although many problems remain unanswered, their work has revealed the general principles of physiological and biochemical adaptations to low temperatures.

The first account of insect cold hardiness was published by Réaumur (1736), who found that insect larvae from the forest were able to withstand -9 °Réaumur without perishing. Early work on insect cold hardiness was summarized by Bachmetjew (1901) and Payne (1926). The studies of R.W. Salt introduced a modern approach to insect cold hardiness, most of which was summarized by himself (Salt 1961). Other important general reviews in this field were presented by Ushatinskaya (1957) and Lozina-Lozinskii (1974), and a fundamental introduction to frost resistance in insects was given by Asahina (1969).

During later years, different aspects of insect cold hardiness have been discussed in a number of review articles (e.g. Merivee 1978; Ring 1980; Sømme 1982; Baust and Rojas 1985; Zachariassen 1985; Storey and Storey 1988; Ushatinskaya 1990; Duman et al. 1991a; Lee 1991), as well as in two recent books (Lee and Denlinger 1991; Leather et al. 1993).

Although the general aspects of cold hardiness in terrestrial arthropods are discussed, the emphasis of the present chapter is on species inhabiting alpine and polar areas, including cold deserts. Reviews related to this subject have been published by Sømme (1981, 1989), Block (1984a), Cannon and Block (1988) and Sømme and Block (1991).

9.2 Concepts of Cold Hardiness

With regard to their cold hardiness, Salt (1961) divided the insects in two principle categories which he termed 'freezing-tolerant' and 'freezing--susceptible'. The first group comprises insects that survive the actual formation of extracellular ice within their tissues. Most insects, however, do not survive freezing, and freeze susceptible species depend on their ability to avoid ice formation by supercooling.

In addition to freeze tolerance and freeze avoidance, the ability to survive long exposures at low temperatures is an important aspect of insect cold hardiness. Biological processes slow down at low temperatures, but some level of activity is required to maintain vital functions. Terrestrial arthropods that are not adapted to low temperature will suffer from lethal *chill injuries*. This applies to temperatures above as well as below haemolymph melting points. While tropical insects and summer-acclimatized temperate species (Bale et al. 1988; Pullin and Bale 1988) may be killed within hours at low temperatures, overwintering species can survive months of exposure at 0 °C or lower (Zachariassen 1985; Sømme and Block 1991). As pointed out by Storey and Storey (1988), chilling injuries are caused by the disruption of metabolic regulation, particularly cellular energetics, which, when too severe or prolonged, lead to irreversible lethal metabolic imbalances. As a consequence, both freeze avoiding and freeze tolerant species must also be adapted to avoid the risks of chilling (Duman et al. 1991a).

Although the terms have been criticized by some authors (Baust and Rojas 1985; Bale 1987), the division into freeze tolerance and freeze susceptibility is a practical way of describing the strategies for cold hardiness of different species. Although they are similar in some respects, freeze tolerance and freeze susceptibilty represent two principally different strategies of adaptation. It must be stressed, however, that all kinds of intermediate situations exsist (Zachariassen 1985; Lee 1991; Leather et al. 1993). Many insects survive freezing at temperatures slightly below their haemolymph melting points (e.g. -1 °C), but as a result of increasing ice formation, mortality increases rapidly with decreasing temperatures. Freeze tolerance in many species shows great seasonal variations, and survival is related to time spent in the frozen state. Overwintering larvae of two beetles, *Dendroides canadensis* and *Cucujus clavipes*, are even known to be freeze susceptible in most years but freeze tolerant in unusually severe winters (Duman et al. 1991a).

In spite of the diffuse differences between freeze tolerance and freeze susceptibility in many species, a number of clearcut cases are known. Collembolans and mites, even those that overwinter in exposed alpine habitats, all depend on supercooling for survival and are classified as freeze susceptible (Sømme and Block 1991). At the other end are a number of beetles and other insects that survive the entire winter in the frozen state (Zachariassen 1985), and may safely be termed freeze tolerant. To avoid confusion, it is important that details of the experimental procedures and results are described to give a documentation of the tolerance or sensitivity of the species in question.

The increasing literature within the field of cold hardiness in terrestrial arthropods has resulted in an expanding nomenclature. According to Leather et al. (1993), certain terms have been used synonymously, and the inconsistent use of some terms by different authors is confusing. A standard terminology is required. A list of definitions of important concepts is given below, partly based on terminology suggested by Zachariassen (1985) with some modifications of Leather et al. (1993).

9.2.1 Definition of Important Concepts

Cold hardiness (*cold tolerance*) is the general ability of an organism to survive at low temperatures, including prolonged exposure in both freeze tolerant and freeze avoiding species.

Chilling injuries are caused by metabolic imbalances in species that are not adapted to low temperatures.

Freeze tolerance is the ability to survive the formation of ice in the body fluids. The term is most commonly used in connection with insects that survive freezing below their supercooling point.

Freeze susceptibility (*freezing sensitivity*, *freezing intolerance*) is the lack of tolerance of ice formation in the body fluids.

Freeze avoidance is the survival strategy of freeze sensitive species that are exposed to low temperatures.

Supercooling is the unfrozen state of water or aqueous solutions (or other liquids) below the melting point.

Supercooling point (*SCP, crystallization temperature*) is the temperature at which spontaneous freezing occurs in a supercooled system. Under experimental conditions the rate of cooling (e.g. 1 °C min^{-1}) is important to obtain comparable SCPs.

Supercooling capacity is the difference between the SCP of an intact insect and the melting point of its haemolymph.

Melting point (MP) is the temperature of a fluid at which the last tiny ice crystal of a frozen sample disappears during slow heating, whereas a slight decrease in temperature results in crystal growth.

Freezing point (FP) is the temperature at which tiny ice crystals in a solution begin to grow. In most solutions, the FP is equal to the MP, except when thermal hysteresis proteins are present.

Hysteresis freezing point is the temperature, lower than the MP, where a tiny ice crystal will start growing in a solution containing thermal hysteresis proteins (proteinaceous antifreeze agents, antifreeze proteins).

Thermal hysteresis describes the difference between the MP and the hysteresis freezing point of a solution.

Ice-nucleating agents (INAs) are substances that cause nucleation or initation of freezing in a supercooled system. INAs may be present in the haemolymph, cells and other body compartments.

9.3 Freeze Susceptible Species

9.3.1 The Importance of Supercooling

When the temperature of water is lowered, freezing will typically not take place at 0 °C unless foreign substances causing heterogeneous nucleation are present. Normally, water will supercool to some temperature below its melting point. If a small ice crystal is added to a container of supercooled water, freezing is immediately intiated. Small volumes of highly purified water, e.g. droplets suspended in mineral oil, supercool to temperatures close to -40 °C (MacKenzie 1977). This is the temperature of homogeneous nucleation of water. As the temperature is lowered below the melting point, the structure of ice-like clusters of water molecules increases in size. When they reach a critical size, the cluster acts as a small embryo ice crystal, nucleating the supercooled water.

The majority of insects and other terrestrial arthropods are killed by freezing, and belong to the freeze sensitive group. For these species the avoidance of freezing is essential, and this is achieved by their greater or lesser ability to supercool. In the same way as small volumes of highly purified water will supercool to temperatures far below the melting point, insects may be considered as liquid containers in which the conditions for supercooling are exceptionally favourable (Sømme 1982). The SCPs of insects seem to correspond well with SCPs of physical solutions with similar volume and osmolality, when INAs are lacking.

A number of factors contribute to the cold hardiness of freeze avoiding species. Most insects overwinter in a specific stage, which must be reached before development and activity are retarded by low ambient temperatures in the autumn. Many collembolans, mites and spiders are known to overwinter in several instars.

Whatever their overwintering stage, terrestrial arthropods must be prepared for the winter. In most species, physiological adjustment takes place to increase their cold hardiness. Considerable seasonal changes may be observed from an increase in cold hardiness in early autumn and winter to the loss of winter adaptations in the spring. Seasonal adaptations which increase freeze avoidance include the removal of ice nucleating agents and the accumulation of cryprotective substances.

Dependence upon supercooling as a way of winter survival has been demonstrated in many taxa of terrestrial arthropods, and Sømme (1982) presented a list of several hundred species. The list comprises a number of Acari, Araneae, Collembola, and insects in the egg, larval, pupal and adult stages. Furthermore, supercooling has been demonstrated in desert scorpions and millipeds (Cloudsley-Thompson 1973, 1991; Crawford 1981). Kirchner (1973) listed 50 species of spiders collected during the winter, and which were all killed by freezing.

Avoidance of freezing is not unlimited, and the capacity for supercooling is affected by a number of factors. The limit of supercooling in terrestrial arthropods is conveniently measured as the supercooling point (SCP), which also represents the lower limit of survival in freeze sensitive species. The SCP is the temperature at which spontaneous freezing occurs during gradual cooling (Fig. 9.1).

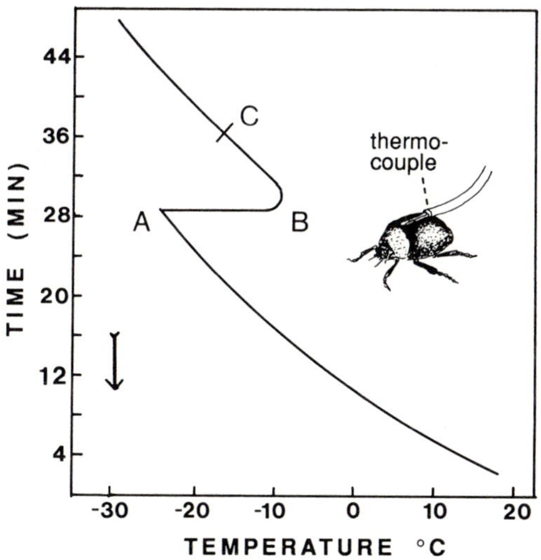

Fig. 9.1. Measurement of the supercooling point of an insect in contact with a copper-constantan thermocouple connected to a recording potentiometer. When spontaneous freezing occurs (*A*), a rapid increase in temperature is recorded (*B*). Most insects will survive if removed to non-freezing temperatures at *B* , but freeze avoiding species are killed when the temperature again decreases, e.g. to *C* or lower. The *arrow* indicates the direction of movement of the paper of the recording potentiometer

For comparative purposes, a constant cooling rate should be applied, since the rate will affect the temperature of crystallization (Salt 1966). A cooling rate of 1 °C min^{-1} has commonly been used, but other velocities may be more appropriate according to the size of the animal. When a supercooled insect freezes, the released heat of fusion will cause the temperature to rise, theoretically close to the melting point of the haemolymph (Zachariassen 1991b). When the thermocouple is not penetrating the insects, as in Fig. 9.1, the temperature of the melting point is not recorded, and the size of the rebound is arbitrary, depending on the degree of external contact between the thermocouple and the insects. The technique for recording the SCPs of insects, based on the use of copper-constantan thermocouples, was devised by Bachmetjew (1901), and further developed by Payne (1926) and Salt (1936). Robinson (1928) observed that piercing the insect with the thermocouple raised the SCP considerably, and Salt (1936) pointed out that the thermocouple must be kept in contact with the surface of the animal if correct measurements are to be obtained. Since piercing was a common procedure of early studies, some of the older data on supercooling are presently mainly of historical interest.

Related to their small size, insect eggs often have low SCPs, commonly in the range of -30 to -40 °C (Sømme 1982). A number of them overwinter in exposed situations such as on twigs and branches of trees. Freeze avoiding overwintering larvae present a wide range of mean SCP values, ranging from -16 to below -60 °C depending on their microhabitat. Most pupae have mean SCPs in the range -20 to -30 °C. Values are frequently still higher in adult insects, although several SCPs in the range -25 to -35 °C have been recorded. Mites have generally low SCPs, while many spiders have poor supercooling capacity, suggesting that they depend on more protected sites for overwintering. Mean SCPs in the range of -20 to -35 °C have been recorded in alpine and polar Collembola (Sømme 1981; Cannon and Block 1988).

9.3.2 Supercooling Capacity in Microarthropods

9.3.2.1 Collembolans

The cold tolerance of microarthropods, with special emphasis on species from south polar areas and other extreme environments, has been reviewed by Cannon and Block (1988). A list of the lowest recorded SCPs in 16 species of Collembola was presented by Sømme (1989), based on field-fresh or laboratory-acclimated specimens. Some species from the Norwegian mountains and the Tyrolean Alps inhabit vegetation and soil of windswept ridges or lichens on exposed rocks. Since there is no protective snow cover, these animals experience very low ambient temperatures during winter. Lowest mean SCPs of -33 °C were recorded in *Vertagopus sarekensis*, while individual specimens of *Tetracanthella wahlgreni* supercooled to -39 °C. Higher mean SCPs of -17 °C were recorded in *Onychiurus vontoerni* and of -19 °C in *Isotoma* sp. from the Alps. Both species probably seek protection in the deeper layers of the soil. In *Entomobrya nivalis*, which is found on glaciers in the Alps, Zettel (1982) showed that juveniles had lower SCPs than adults. This is probably related to the reduced chances of nucleation with smaller size. Two species of entomobryid Collembola from 4500 m on Mount Kenya, with mean SCPs of ca. -20 °C, avoid freezing at the low night temperatures typical of high altitudes in tropical mountains (Sømme and Zachariassen 1981).

Among antarctic species, *Isotoma klovstadi* has been reported to survive temperatures down to -50 or -60 °C (Pryor 1962), and confirmation of this result would be interesting. In *Gomphiocephalus hogdsoni* from South Victoria Land, Block (1985b) found mean SCPs of -30 °C during summer, and Sømme (1986a) reported SCPs of -36 °C in *Cryptopygus sverdrupi* from Dronning Maud Land following gradual acclimation at low temperature. The common antarctic collembolan *Cryptopygus antarcticus* has a mean SCP of -26 °C (Cannon and Block 1988).

Folsomia quadrioculata from Arctic Alaska survived temperatures down to -23 °C (Tanno 1975), while summer specimens of *Onychiurus arcticus* from Spitsbergen had mean SCPs of -6 to -8 °C (Block et al. 1994).

9.3.2.2 Mites

Lowest mean SCPs were in the range -22 to -33 °C in 14 species of alpine cryptostigmatid mites (Sømme 1989). Species from windswept mountain ridges, such as *Carabodes labyrinthicus* from Norwegian mountains and *Lepidozetes singularis* from the Tyrolean Alps, had the greatest capacity for supercooling. The cryptostigmatid mite *Alaskozetes antarcticus*, which is widely distributed in the maritime Antarctic, has mean SCPs of -30 to -31 °C during the austral winter (Cannon and Block 1988). Another cryptostigmatid, *Halozetes belgicae*, from Signy Island has a mean SCP of -34 °C.

9.3.3 Removal of Ice-Nucleating Agents

As pointed out by Zachariassen (1985), most tropical insects as well as summer- and winter-active temperate insects have SCPs in the range -7 to -12 °C. The reason why crystallization takes place at these temperatures and not closer to the temperature of homogeneous nucleation in pure water (-40 °C), is due to the presence of ice-nucleating agents (INAs).

The cessation of feeding and clearing of the gut at the onset of winter in many freeze avoiding insects increases supercooling capacity (Sømme 1982). The ice-nucleating effect of gut contents in insects appears to be well documented (Zachariassen 1985; Shimada 1989). Components causing nucleation are probably mineral dust and the food particles themselves. It has recently been demonstrated that the ingestion of ice-nucleating bacteria raises the SCPs of feeding insects dramatically (Lee 1991). Although evidence in favour of gut nucleation is overwhelming, there are also exceptions that require further study, as pointed out by Baust and Rojas (1985) and Leather et al. (1993). As shown below, nucleation may also take place in other body compartments or by surface contact with ice.

INAs are also likely to be present in other compartments of the insect body (Zachariassen 1985; Duman et al. 1991a). Different proteins and lipoproteins may have ice-nucleating activity. The haemolymph is apparently not the site of nucleation in summer active insects, but INAs may presumably be present intracellularly. The freeze avoiding tenebrionid beetle *Bolitophagus reticulatus* has the ability to remove or mask INAs in winter, but during warm acclimation of starved specimens, the SCP rises abruptly from -20 to -8 °C (Zachariassen 1985), suggesting reappearance of intracellular INAs. The removal of haemolymph lipoproteins with ice-nucleating activity, however, is associated with a lowering of SCPs in the stag beetle *Ceruchus piceus* (Neven et al. 1986). In this way the presence of gut content and endogeneous INAs limits the supercooling capacity of active insects (Fig. 9.2), while the removal or inactivation of these substances increases the chances of avoiding freezing.

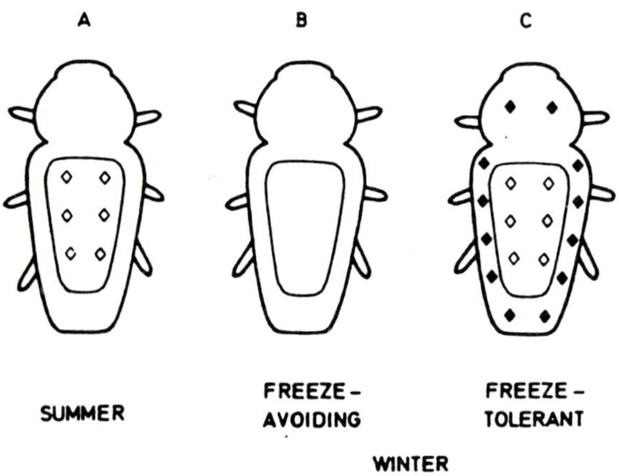

Fig. 9.2A-C. The distribution of ice-nucleating agents in body fluids of various categories of insects. *Open symbols* Ice-nucleating agents in intracellular or intraintestinal fluid; *closed symbols* ice-nucleating agents in haemolymph. (Zachariassen 1980)

As pointed out by Duman et al. (1991a), the removal of endogeneous INAs, on either evolutionary or seasonal time scales, could be the preferred overwintering mechanism for some species, since energy need not be expended in the production of antifreezes. Complete removal of proteinous

INAs, however, may not be feasible if their function is required over the winter. Some of the structures of proteins and lipoproteins that impart ice-nucleator activity may also be required for their normal functioning. Selection against nucleating proteins may therefore not be possible.

9.3.4 Seasonal Changes

9.3.4.1 Microarthropods

While the previous section (9.3.2) presented the lowest values of mean SCPs that have been recorded, considerable seasonal variations in supercooling capacity have been monitored in a number of alpine and polar microarthropods (Cannon and Block 1988; Sømme 1989).

According to Sømme and Conradi-Larsen (1977), the mean SCPs of the collembolan *Tetracanthella wahlgreni* and the cryptostigmatid mite *Carabodes labyrinthicus* were in the range of -6 to -8 °C (Fig. 9.3). Both species inhabit windswept mountain ridges at Finse, Norway. Greater ranges of SCPs were found during the autumn with several individuals tolerating temperatures far below the summer means. SCPs of both species were in the range of -20 to -39 °C in mid-winter, while an increase in values was observed in *Tetracanthella wahlgreni* during the month of May. Winter conditions are still prevailing at this time of the year, and the habitat of these species is mostly covered by ice. Due to high insolation in late May and early June, rapid melting took place and the mean SCPs of the Collembola returned to a summer value of -6 °C. This increase was slightly delayed in *Carabodes labyrinthicus*, which exhibited a large range: a number of specimens had high SCPs in early June.

Seasonal variations have also been reported in *Anurophorus laricis* and *Xenylla maritima* (Leinaas and Sømme 1984) and in *Vertagopus sarekensis* and *V. westerlundi* (Leinaas and Fjellberg 1985) from Norwegian mountains. In *Entomobrya nivalis* from the Alps, von Almen and Zettel (1984) found that the annual cycle of cold hardiness is influenced by both temperature and photoperiod, while the supercooling capacity of *Isotoma hiemalis* is directly influenced by temperatures below 0 °C.

Schenker (1983) demonstrated that SCPs in cryptostigmatid mites from the Alps were lowered during acclimation at temperatures from 10 to -5 °C, suggesting that seasonal changes will be induced by decreasing temperatures during the autumn. The antarctic mite *Alaskozetes antarcticus* has a median SCP of -30 °C in mid-winter, a larger range in the spring and a median of -4 °C in the summer (Cannon and Block 1988). Even in summer a few individuals have SCPs between -10 and -30 °C, giving a skewed distribution of individual measurements.

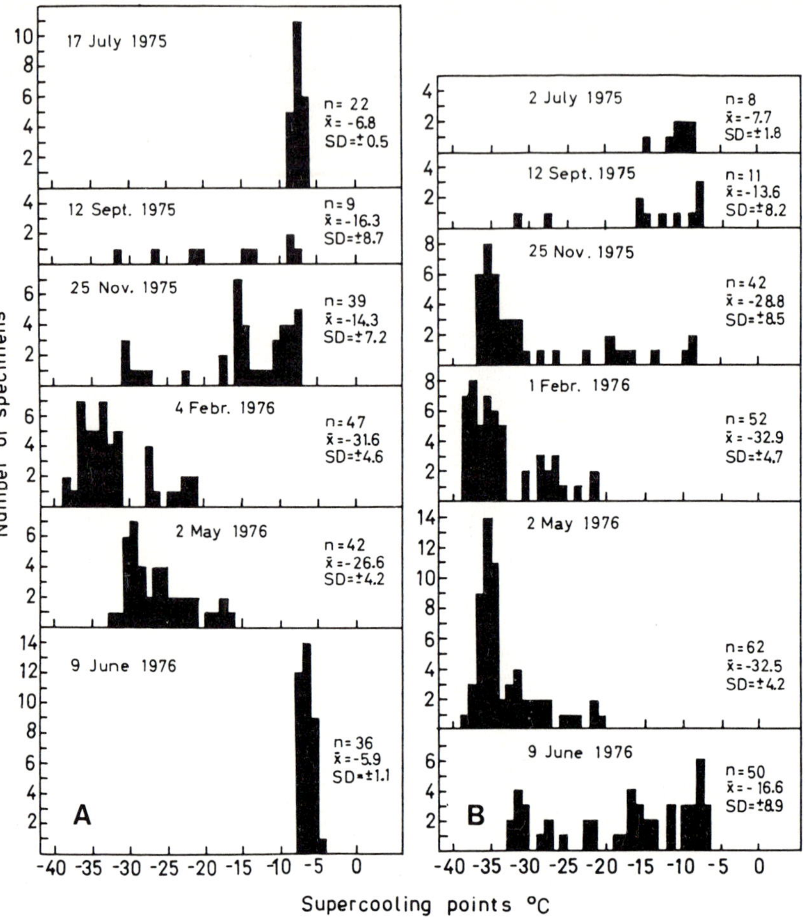

Fig. 9.3A,B. Supercooling point distribution histograms of the springtail *Tetracanthella wahlgreni* (**A**) and the oribatid mite *Carabodes labyrinthicus* (**B**) at different times of the year. (Sømme and Conradi-Larsen 1977)

9.3.4.2 Beetles

Seasonal changes in cold hardiness have also been demonstrated in alpine beetles. At Finse, Norway, several carabid beetles overwinter in habitats covered by layers of deep snow (Østbye and Sømme 1972; Bakken 1985). One of them, *Pelophila borealis*, depends on supercooling to survive ambient temperatures below -5 °C; during the autumn the SCPs drop to within the range -12 to -20 °C (Fig. 9.4). In *Patrobus septentrionis* and *Calathus melanocephalus*, the mean SCPs dropped from -6 °C in the summer to -14 °C in mid-winter (Bakken 1985).

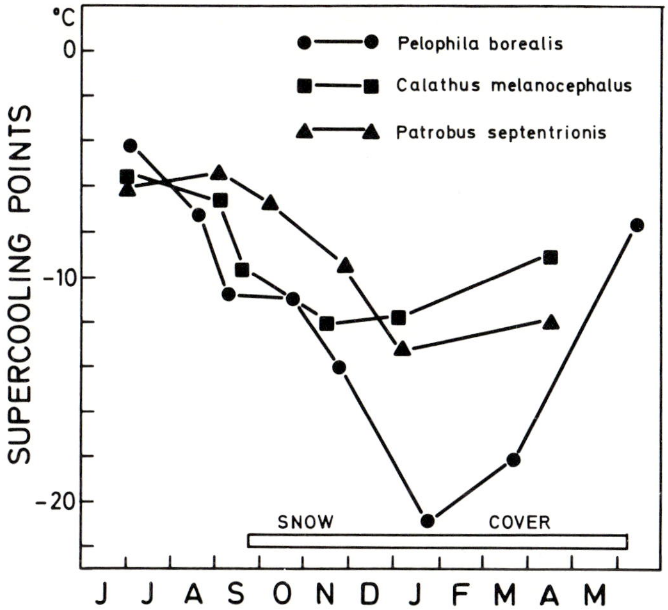

Fig. 9.4. Seasonal variation in mean SCPs of three alpine species of carabid beetles from Finse, Norway.

Apparently, their supercooling capacity is greater than required, since the deep snow cover protects them from temperatures as low as this. The critical time for the beetles is during spells of early frost in the autumn, before sufficient snow has accumulated.

Bakken (1985) found that samples of the haemolymph of *Patrobus septentrionis* and *Calathus melanocephalus* were free of potent INAs and supercooled to values far below the SCPs of intact beetles. The site of nucleation in *C. melanocephalus* did not lie in the gut, but in some other unidentified part of the body. In *P. septentrionis* the site of nucleation was in the hindgut. In both species the action of INAs was reduced by low temperature acclimation and the accumulation of low molecular weight cryoprotectant substances, of which myo-inositol was an important component.

These results reveal a pattern of adaptations in the cold hardiness of beetles that is different from that of microarthropods in several respects. Alpine freeze avoiding beetles would not be able to survive in the exposed windswept habitats of springtails and mites. They are also less cold-hardy than many freeze tolerant species of beetles from alpine and arctic environments (Chap. 9.4.3).

9.3.5. Gut Content and Ice-Nucleating Agents

Populations of terrestrial microarthropods frequently show a marked bimodality in the distribution of their SCPs, with one group having high values and another low. The border between the high and low groups usually lies between -15 and -20 °C (Cannon and Block 1988). In *Tetracanthella wahlgreni* this bimodality was closely related to the presence or lack of gut contents (Sømme and Conradi-Larsen 1977). Where microscopic examination revealed remnants of food in the gut, the specimens had SCPs above -15 °C, while values for most of those with empty guts were in the range -15 to -32 °C. Due to this correlation it was postulated that some microarthropods may have to eliminate their gut contents in the autumn in order to avoid heterogeneous ice nucleation during the winter. In addition to the elimination of gut contents, supercooling capacity is often increased by the accumulation of cryoprotective substances (Chap. 9.3.6) The extent of the increase in supercooling capacity by gut content elimination, however, is variable. Specimens of the prostigmatid mite *Stereotydeus mollis* and the collembolan *Gomphiocephalus hodgsoni* succumbed more quickly to cold when they were starved than when well-fed (Fitzsimons 1971).

In experiments with *Cryptopygus antarcticus*, Sømme and Block (1982) demonstrated that a large proportion of the specimens lowered their SCPs considerably during starvation at 5 °C (Fig. 9.5). The experiments were designed to avoid the effect of possible ice nucleators in the exuviae and faecal pellets that the springtails are likely to eat. When the collembolans were kept floating on distilled water in small vials, their faecal pellets sunk to the bottom. Exuviae, however, floated and were available for consumption. The effect of starvation at 5 °C on the capacity for supercooling was considerably greater, and occurred much faster, in individuals kept in separate vials than in those maintained in bulk on distilled water. The reason for this difference is probably that more individuals had a chance to feed on exuviae when they were together than when isolated in the vials.

Feeding on different diets produced different effects on the distribution of SCPs in *Cryptopygus antarcticus* (Sømme and Block 1982). During the experiments, the springtails were initially starved individually on distilled water. Subsequently, they were fed on a homogenate made from moss turf, or given a diet of purified unicellular green algae (a mixture of *Stichococcus* and *Monodus*). Of those given the moss homogenate, 87 % of specimens had SCPs above -15 °C (Fig. 9.5), while in those fed on green algae, 65 % had values below this temperature, with a mean of -24 °C. The results suggest that the moss homogenate had a high content of ice-nucleating agents, while the green algae were more or less free from such substances.

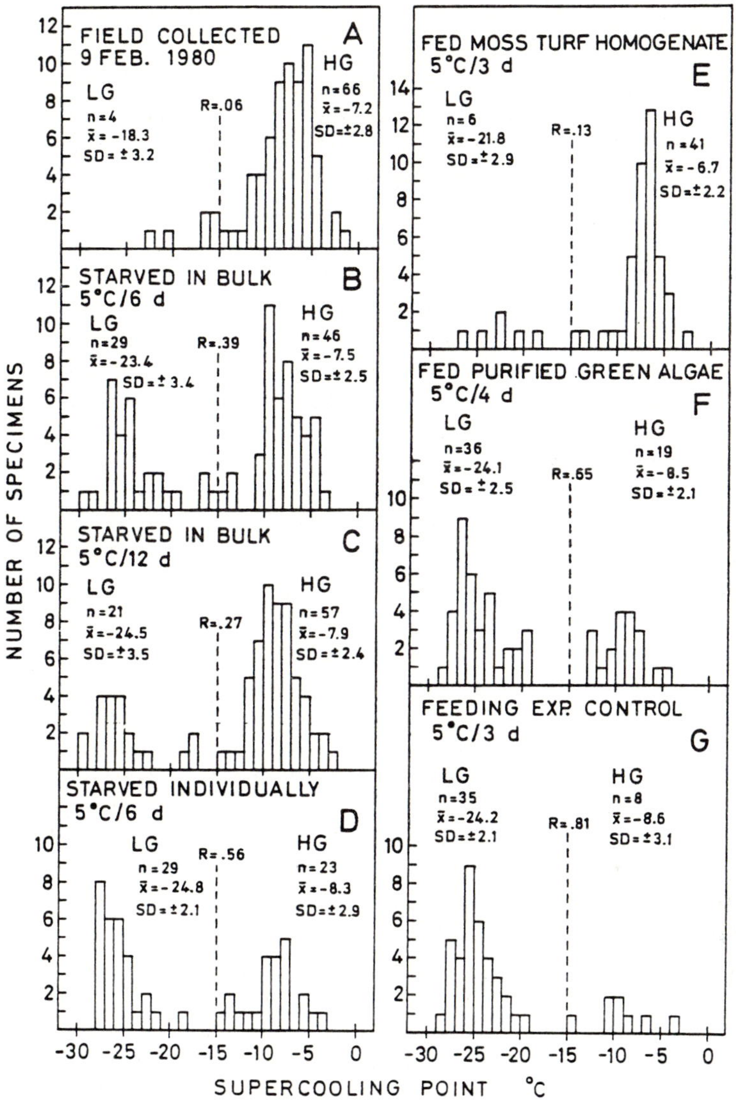

Fig. 9.5A-G. Supercooling point distribution histogram for *Cryptopygus antarcticus*. **A** Field-collected specimens. **B,C** Specimens starved in bulk at 5 °C. **D** Specimens starved individually at 5 °C for 6 days, following starvation in bulk at 5 °C for 12 days. **E,F** Specimens fed moss turf homogenate and purified green algae, respectively, following starvation as in **D**. **G** Feeding experiment control. *LG* Low group SCPs; *HG* high group SCPs; *R* LG/(LG+HG) ratio; *n* number of specimens; mean (± SD) is given for each LG and HG. (Sømme and Block 1982)

This observation is important if one is to understand how microarthropods living under the most extreme continental antarctic conditions are able to survive. Summer temperatures 3 cm below the surface of moss patches and *Prasiola* algae at Cape Bird, Ross Island ranged from -7 to 14 °C (Block 1985b). Similarly, recordings from the microhabitat of *Cryptopygus sverdrupi* in the Mühlig-Hofmannfjella, Dronning Maud Land (Fig. 1.8) show diel fluctuations between -10 and 9 °C (Sømme 1986a). Under these conditions, feeding during daytime would result in freezing during the night if the food contained high concentrations of INAs. The SCPs of *Cryptopygus sverdrupi* range from -25 to -32 °C even in the middle of the summer. Their diet probably consists of tallous *Praesiola* algae and of unicellular green algae growing on its surface. Apparently, this diet is free of strong INAs.

The prostigmatid mites *Eupodes tottanfjella*, *Nanorchestes bifurcata* and *Nanorchestes brekkerista* from Vestfjella, Dronning Maud Land, had summer SCPs in the range -20 to -30 °C (Sømme 1978). The springtail *Gomphiocephalus hodgsoni* and the prostigmatid mite *Stereotydeus mollis* showed considerable supercooling capacity at similar levels (Block 1985b). Observations like these demonstrate that several microarthropods are able to feed during the summer months without consuming INAs and running the risk of heterogeneous nucleation at low ambient temperatures.

Although Sømme and Block (1982) found that a large proportion of individual *Cryptopygus antarcticus* lowered their SCPs during starvation (Fig. 9.5), some specimens remained in the high group. These experiments were performed on animals in the summer. Among winter-acclimated specimens of *C. antarcticus*, a large proportion exhibited a considerable loss in supercooling capacity when allowed to take up distilled water at 5 °C (Cannon et al. 1985). Control specimens maintained at the same temperature, but without access to water, showed no loss of supercooling potential (Fig. 9.6).

The different effects of access to distilled water in summer-acclimated and winter-acclimated animals is not completely understood. One possible explanation is that the loss of supercooling capacity in winter specimens is related to the reactivation of potential INAs within the animals (Cannon et al. 1985; Cannon and Block 1988). The water content of specimens collected during the winter was as low as 37 %, but increased to 68 % during uptake of distilled water. The increase in water content may in some way result in the reactivation or unmasking of internal INAs. Another possibility is the presence of INAs in the distilled water itself or from contamination by ice-nucleating bacteria, but this seems unlikely since more animals should then have been affected in the experiments conducted during the summer.

A clear relation between feeding and supercooling has also been demonstrated in *Alaskozetes antarcticus* (Block 1984b). A number of specimens from laboratory cultures had individual SCPs between -2 and-18 °C but,

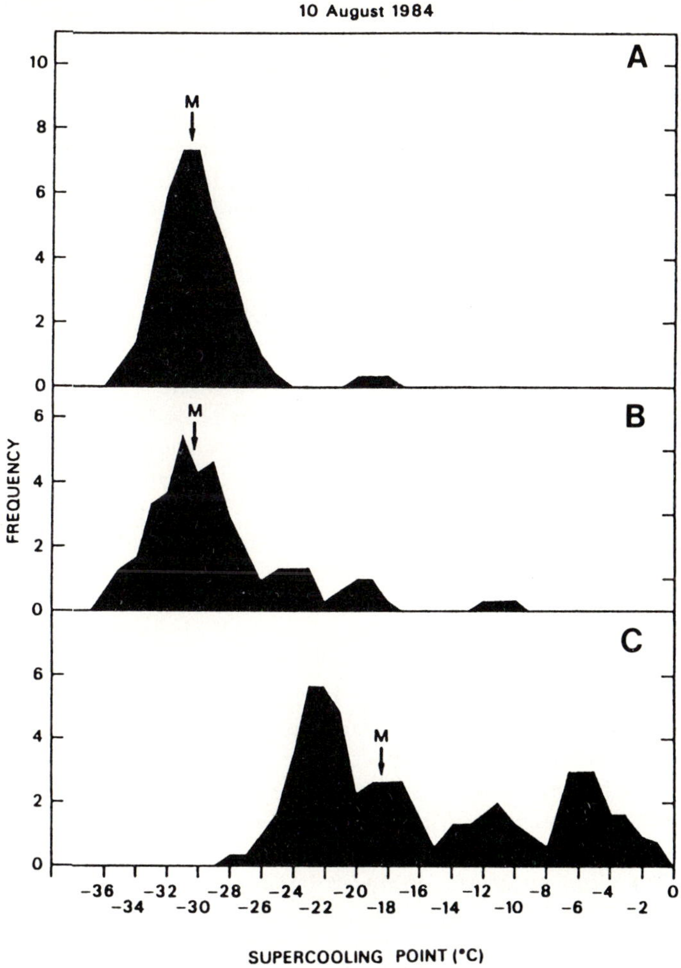

Fig. 9.6A-C. Supercooling point distributions for *Cryptopygus antarcticus* collected in the field 10 August 1984 and given access to distilled water. **A** Field-collected. **B** Dry. **C** Distilled water. *M* Median SCP. (Cannon et al. 1985)

during starvation over 4 weeks, the number of specimens with SCPs below -20 °C increased considerably. Shimada et al. (1992) showed that SCPs of tritonymphs and adults of this species on King George Island varied considerably during the summer. Means ranging from -22.4 to -3.8 °C were recorded from active mites, and were related to different feeding activity and habitat humidity. A lowest mean SCP of -28.3 °C was measured in inactive premoult tritonymphs collected under rocks. Overwintering *Alaskozetes antarcticus*, however, showed loss of supercooling capacity when acclimated at 4 °C without food but with access to distilled water (Cannon 1986). Control animals without water showed only a small loss of their ability to supercool. This again suggests the presence of INAs either associated with water or derived from internal sources within the mites.

Whatever the source of these INAs, their presence may have great ecological consequences. When water is taken up by partly dehydrated overwintering microarthropods, their supercooling capacity is reduced and mortality at lower temperatures increases. This situation is likely to arise in the spring, when mild weather followed by a cold spell could be fatal.

9.3.6 Low Molecular Weight Cryoprotectant Substances

A striking characteristic of overwintering insects is their capacity to accumulate high concentrations of low molecular weight polyols and sugars (Lee 1991). Due to their effect on supercooling or on survival in freeze tolerant species, these substances are often referred to as cryoprotectants. The most widely distributed cryoprotectant in freeze avoiding terrestrial arthropods is glycerol (Sømme 1982; Duman et al. 1991a). As a non-ionizing solute, one molal concentration of glycerol depresses the MP of water by about 1.86 °C.

Other colligatively acting, low molecular weight cryoprotectants include polyols like mannitol and sorbitol, and in some instances ribitol, erythritol and inositol (Sømme 1982; Lee 1991). In the bark beetle *Ips acuminatus*, Gehrken (1984) found up to 2.7 M concentrations of ethylene glycol. Accumulation of sugars such as glucose, fructose and trehalose also lowers the haemolymph MPs of many species. In some insects, a multicomponent cryoprotective system has been described, including the combination of glycerol, sorbitol, glucose and trehalose found in the larvae of *Scolytus ratzeburgi* (Ring 1977). Concentrations of glycerol as high as 5 M have been found in overwintering larvae of the parasitoid wasp *Bracon cephi* (Salt 1961), while Miller (1982) recorded 4.8 M concentrations in willow cone gall larvae *Rhabdophaga strobiloides* from Alaska. Such high concentrations are exceptional, but many overwintering insects accumulate 1 to 3 M levels of cryoprotectants.

As reviewed by Sømme (1982), seasonal changes in the supercooling capacity of many species are related to parallel changes in the concentration of glycerol and other cryprotectant substances. An example is shown in Fig. 9.7, illustrating seasonal changes in larvae of the moth *Petrobia resinella* in Estonia (Hansen 1973). Low mid-winter SCPs are concurrent with high concentrations of glycerol, while specimens without glycerol in the autumn and the spring have high SCPs.

As pointed out by Ring (1980), high concentrations of glycerol and other low molecular weight cryoprotectants may be toxic. Excessive concentrations of each solute may be avoided in multicomponent systems, since protection against cold is offered by the combination of several substances (Morrissey and Baust 1976; Ring 1980). Ethylene glycol, as found in *Ips acuminatus* by

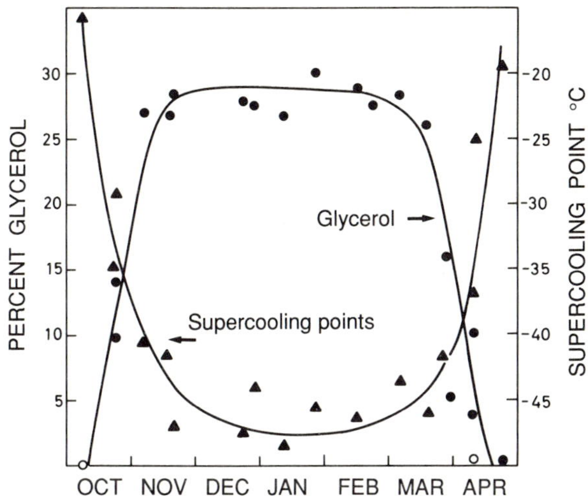

Fig. 9.7. Seasonal changes in glycerol content (% of fresh weight) and SCP in larvae of *Petrobia resinella. Closed symbols* Second instar; *open symbols* last instar. (After Hansen 1973)

Gehrken (1984), is toxic to humans (Duman et al. 1991a) and probably to a number of other organisms. The toxicity of low molecular weight cryoprotectant substances requires further study, but it has been suggested that increased cold hardiness in some terrestrial arthropods may be associated with increased tolerance to the toxicity of high polyol concentrations.

Several authors (e.g. Salt 1961; Sømme 1982; Zachariassen 1985) have shown that the SCPs of freeze avoiding insects are lowered more with increasing polyol concentrations than the MPs (Fig. 9.8, line 4). The SCPs also decrease more than the MPs in experiments with small drops of pure aquaous solutions (MacKenzie 1977) (Fig. 9.8, lines 5 and 6), while the effect is depressed by INAs from freeze tolerant beetles (Fig. 9.8, lines 2 and 3). In reviewing the literature, Zachariassen (1985) concluded that a value of ca. 2 is the most correct quotient between SCP and MP depressions caused by accumulation of polyols in freeze avoiding insects. The factor of 2 also applies to highly purified solutions produced by double distillation and microfiltering (MacKenzie 1977). As pointed out by Zachariassen (1985), these results suggest that the enhanced depressive effect of polyols on SCPs in freeze avoiding insects, compared with the MP depression, can be explained from the physical effects of solutes in pure INA- free solutions.

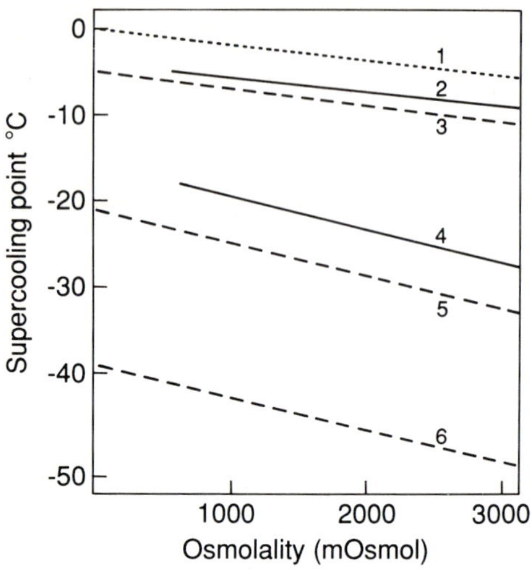

Fig. 9.8. The effect of increasing solute concentration on SCPs of intact insects and aqueous solutions. *1* Melting point depression; *2* SCPs from four species of freeze tolerant beetles; *3* glycerol solutions with haemolymph INAs from freeze tolerant insects; *4* nine freeze avoiding beetles; *5* ordinary solutions of glycerol; *6* highly purified solutions of glycerol. (Based on data from Zachariassen 1985)

9.3.7 Thermal Hysteresis Proteins

In addition to low molecular weight cryoprotectant substances, several terrestrial freeze-avoiding arthropods produce proteins or glycoproteins of importance for their cold hardiness (Zachariassen 1985; Duman et al. 1991a,b). These "antifreeze" proteins depress the freezing point of the haemolymph but not the melting point. The difference between the FP and the MP is termed thermal hysteresis. Apparently, the thermal hysteresis proteins (THPs) adhere to the surface of seedling ice crystals, inhibiting further growth. Although their presence in insects was discovered a long time ago, the role of THPs in cold hardiness has came known only during the last two decades. The FP is usually depressed by 2 to 5 °C below the MP (Duman et al. 1991a,b).

In insects producing antifreeze proteins, seasonal lowering of SCPs is generally associated with increased THP concentration. Thermal hysteresis reaches a level of 4 °C in mid-winter, declines rapidly during the spring and is lost in the summer in larvae of the tenebrionid beetle *Meracantha contracta* (Duman et al. 1991b). The seasonal changes are affected by environmental cues; both low temperature and short photoperiod may induce production of THPs.

THPs may even have more important effects on the cold hardiness of insects than just lowering the SCPs. Based on experiments with the beetle *Rhagium inquisitor*, Zachariassen and Husby (1982) suggested that THPs are able to stabilize the supercooled state over the entire supercooled range. An inverse relationship was found between the thermal hysteresis of a sample containing THPs and the size of the seed crystals used in the measurements. By extrapolation of the regression line, although hypothetical, the hysteresis freezing point should be -25 °C with ice crystals at the size critical for homogenous nucleation. This corresponds exactly to the mean SCPs of the beetles at -25 °C. Supercooling is a "metastable" state, and there is always the risk of spontaneous freezing during prolonged periods of supercooling. For such insects, the stabilizing effect of THPs would be an important adaptation. Furthermore, it has been suggested that THPs may protect against inoculative freezing from contact with ice crystals on the body surface (Zachariassen 1985; Gehrken 1992).

9.3.8 Cryoprotectants in Alpine and Polar Microarthropods

Glycerol has been found to accumulate in several overwintering alpine microarthropods (Sømme 1989). In mid-winter samples of the collembolan *Tetracanthella wahlgreni* the mean content of glycerol was 36 $\mu g\ mg^{-1}$ fresh weight. Corresponding values for the oribatid mites *Calyptozetes sarekensis* and *Carabodes labyrinthicus* were 59 and 40 $\mu g\ mg^{-1}$ fresh weight, respectively. Accumulation of glycerol is probably induced by low environmental temperatures, and all three species produced glycerol in the laboratory during acclimation at -5 °C. The increased concentrations of glycerol probably explain the additional lowering of mid-winter SCPs in these species (Fig 9.3). Thus, the preparation for overwintering is a two-step process in some microarthropods. Initially, evacuation of the gut content is required, and secondly, accumulation of low molecular weight cryprotectants must take place. Several species of oribatid mites from the Alps also produce glycerol during low temperature acclimation (Sømme 1989), while some species of collembolans fail to do so.

The antarctic springtail *Cryptopygus antarcticus* possesses a multicomponent cryoprotective system (Cannon and Block 1988). Sorbitol and mannitol are the principle substances, but glycerol and myo-inositol are also accumulated. In the closely related *Cryptopygus sverdrupi*, glycerol was the main cryoprotectant accumulated during low-temperature acclimation (Sømme 1986a), but glucose and fructose also increased in concentration.

According to Block (1984b), *Alaskozetes antarcticus* can supercool to as low as -26.5 °C without the presence of ice nucleators and glycerol. This degree of cold hardiness may be insufficient for the severe austral autumns and winters at Signy Island, where surface temperatures of -26 °C have been recorded. Accumulation of glycerol is initiated at low temperature and a linear relation between glycerol concentration and SCP has been demonstrated. Concentration increases up to 50 $\mu g\ mg^{-1}$ of body water and individual SCPs as low as -31 °C have been recorded. In comparison, the oribatid mite *Halozetes belgicae* from Signy Island is even better prepared for the winter with glycerol concentrations up to 91.9 μg mg^{-1} fresh weight and individual SCPs down to -38 °C (Cannon and Schenker 1985).

Few investigations have been carried out on THPs in alpine and polar microarthropods. Their presence, however, was demonstrated in the collembolan *Isotoma* sp. from Spitsbergen (Aunaas et al. 1983) and in several springtails from the Swiss Alps (Zettel 1984). Block and Duman (1989) found THPs in *Alaskozetes antarcticus*, where thermal hysteresis reached a level of 1.9 °C in nymphs and 1.1 °C in adults.

As pointed out above (Chap. 8.3), microarthropods overwintering in the soil are apparently at a great risk from freezing as a result of contact with ice crystals. If THPs protect against inoculative freezing, as suggested by Zachariassen (1985), this would be an adaptation of great importance. Experimental evidence for this hypothesis has recently been presented in beetles by Gehrken (1992) (Chap. 8.4.2).

In conclusion, it appears that alpine and polar microarthropods base their cold hardiness on freeze avoidance. No freeze tolerant collembolans or mites have been found. In principle, the strategies of cold hardiness are similar to those of freeze avoiding terrestrial arthropods from other geographical regions. The supercooling capacity of alpine and polar species is increased by the elimination of INAs and further increased by accumulation of low molecular weight cryoprotectant substances. So far, thermal hysteresis proteins have been demonstrated in only a few species, but are likely to be more commonly distributed, contributing to the stabilization of the supercooled state.

9.4 Freeze Tolerant Species

9.4.1 Why is Freezing Injurious?

Insects and other invertebrates that tolerate freezing possess adaptations that are exceptional in the animal kingdom. In freeze tolerant insects protective physiological mechanism have evolved, such as the production of ice nucleators and cryoprotectant substances. These characteristics will be further discussed in the following sections.

There is general agreement that freezing is lethal when it occurs within the cells - intracellular freezing - both in whole insects and in cryopreserved biological material (e.g. Ring 1980; Zachariassen 1985; Duman et al. 1991a). The reason for this is twofold. Initially, removal of water into ice results in concentration of solutes in the remaining liquid, and this will have consequences for metabolic processes (Storey and Storey 1988). Secondly, the increased osmolality in the unfrozen fraction leads to osmotic swelling of the cells and subsequent rupture or physical damage to the cytoskeleton and cell membrane.

The formation of extracellular ice is also potentially injurious, but this danger is more or less controlled in freeze tolerant insects. Mechanical injury may increase from recrystallization of ice. During slow warming or during prolonged exposure to freezing temperatures, large ice crystals grow at the expense of smaller ones (Luyet 1966). As pointed out by Lee (1991), this form of injury may be of importance to overwintering insects that experience extended periods of exposure at temperatures where recrystallization may take place.

The main reasons for lethality during extracellular freezing are still being debated and various hypotheses have been proposed (Storey and Storey 1988; Duman et al. 1991a). As with intracellular freezing, extracellular freezing may result in physical disruption of finer structures. Structural damage from ice includes distortion and deformation of cells, cell to cell connections disrupted and capillaries damaged (Mazur 1984).

The osmotic gradient formed by the concentration of solutes of the extracellular water causes the cells to shrink. The ice continues to grow until the melting point of the unfrozen water fraction is equal to the surrounding temperature. If the concentration gradient is large, an osmotic shock may result from rapid fluxes of water and solutes across cell membranes. Continued temperature decrease results in the formation of more ice and increased solute concentrations. Lovelock (1953) proposed that lethal injury is caused by extremely high levels of solute, especially inorganic ions, produced by freeze concentrations. The dehydration caused by elevated levels of solutes in remaining liquid compartments may have irreversible injurious effects on the structure and function of enzymes and subcellular organization (Mazur 1984).

Extracellular freezing may also cause the cells to shrink to such an extent that their external membranes rest on a matrix of intracellular membrane structures (Meryman 1974). According to this theory, a minimum critical cell volume exists, beyond which further shrinkage results in irreparable damage. As pointed out by Storey and Storey (1988), the observed lethal limit of approximately 65 % of the water transformed to ice in freeze tolerant invertebrates and cryopreserved mammalian cells supports this view.

According to Mazur (1984), the crucial factor is the extracellular unfrozen fraction itself due to the deformation of cells trapped in minute channels of unfrozen solution.

The injuries of intracellular freezing have rarely been demonstrated in insects. The observations by Tanno (1967, 1968), as reviewed by Asahina (1969), are interesting in this respect. Overwintering prepupae of the Japanese poplar sawfly *Trichiocampus populi* are extremely freeze tolerant, and may even survive immersion in liquid nitrogen. In this species clear morphological differences were observed between the fat body cells in the visceral layer (diameter 230 μ) and those of the parietal layer (diameter 170 μ). The large size of these cells causes ice to be formed internally, and most easily in the visceral ones (Fig. 9.9).

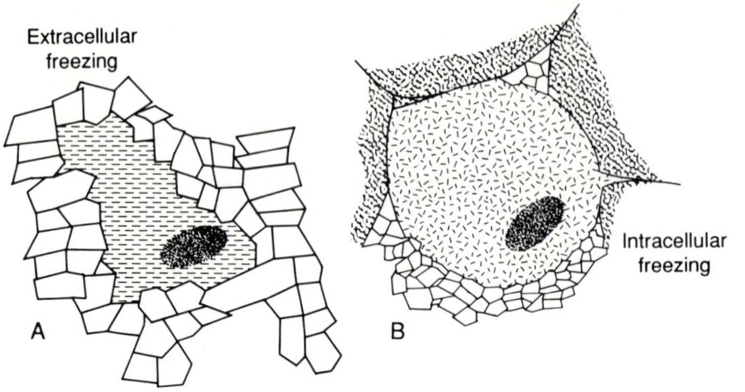

Fig. 9.9A,B. Extracellular and intracellular freezing of fat body cells from *Trichiocampus populi*. **A** Cell surrounded by ice crystals, cell content unfrozen; **B** Cell frozen intracellularly; small ice crystals are also seen in the extracellular liquid. (Drawn from a microphotograph by Tanno 1967)

By microscopic examination of sectioned frozen prepupae, Tanno (1968) demonstrated that different numbers of fat body cells froze at different cooling rates, and this had different effects on the survival of intact specimens treated in the same way. At a cooling rate of 0.4 °C min^{-1} no intracellular freezing took place, and all prepupae completed normal metamorphosis after thawing. When the rate of cooling was increased to 0.8 °C min^{-1}, approximately one-sixth of the visceral fat body cells froze intracellularly. Following this treatment the prepupae resumed development to pupae, but only half of the pupae moulted to adults. At a cooling rate of 4 °C min^{-1} all visceral cells and one fifth of the parietal cells froze. In this case, the prepupae were unable to resume development after thawing, but lived for up to 50 days. By extremely rapid cooling at 327 °C min^{-1}, all fat body cells froze and the prepupae died shortly after thawing. The results corresponded to the use of resources from different parts of the fat body during metamorphoses. Most of the parietal cells are used during transformation from prepupae to pupae, while those of the visceral layer are consumed throughout the metamorphosis from prepupae to adult (Tanno 1967).

Cooling rates also affect survival in larvae of *Chymomyza costata* (Diptera, Drosophilidae) from the mountains of Hokkaido, Japan (Shimado and Riihimaa 1988). High survival rates were obtained by a cooling rate of 0.1 °C min^{-1}, while almost all larvae were killed at a rate of 1 °C min^{-1}. Miller (1978) demonstrated that a slow cooling rate of 0.1 to

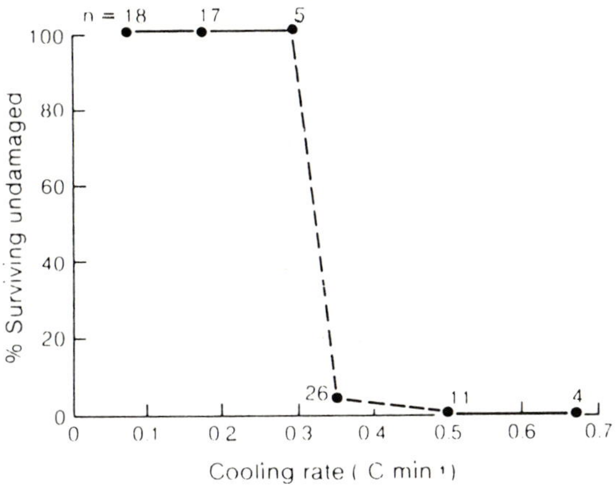

Fig. 9.10. Relationship between cooling rate and freezing survival in adults of the tenebrionid beetle *Upis ceramboides*, following exposure to ca. -50 °C for 1 h. *n* Number of specimens tested at each cooling rate. (Miller 1978)

0.3 °C min^{-1} is essential for maximum survival at -50 °C in the tenebrionid *Upis ceramboides*. (Fig. 9.10). Apparently, a low cooling rate is important for survival in freeze tolerant insects. Under natural conditions, cooling will usually be slow but this may frequently have been overlooked in laboratory studies.

In spite of the general impression that freeze tolerant insects can survive only if the ice formation is extracellular, there are exceptions. Salt (1962) reported that the fat body cells of larvae of the goldenrod gall fly *Eurosta solidaginis* survived freezing, and this observation has recently been confirmed by Lee et al. (1993). They showed that a high percentage of isolated fat body cells of *E. solidaginis* survived freezing in Grace's insect medium at -5 and -10 °C. When 1 M glycerol was added, most of the cells also survived at -25 and -80 °C. Viability of the cells following freezing was assessed by cytochemical techniques. Although the cells survived by these criteria, it is of course not possible to know if freezing would have affected their function in later stages of metamorphosis. Only cells from winter-acclimated larvae survived, and Morason et al. (1994), using a high pressure cryofixation procedure, found that the ultrastructure of cytoplasmatic organelles and nucleoplasm remained intact when cryoprotectants were present.

9.4.2 The Occurence of Freeze Tolerance in Insects

In his studies on insects, Réaumur (1736) was aware that some species survive the winter in the frozen state. In more recent times, although numerous examples of freeze tolerance have been recorded, it has been realized that most species do not freeze but remain supercooled. Apart from insects, freeze tolerance has not been documented in other terrestrial arthropods with the exception of one species of scorpions and one of millipeds. Up to now, freeze tolerance has not been found in spiders, mites or springtails, in spite of the fact that many species overwinter under severe temperature conditions.

Normally, the overwintering stage only is freeze tolerant. In arctic insects with prolonged life cycles several stages may be able to survive freezing. As explained below, the degree of freeze tolerance varies greatly from species to species, and there are also great seasonal variations within species.

Among the first investigations on freeze tolerance in insects, Salt (1961) showed that larvae of the parasitoid wasp *Bracon cephi* tolerated temperatures below -40 °C and accumulated glycerol up to 25 % of fresh weight as a cryoprotectant. The larvae overwinter in the frozen state on the dry prairies of western Canada, where temperatures are extremely low during the winter. As reviewed by Asahina (1969), prepupae of the slug caterpillar *Monema flavescens* and pupae of the swallowtail *Papilio machaon* from Hokkaido, Japan, tolerate freezing at temperatures of -40 and -50 °C, respectively. Provided initial freezing was at -30 to -40 °C for 1 h, both species survived immersion in liquid oxygen for 2 days. Development to adults was resumed, but the moths were unable to shed their pupal cuticle. Some of them emerged as abnormal adults. The overwintering third instar larva of the butterfly *Aporia crataegi*, however, survived immersion in liquid nitrogen and developed normally if they were pre-frozen for at least 1 h at -30 °C. Although the temperature of liquid oxygen and nitrogen is of little ecological interest, the experiments demonstrate at least the great potential of freeze tolerance in certain insects.

Freeze tolerant insects are found among species from several orders such as Coleoptera, Diptera, Lepidoptera and Hymenoptera. All investigations prior to 1968 were concerned with immature stages and the discovery of freeze tolerance in the carabid beetle *Pterostichus brevicornis* was the first example of an adult insect (Miller 1982). Freeze tolerance is well documented in larval and pupal stages but not from the egg stage of any insect.

9.4.3 Polar and alpine insects

A list of freeze tolerant insect from alpine and polar areas is presented in Table 9.1, indicating their ability to supercool, as well as the lowest temperature they are able to survive in the frozen state. Among adult beetles, the small carabid *Pterostichus brevicornis* overwinters in decaying stumps in Alaskan forests and is highly resistant to freezing (Baust and Miller 1970; Miller 1982). Although partly protected by snow, part of the population is exposed to the full impacts of winter cold. Temperatures down to -70 °C were tolerated in the laboratory. Similarly, the large tenebrionid *Upis ceramboides*, found under loose bark of dead trees in the boreal forest of interior Alaska, survives freezing to approximately -60 °C. In the same region, a small chrysomelid beetle, *Phratora* sp., overwinters in large numbers beneath loose bark of aspens. Field populations of this species showed almost complete recovery after exposure to -45 °C. In *Pytho americanus* from the North Western Territories in Canada, both larva and adult are freeze tolerant (Ring 1982).

Among alpine beetles, adults of the chrysomelid *Melasoma collaris* from the alpine zone of the Hardangervidda in Norway tolerated freezing at -35 °C (Sømme and Conradi-Larsen 1979). A related species, *Phyllodecta laticollis*, overwintering beneath loose bark of willows in subalpine areas of southern Norway, tolerated -42 °C (van der Laak 1982). This species is exceptional in that it maintains freeze tolerance during summer, but at high altitudes in the inland areas of its distribution, temperatures may drop below zero even at this time of the year.

At 4000 m on Mount Kenya, freezing temperatures down to -7 °C are common during the night, and two adult curculionid beetles were found to be freeze tolerant at this level (Sømme and Zachariassen 1981). Similarly, a caterpillar of the genus *Agrotis*, endemic to the summits of tropical Mauna Loa and Mauna Kea on Hawaii, are freeze tolerant (Duman and Montgomery 1991). Individuals collected near the summit at 3800 m had full guts and SCPs of only -4.6 °C. In this cold and xeric habitat there is little food, but the caterpillars are known to feed on lichens.

Adult grasshoppers of the species *Meridacris subaptera* at 4200 m in the Venezuelean Andes Mountains survived short periods of freezing down to -9 °C (Table 9.1), corresponding to the low night temperatures of their surroundings (Sømme 1986c). Due to the great temperature fluctuations in alpine habitats close to the equator, the grasshoppers experience daily periods of high temperatures. Their limited survival on freezing is in accordance with the short duration of frost during the night. In New Zealand, the endemic mountain 'weta' *Hemideina maori* overwinters in several stages under rocks above the timber line (Ramløv et al. 1992). Ambient temperatures periodically drop below the SCPs of this large grasshopper (3-7 g), which apparently remains in the frozen state for large parts of the winter. In the laboratory the weta tolerated temperatures down to -10 °C, while temperatures in their microhabitat rarely fall as low as this.

Table 9.1. Freeze-tolerant insects from polar and alpine areas

Species	Stage	Location	Overwintering habitats	Lowest mean SCP, ° C	Lower lethal temp., ° C	Reference
Orthoptera						
Meridacris subaptera	A	Andes Mts	Vegetation[a]	-4	-9	Sømme (1986c)
Hemideina maori	A	New Zealand	Under rocks	-4	-10	Ramløv et al. (1992)
Neuroptera						
Hemerobius simulans	A	Alaska	Bark of aspen	-36	-50	Miller (1982)
Homoptera						
Typhlocyba sp.	A	Alaska	Bark of aspen	-32	< -40	Miller (1982)
Coleoptera						
Pterostichus brevicornis	A	Alaska	Stubs, timber	-11	-70	Miller (1982)
Pelophila borealis	A	Norway	Soil hibernacula	-21	-5	Sømme (1974b)
Upis ceramboides	A	Alaska	Under bark	-6	-60	Miller (1982)
Pytho americanus	A	NW Territories	Under bark	-8	< -45	Ring and Tesar (1980)
Pytho americanus	L	Rocky Mts.	Under bark	-5	-44	Ring (1982)
Pytho deplenatus	L	Rocky Mts.	Under bark	-54	-55	Ring (1982)
Phratoria sp.	A	Alaska	Bark of aspen	-8	< -45	Miller (1982)
Melasoma collaris	A	Norway	In litter	-6	-35	Sømme (1989)
Phyllodecta laticollis	A	Norway	Bark of willows	-5	-42	van der Laak (1982)
Ostoma ferruginea	A	Alaska	Under bark	-12	< -55	Miller (1982)
Cucujus clavipes	A	Alaska	Under bark	-	< -55	Miller (1982)
Seneciobius kenyanus	A	Mt. Kenya	Vegetation[a]	-5	-7	Sømme and Zachariassen (1981)
Parasystates elongatus	A	Mt. Kenya	Vegetation[a]	-5	-7	Sømme and Zachariassen (1981)
Lepidoptera						
Nymphalis antiopa	A	Alaska	Artificial hibernacula	-20	-34	Miller (1982)
Polygomia sp.	A	Alaska	Artificial hibernacula	-25	-34	Miller (1982)

Martyrhilda ciniflonella	A	Alaska	Bark of spruce	-2	< -40	Miller (1982)
Rheumaptera hastata	P	Alaska	Leaf litter	-25	< -25	Miller (1982)
Gynaephora groenlandica	L	Ellesmere Isl.	Litter	-7	-70	Kukal (1991)
Agrotis sp.	L	Mt. Loa, Hawaii	Snowfields	-5	-6	Duman and Montgomery (1991)
Diptera						
Mycetophila sp.	A	Alaska	Bark of aspen	-33	< -40	Miller (1982)
Belgica antarctica	L	Antarctica	Soil	-6	-10	Baust and Edwards (1979)
Eretmoptera murphyi	L	Signy Island	Soil	-6	-16	Block et al. (1984)
Metioctnimus sp.	L	Mt. Kenya	Frozen pools[a]	-	-7	Sømme and Zachariassen (1981)
Xylophagus sp.	L	NW Territories	Under bark	-6	-43	Ring (1982)
Hymenoptera						
Pontania sp.	L	NW Territories	Galls on willow	-8	-	Ring (1981)
Amauronematus sp.	L	NW Territories	Catkins of willow	-7	-40	Humble and Ring (1985)
Syndipnus sp.	L	NW Territories	Parasite in host	-7	-40	Humble and Ring (1985)

[a]No winter, but diel temperature fluctuations.

According to Miller (1982), a variety of adult insects from the interior of Alaska overwinter in the frozen state (Table 9.1). In addition to the beetles mentioned above, his list includes two nymphalid butterflies, one moth, one mycetophilid midge, one neuropteran and one cicada. A striking feature of all these insects is that their lower temperature limits are only a little below the record minimum temperature of interior Alaska.

As will be discussed below, most freeze tolerant insects have high SCPs (Table 9.1) due to the presence of ice-nucleating agents (INAs) in their haemolymph (Table 9.1). The larvae of the arctic and alpine beetle *Pytho americanus* with a mean SCP of -4.4 °C is a typical example (Ring 1982). More unusual is a mean SCP of -54 °C in *Pytho deplanatus*, corresponding to the lower lethal temperature of the frozen larvae. Both *Pytho* species live under the bark of fallen spruce in the Rocky Mountains.

Freeze tolerance in arctic insects was first documented in larvae of chironomid midges (Scholander et al. 1953). In the Antarctic, the chironomid midge *Belgia antarctica* does not survive freezing at temperatures below -10 °C (Baust and Edwards 1979). Air temperatures at the Antarctic Peninsula are much lower, but the larvae inhabit soil microhabitats along the coast that are warmed by the temperature of the surrounding sea water. A population of the midge *Eretmoptera murphyi* was accidentally introduced from South Georgia to Signy Island more than 20 years ago, and has survived in sheltered microhabitats in the soil (Block et al. 1984). In this midge as well, the larva is freeze tolerant.

The lymantriid moth *Gynaephora groenlandica* has an exceptionally long life cycle and overwinters in several different larval instars (Kukal 1991). The species inhabits some of the most adverse dry and cold environments of Greenland and the Canadian arctic archipelago. Unlike most cold-adapted insects, *G. groenlandica* does not lose its freeze tolerance capability during active growth and development. In the summer, it survives freezing at -15 °C, but tolerates temperatures down to -70 °C in the winter. As pointed out by Kukal (1991), the accumulation of low glycerol levels probably contribute to its survival for prolonged periods at temperatures below -30 °C, and even lower when air temperatures fall to -60 °C.

9.4.4 Variations in the Degree of Freeze Tolerance

From the examples discussed in the preceding sections, it appears that great differences are found in the freeze tolerance of different species (Table 9.1). Naturally, each species is adapted to the particular conditions of its own overwintering site.

Most of the species that are freeze tolerant during the winter cannot survive freezing during the summer, although exceptions like *Gynaephora groenlandica* and *Phyllodecta laticollis* have been cited previously.

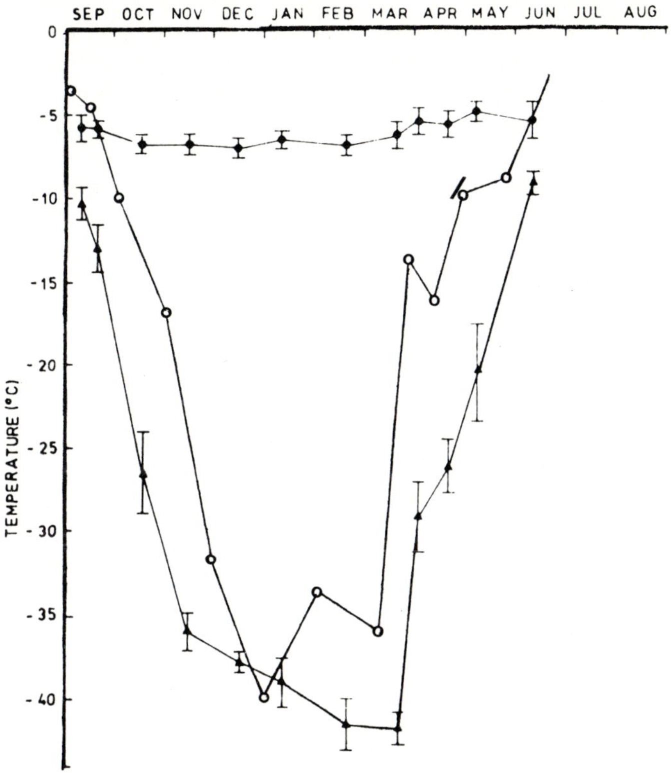

Fig. 9.11. SCPs (*closed circles*) and low temperature tolerance limit (*closed triangles*) of *Phyllodecta laticollis*, and minimum environmental temperatures (*open circles*) at the collecting site during the year. (van der Laak 1982)

Typically, freezing tolerance is gradually built up during the autumn and reaches a maximum in the middle of the winter (e.g. Baust and Miller 1970; Sømme and Conradi-Larsen 1979; van der Laak 1982). The increase in freeze tolerance is usually associated with the formation of glycerol and other polyols (e.g. Asahina 1969; Storey and Storey 1988). Both in the laboratory and outdoors, synthesis of polyol may be triggered by low temperatures.

Figure 9.11 shows the seasonal variations in freeze tolerance of *Phyllodecta laticollis* from the alpine region at Oppdal, Norway. The freeze tolerance generally follows the mean air temperature at the site of collection, while the SCPs remain high due to haemolymph INAs (Chap. 9.4.5). The seasonal changes in freeze tolerance of this species is related to changes in the concentration of glycerol and other cryoprotectant substances.

In the literature, the freeze tolerance of insects is often given as the lower lethal temperature. This value is of interest in comparing different species and seasonal variations within single species. Under experimental conditions, the lower lethal temperature of some insects is much lower that those of their natural environment, e.g. in species that tolerate temperatures of liquid gases (Asahina 1969; Miller 1982; Shimada and Riihimaa 1988). Although the experiments illustrate that some insects have a great potential of survival, this degree of freeze tolerance may be of little ecological relevance.

To understand the survival of freeze tolerant species under natural conditions, the time aspect must also be considered. The fact that freeze tolerant insects survive the winter in their overwintering sites clearly shows that they are adapted to tolerate long exposures in the frozen state. In spite of this, they may not survive for prolonged periods at temperatures below those of their natural environment.

Relatively few studies have been carried out on the relation between time and temperature in freeze tolerant insects. When long-term experiments are carried out, decreasing survival is often found with increasing exposure times. Thus, Asahina (1969) showed that all prepupae of the slug caterpillar *Monema flavescens* survived 100 days at -10 or-20 °C, but increasing mortalities were recorded with storage up to 250 days.

Table 9.2. Mortality rates in *the* alpine carabid beetle *Pelophila borealis* during exposure to different temperatures and for different time intervals in the frozen state. N = number of specimens (Sømme 1974b)

Temp. °C	Time interval	N	Percent mortality
-10	10 min	5	100
	1 h	19	100
-5	4 days	6	0
	8 days	13	1
	12 days	4	20
	16 days	24	46
	32 days	12	75
	42 days	19	95
-3	16 days	7	0
	32 days	12	0
	64 days	8	25

Adults of the alpine carabid beetle *Pelophila borealis* offer another example of the relation of temperature and time on rates of mortality on frozen insects (Østbye and Sømme 1972; Sømme 1974b). The beetles are killed when freezing takes place at their SCPs in the range -10 to -21 °C. In the laboratory, freezing at higher temperatures can be inoculated by contact with ice. At -5 °C, mortality rates increased gradually with from 0 % at 4 days to 95 % at 42 days (Table 9.2), while more than 75 % of the beetles survived 64 days at -3 °C. In their overwintering habitats, inoculative freezing takes place when the beetles are enclosed by ice, but temperatures seldom fall below -2 or -3 °C due to a deep protective layer of snow.

9.4.5 Ice Nucleation in Freeze Tolerant Species

9.4.5.1 Haemolymph Ice-Nucleating Agents

As described in Chapter 9.3.1, purified water will supercool far below its melting point unless heterogeneous nucleation is caused by the presence of impurities. In biological systems ice-nucleating agents (INAs) are always present. For freeze susceptible insects it is important to mask or get rid of the nucleators before the winter sets in.

In contrast, INAs are produced rather than removed in many freeze tolerant insects. Zachariassen and Hammel (1976) found that the freeze tolerant tenebrionid beetles *Eleodes* and *Coelocnemis* have SCPs in the range -6 to -7 °C. The beetles live in mountain ranges of southern California and are exposed to freezing temperatures during the winter. The nucleating activity of the beetles was located in the haemolymph. To test this, samples of haemolymph were placed in capillary tubes sealed with liquid paraffin at both ends. Freezing in the capillaries were observed to take place at exactly the same range (-6 to -7 °C) as the SCPs of intact beetles. Control capillaries with 0.9 % NaCl solution will normally supercool to between -17 and -20 °C. The ice-nucleating activity of haemolymph from the beetles appeared to be seasonal and was lacking in specimens collected during the summer, at a time when they are freeze susceptible.

Several later studies have demonstrated that most freeze tolerant insects have high supercooling points, as reviewed by Zachariassen (1985), Duman et al. (1991a,b) and others. Examination of older literature reveals that high supercooling points have previously been recorded in several freeze tolerant species. From Table 9.1 it also appears that SCPs in the range -6 to -12 °C are typical for alpine and polar freeze tolerant insects, notably beetles.

The primary function of haemolymph INAs is apparently to inhibit lethal intracellular ice formation (Zachariassen 1985; Duman et al.

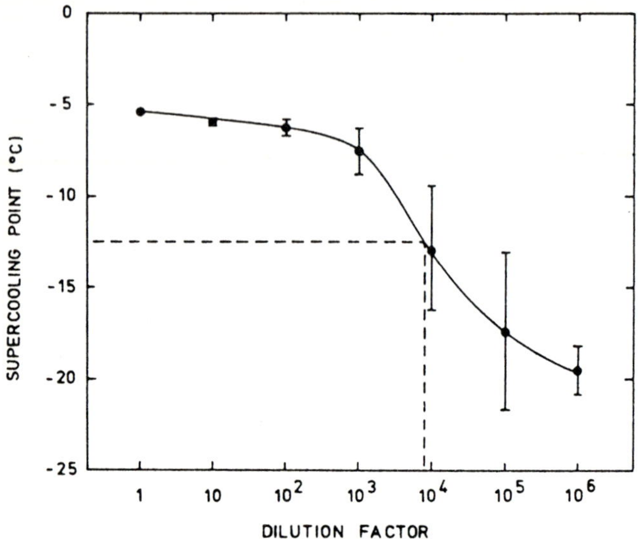

Fig. 9.12. Effect of dilution on nucleator activity of 5-μl samples of haemolymph from the tenebrionid beetle *Eleodes blanchardi*. Each subsequent step involved a tenfold dilution in 0.9 % NaCl solution. (Zachariassen et al. 1982)

1991a, b). When an insect is highly supercooled, the risk of intracellular freezing is much greater. Injurious freezing and subsequent swelling due to increased osmotic pressure can also take place in the intestines, where INAs frequently are contained (Sect. 9.3.2).

The preceding paragraphs show that freezing is usually initiated at temperatures above -10 °C when haemolymph nucleators are present. When ice is formed, the extracellular fluid becomes concentrated and an outflux of water from the cells takes place due to the osmotic gradient that arises. The dehydration of the cells lowers the melting points and SCPs of their intracellular water, and chances of intracellular freezing are effectively inhibited. In addition, the cell membrane probably represents a barrier to penetration of ice from the haemolymph (Duman et al. 1991a).

In freeze tolerant insects, INAs are present in both the haemolymph and other compartments (Fig 9.2). An important condition for this system to function is that haemolymph INAs are more effective in initiating freezing at higher temperatures than intracellular or intestinal INAs. Zachariassen et al. (1982) showed that the haemolymph INAs of *Eleodes blanchardi* are very effective (Fig. 9.12).

The nucleating activity was only slightly reduced when the haemolymph was diluted by a factor of 10^3 with 0.9 % NaCl solution, while further dilution resulted in decreased activity. The biological importance of this apparently superfluous concentration in the haemolymph is not clearly understood.

Haemolymph INAs are inactivated by heat and the enzymatic action of protease (Zachariassen 1985; Duman et al. 1991a), which indicates that they are proteinaceous macromolecules. Some of these molecules have been isolated and characterized, such as INAs from the cranefly *Tipula trivittata* (Duman et al. 1991a,b). This molecule is a globular 800-kDa lipoprotein composed of 45 % protein, 51 % lipid and 4 % carbohydrate. It seems likely that hydroxyl groups of the carbohydrate and certain side chains of the protein fraction are able to bind water molecules and structure them into an embryo ice crystal, thus initiating ice formation in the haemolymph.

9.4.5.2 Inoculative Freezing

As pointed out in Chapter 8.4 the avoidance of inoculative freezing is important to freeze susceptible species of terrestrial arthropods. Various adaptations to escape contact with ice are found in species inhabiting moist overwintering sites.

In freeze tolerant species the situation again is the contrary. Several species depend on freezing at high temperatures to survive when the temperature is further decreased. In principle, this is a situation similar to that in insects in which freezing is initiated by haemolymph INAs. It confirms the general theory that freezing at high temperatures reduces the risk of injurious intracellular freezing.

According to Fields and McNeil (1986), inoculative freezing is essential for freeze tolerance in the arctiid moth *Cisseps fulvicollis*. Under dry cooling conditions, the larvae have SCPs of -12 to -17 °C and are killed by freezing. When cooled in contact with external ice, they freeze above -4 °C and survive. As mentioned previously, adults of the carabid *Pelophila borealis* are killed when freezing takes place at their SCPs, but survive inoculative freezing down to -5 °C. Similarly, prepupae of *Sciara* sp. supercool to -18 °C in dry surroundings but are killed by freezing at this temperature (Tanno 1977). In their moist natural environment freezing takes place at -4 °C and the prepupae subsequently survive freezing down to -12 °C.

Even more spectacular is the freeze tolerance of diapausing larvae of the drosophilid fly *Chymomyza costata* (Shimada and Riihimaa 1988). In contact with ice, freezing is initiated at -2 °C. Provided that a slow cooling rate is used, the larvae survive freezing to -40 and -80 °C, after which they develop normally to pupae. If the larvae are cooled in dry surroundings, they freeze at approximately -20 °C and are unable to resume development.

The ichneumonid wasp *Syndipnus* sp. is an endoparasitoid on the sawfly *Amauronematus* sp. in the Northwestern Territories of Canada (Humble and Ring 1985). The freeze tolerant sawfly larvae overwinter in their cocoons on the host plant above the snow. They have a mean SCP of -6.5 °C (Table 9.1), and the larvae of the endoparasitoid freeze by inoculation within their host. When the parasitoid is removed from the host, their SCPs drop to -17 °C, but the larvae also survive freezing at this temperature.

The North American centipede *Lithobius forficatus* can survive freezing at-5 to -7 °C, provided that it is cooled in contact with water and freezes by inoculation at -1 to -2 °C (Duman et al. 1991b). In contrast, all specimens are killed when frozen in dry condition at temperatures of their supercooling points in the range -3 to -6 °C. The moderate freeze tolerance of this species makes it possible for it to survive in its winter habitat among leaf litter, usually under logs.

9.4.5.3 Other Sites of Nucleation

In some species, nucleation at high temperatures is not caused by haemolymph INAs or by inoculative freezing, but by nucleators in other body compartments. The arctic lymantriid moth *Gynaephora groenlandica* is an example (Kukal 1991). In the winter, the larvae have a mean SCP of -7 °C but samples of the haemolymph supercool to -20 °C. The site of nucleation is not known, but may possibly take place in the gut content. Similarly, the freeze tolerant larvae of the goldenrod gallfly *Eurosta solidaginis* have SCPs of approximately -8 °C (Baust and Nishino 1991). Investigations on some populations have indicated the presence of haemolymph INAs, while the haemolymph of larvae from other populations supercool to -20 °C, in which case the site of nucleation presumably must be the gut. Freezing is inoculated at -5 to -6 °C in early autumn when the galls have high moisture content (Layne et al. 1989). Later in the season, the water content of the galls is greatly reduced and the inoculative effect is lost. In this case, the larvae do not depend on inoculative freezing to survive, since endogeneous INAs prevent excessive supercooling in late autumn and winter.

Several exceptions to the rule of high SCPs in freeze tolerant insects are known. Some of them have been reported from arctic insects (Miller 1982; Ring 1982) and are included in Table 9.1. As already noted, the low SCP of *Pytho deplanatus* at -54 °C is remarkable, and several other species have SCP in the range -20 to -35 °C. Other examples include the freeze tolerant larvae of the parasitoid wasp *Bracon cephi* which have individual SCPs as low as -47 °C (Salt 1961). Similarly, Sømme (1964) reported a mean SCP of -49 °C in larvae of the parasitoid *Eurytoma gigantea* and Baust et al. (1979) found SCP below -30 °C in pupae of *Eurytoma obtusiventris* and the beetle *Mordellistena unicolor*.

All these species clearly lack efficient nucleators in their haemolymph as well as in cells and other body compartments. More studies are required to understand where freezing is initiated in freeze tolerant insects with low SCPs.

9.4.6 Cryoprotectant Substances

Increasing concentrations of glycerol and other polyols increase the cold hardiness of freeze tolerant insects (Zachariassen 1985). Glycerol is often present in molar concentrations and the separate substances of multiple cryoprotectant systems may similarly add to high levels (Duman et al. 1991a). As summarized by Storey and Storey (1988), other cryoprotectant substances reported in freeze tolerant insects include the polyols sorbitol, ribitol, erythritol and threitol. Furthermore, the disaccharides trehalose and sucrose may occur in high concentrations, while the role of glucose and fructose as cryoprotectants is questionable.

Glycerol and other low molecular weight cryoprotectant substances reduce the amount of ice in a colligative manner (Zachariassen 1985; Duman et al. 1991a). Thus, the percentage of frozen body water at any given temperature is less than in the absence of the cryoprotectant. As a consequence, problems associated with freezing (Sect. 9.4.1.), such as shrinkage to critical cell volume, high ion concentration and reduced volumes of unfrozen water, are lessened.

Substances other than polyols and sugars may also act as cryoprotectants but few studies have been carried out with regard to insects. Sømme (1964) showed that some species of freeze tolerant insects apparently lack high concentrations of polyols and sugars. As pointed out by Zachariassen (1985), however, cold hardening in freeze tolerant insects is accompanied by an increase in the concentration of free amino acids. Van der Laak (1982) found that the concentration of free amino acids in the crysomelid beetle *Phyllodecta laticollis* may exceed 300 mmolal in the winter. The accumulation of free amino acids may depress the lethal temperatures and represent a significant contribution to winter survival of freeze tolerant insects.

Another interesting aspect of the protective effect of solutes such as polyols, sugars and amino acids is their ability to increase the availabilty of water at the surface of proteins (Carpenter and Crowe 1988). Since such substances are preferentially excluded from contact with the surface of proteins, they maintain a layer of water around the macromolecules and inhibit their denaturation during freezing and thawing.

9.5 Freeze Tolerance Versus Freeze Avoidance

Several authors (e.g. Zachariassen 1985; Block 1991; Duman et al. 1991a) have discussed why some insects are freeze tolerant and some are freeze susceptible. Is the first strategy of adaptation better than the other or vice versa? This question has no general answer since each species of arthropods may have its own evolutionary and natural history leading to cold hardiness. A list of advantages and disadvantages appears to be meaningless as long as survival strategies probably have evolved independently in a number of species. No choice of strategy has been offered during the course of evolution.

Table 9.3. Characteristics of cold tolerance exhibited by cold hardy terrestrial arthropods (After Block 1991)

Freeze susceptibility	Freeze tolerance
High supercooling capacity	Low supercooling capacity
Supercooling point approximates to lower lethal temperature	Supercooling points is not lower lethal temperature
Ice nucleating agents absent or masked	Ice-nucleating agents present or activated
Polyols function as antifreezes	Polyols function as cryoprotectants
Therma l hysteresis proteins often present (stabilize supercooled state)	Thermal hysteresis proteins act as antifreezes
Ice formation lethal	Survive formation of extracellular ice

A comparison between the characteristics of freeze susceptible and freeze tolerant terrestrial arthropods shows substantial differences (Block 1991) (Table 9.3). It is interesting to note that glycerol and other polyols appear as antifreezes in freeze susceptible species, while they protect against freezing injuries in freeze tolerant species.

As pointed out previously, all species of collembolans, mites and spiders are freeze susceptible, suggesting that freeze tolerance for some reason is impossible in these taxa. Small size, in general, favours supercooling for physical reasons and may subsequently affect the evolution of survival by freeze avoidance. Among insects, there is presently no clear phylogenetic trend regarding freeze tolerance or freeze avoidance. The first strategy appears to be more common in higher insect orders such as Coleoptera, Lepidoptera, Diptera and Hymenoptera (Cannon and Block 1988; Block 1991), but this may reflect the larger number of species in these orders as well as the preferences of the researchers (Duman et al. 1991a). Regardless of taxa, the strategy of freeze avoidance and supercooling is definitely more widespread than freeze tolerance (Sømme 1982).

In spite of these considerations, a comparison of freeze tolerant and freeze susceptible beetles clearly shows that the protective effect of glycerol and other cryoprotectant substances is higher in the former group (Zachariassen 1985). The relation between the mOsm concentration of the haemolymph and the SCPs in freeze tolerant beetles and the lower lethal temperature in two freeze tolerant species is shown in Fig. 9.13.

In beetles tolerant to freezing, the carbohydrate resources are more effectively utilized to promote cold hardiness than in the freeze susceptible ones. Accordingly, Zachariassen (1985) concluded that freeze tolerance is probably the most successful strategy of cold hardiness in areas with extremely low temperatures. This is supported by the fact that many arctic insects and species from the interior of Alaska and Canada are tolerant to freezing.

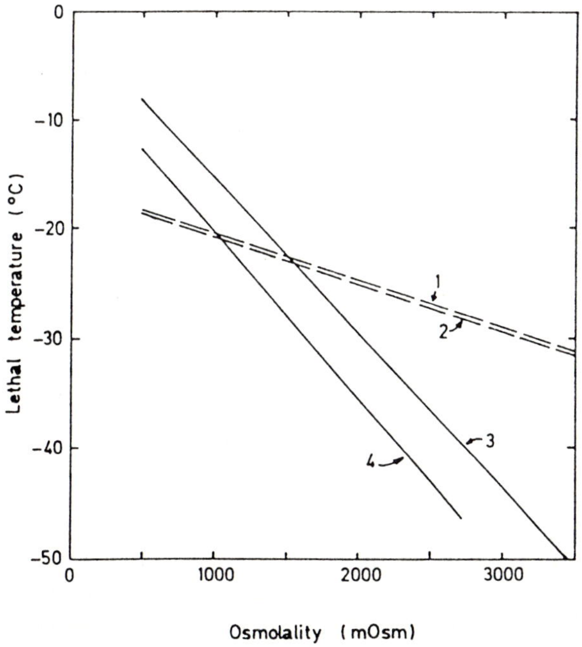

Fig. 9.13. Lethal temperatures of freeze avoiding insects (*dashed lines*) and freeze tolerant insects (*solid lines*) as a function of body fluid osmolality. *1* Regression line of 9 species of freeze avoiding beetles; *2* regression line of *Ips acuminatus; 3* lower lethal temperatures of *Pytho depressus; 4* lower lethal temperatures of *Phyllodecta laticollis*. (Zachariassen 1985)

10 Hot Desert Arthropods at Low Temperatures

10.1 Preadaptation to Cold

As mentioned previously, insects can be considered as containers in which the physical conditions for supercooling are favourable (Sømme 1982). Although they may not be exposed to freezing temperatures, insects from tropical regions and summer-acclimated specimens from temperate regions have SCPs in the range of -7 to -12 °C (Zachariassen 1985). Due to the ability of terrestrial arthropods to supercool, Cloudsley-Thompson (1973) and Constantinou and Cloudsley-Thompson (1986) pointed out that they are partly pre-adapted to inhabit cold environments. For this reason, species living in hot deserts may have the supercooling capacity to tolerate freezing temperatures of their environments. Cold hardiness, however, is not merely a question of supercooling, but also involves the maintenance of vital physiological and biochemical processes at low temperatures (Duman et al. 1991a). In fact, some insects such as the summer-acclimated aphids described by Bale et al. (1988), are killed in a few minutes at subzero temperatures before their SCPs are reached. Very little is known about the harmful effects of exposure to cold above freezing temperatures. Summer--acclimated temperate insects and tropical insects usually succumb within a short time, and offer a striking contrast to overwintering stages of temperate, polar and alpine species that survive for several months at temperatures below zero. Cold hardiness in terrestrial arthropods from hot deserts offers an interesting example of how species primarily adapted to warm climates may also be able to survive periods of freezing conditions in their natural habitats.

10.2 Cold Hardiness of Different Taxa

10.2.1 Scorpions

Cold hardiness has been studied in relatively few species of scorpions, and some of the data available do not really allow classifications of scorpions as freeze susceptible or freeze tolerant (Whitmore et al. 1985). The first report on supercooling in scorpions was published by Cloudsley-Thompson and Crawford (1970a), who found a mean SCP of -6 °C in *Vejovis* sp. from New Mexico. Cloudsley-Thompson (1973) showed that *Leiurus quinquestriatus* (Fig. 10.1) from the Sudan had a mean SCP of -7.5 °C.

Fig. 10.1. *Leiurus quinquestriatus* from Sudan and Egypt in defence posture. (Photo courtesy J.L. Cloudsley-Thompson)

The scorpion *Diplocentrus spitzeri* (=*poloncillus*) lives in shallow burrows beneath rocks in the Peloncillo Mountains of New Mexico. When winter temperatures drop below normal for short periods, part of the population may be killed (Crawford and Riddle 1974,1975). SCPs in the range of -6 to -8.5 °C and haemolymph melting points did not vary significantly with season. The species displayed a variable but generally poor recovery after brief periods of freezing at the temperatures of their SCPs. Crawford and

Riddle (1975) concluded that *D. spitzeri* is freeze susceptible, or possibly moderately freeze resistant since some specimens survived. From general experience with insects, however, it seems likely that all specimens of this species would have been killed by freezing at slightly lower temperatures or by a small increase in time of exposure.

In comparison, SCPs of the scorpion *Paruroctonus aquilonalis* (=*utahensis*) decreased from -8.5 °C in the autumn to -11.9 °C in January (Riddle and Pugach 1976). Freezing following supercooling proved immediately fatal to nearly all specimens tested. The change in SCP was not related to accumulation of glycerol and sorbitol but associated with the cessation of feeding in the autumn.

A different pattern was found in another scorpion, *Centruroides vittatus*, from Texas (Whitmore et al. 1985). SCPs of this species were in the range of -1 to -3 °C during the entire year. Following initiation of crystallization, the scorpions were cooled again to approximately the temperatures of their SCPs, or slightly below. Under these conditions most specimens survived freezing. Mortality increased greatly at lower temperatures, and -6 °C appeared to be a critical limit of survival. The high values of the SCPs were explained by the concentration of INAs, that were substantially higher in the gut content than in the haemolymph. *C. vittatus* displays several characteristics found in freeze tolerant insects such as high SCPs and presence of INAs. In conclusion, Whitmore et al. (1985) considered this as the first reported observation of freeze tolerance in scorpions.

The results from the different studies on scorpions illustrate the difficulties of designing species to definite categories of freeze tolerance or freeze susceptibility. Both freezing temperature and exposure time are important in understanding the survival of frozen animals. If freeze tolerance is to have any ecological relevance, survival must correspond to the freezing conditions likely to occur in the natural habitats of the species.

10.2.2 Centipedes

According to Cloudsley-Thompson and Crawford (1970b) summer--collected specimens of the large centipede *Scolopendra polymorpha* from New Mexico have a mean SCP of -3.1 °C. Crawford et al. (1975) suspected that a seasonal lowering of SCPs might take place, since soil temperatures in this area often fall below freezing in the winter. Specimens of *S. polymorpha* may readily be collected beneath rocks in the Peloncillo Mountains of New Mexico. Crawford et al. (1975) found no clear seasonal variation, but mean SCPs in different months of the year varied from -3.7 to -7.5 °C in freshly collected specimens. Specimens collected in the winter showed good recovery after 10 days from a single cooling to the SCP. In these experiments the centipede was removed from the cooling device immediately after the SCP was reached. More than one such exposure sometimes had adverse effects, which indicates that the centipedes are susceptible to ice formation following crystallization. When the centipedes were cooled to a temperature 1 °C below their SCPs, recovery at room temperature after 24 h was poor, and only few individuals survived after 10 days (Crawford and Riddle 1974). According to these authors, the results suggest that longer exposures in the frozen state might have been fatal. In conclusion, it seems reasonable to classify *S. polymorpha* as a freeze susceptible animal.

Crawford and Riddle (1974) measured temperatures as low as -3 °C in burrows of *Scolopendra polymorpha* under rocks in the Peloncillas Mountains. Higher temperatures were recorded in the soil at depths of 8 and 24 cm. Centipedes confined to burrows in this range will experience a relatively stable microclimate. Based on estimates of population densities above 8 cm, meteorological data from the area and survival of the centipedes in freezing experiments, Crawford and Riddle (1974) suggested an average yearly centipede mortality of 10 % due to cold.

10.2.3 Spiders

According to Lee and Baust (1985), the funnel-web weaving spider *Agelenopsis aperta* did not become more cold-hardy during low temperature acclimation. The species is found in desert grassland and lava beds at 1600 m in southern New Mexico. The SCPs of both males and females were in the range -8 to -12 °C, and remained relatively constant, regardless of previous low temperature exposure. The spiders tolerated temperatures close to their SCPs, but were killed by freezing. The funnels of *A. aperta* may extend more than 25 cm into cracks in the ground. In January, Lee and Baust (1985) measured ambient temperatures of -14 °C, while the temperature at 4 cm in one of the burrows was only -6 °C. Thus, by moving only a few centimetres into the substrate, *A. aperta* may avoid environmental temperatures approaching its SCP.

10.2.4 Insects

10.2.4.1 Grashoppers

Locusts are also able to supercool, and Cloudsley-Thompson (1973) found mean SCPs of -5.4 and -7.5 °C, respectively, in adults and 5th instar nymphs of the desert locust *Schistocerca gregarià*. Corresponding values for the migratory locust *Locusta migratoria migratoroides* were -3.5 and -3.8 °C. The degree of supercooling in locusts depends on several factors. In specimens tested shortly after they were given water to drink, SCPs were elevated. Starved specimens of *S. gregaria* had lower SCPs than fed ones, but the difference was not statistically significant. Acclimation at 5 °C for 18 h lowered the SCPs in *S. gregaria* but not in *L. migratoria migratoroides.*

Locusts that are exposed to diurnal cycles of LD 6:18 are known to elicit a clear circadian periodicity (Cloudsley-Thompson 1978). The mean SCPs of *Locusta migratoria migratoroides* were significantly lower in the dark than in the light phase of the circadian locomotory rhythm, and a similar tendency was observed in *Schistocerca gregaria.* Cloudsley-Thompson (1978) argued that this difference would be of adaptive value since it would enable an insect to withstand greater cold at night when ambient temperatures are lower than during the day.

Since locusts are rarely or never exposed to freezing temperatures, the effect of circadian rhythms on supercooling seems to be of little importance to them. In equatorial alpine species that are exposed to great diel temperature fluctuations, however, including freezing night temperatures, increased cold hardiness during the dark phase would have considerable adaptive value, but this remains to be investigated. As mentioned previously (Chap. 9.2.3), both *Meridacris subaptera* from the Andean Mountain Range (Sømme 1986c) and *Hemideina maori* from New Zealand (Ramløv et al. 1992) have poor supercooling capacity, as is the case with the locusts. The alpine species, however, are freeze tolerant and able to survive degrees of freezing appropriate to their environments.

10.2.4.2 Beetles

The cold hardiness of beetles from tropical high alpine altitudes has been discussed briefly in Chapter 9.4.1. At lower elevations in Kenya, Zachariassen (1980) measured the SCPs of a number of beetles and found that they ranged from -6 to -12 °C. All these species were freeze susceptible. In the tenebrionid beetles *Ocnera hispida* and *Adesmia antiqua* from Karthoum, Sudan, Cloudsley-Thompson (1973) found mean SCPs of -10.4 and -9.0 °C, respectively. In these experiments the animals were removed from the cooling device to room temperature within 1 min after reaching their SCPs. More than half of them recovered from this treatment. As pointed out by Cloudsley-Thompson (1973), this remarkable ability to survive low temperature exposures cannot have any adaptive significance, since winter temperatures in Khartoum seldom fall below 5.6 °C.

Eleodes spp. (Coleoptera, Tenebrionidae) from Arizona and New Mexico have mean SCPs of -11.8 °C and recover from freezing (Cloudsley--Thompson and Crawford 1970a). Tenebrionid beetles of the genera *Eleodes* and *Coelocnemis* from mountains in southern California are freeze tolerant during the winter (Zachariassen and Hammel 1976). They contain highly effective INAs which are responsible for their SCPs in the range of

-6 to -7 °C. The INAs were present only in winter specimens, and SCPs were lower during the summer. Zachariassen et al. (1979) also showed that approximately 25 % of the total body water of the freeze tolerant *Eleodes blanchardi* is unfreezable, but the proportion did not change with warm acclimation and therefore does not explain the freeze tolerance of the beetles. The beetles do not accumulate significant levels of glycerol or other polyols and the unfreezable water is probably bound to ions and macromolecules.

10.3 Conclusions

Studies on the cold hardiness of terrestrial arthropods from hot deserts clearly illustrate how some species are simultaneously adapted to drought and cold. The general mechanism of cold hardiness known from temperate and polar species (Chap. 9), appears to be similar in species from arid and hot areas. The physical potential of all insects and other terrestrial arthropods for supercooling is reflected in desert species, although other factors are also of importance for their cold hardiness. Freezing winter temperatures are common in temperate deserts around the world. In African and North American deserts, many terrestrial arthropods have sufficient supercooling capacity to avoid freezing at the ambient temperatures of their natural habitats. Moderate degrees of freeze tolerance have been demonstrated in one species of scorpions and in several tenebrionid beetles from hot deserts as well as in grasshoppers and curculionid beetles from high altitudes in tropical mountains.

11 Cold Tolerance in "Lower" Invertebrates

11.1 Nematodes

Nematodes are found in a wide variety of habitats that are exposed to subzero temperatures during winter. This applies to free-living soil-dwelling nematodes, many plant parasites and the free-living stages of nematodes parasitic on vertebrates and invertebrates. Nematodes are important components of the terrestrial alpine and polar fauna. They occur in enormous numbers both in the soil and associated with vegetation. In hot desert soils they are most numerous close to shrubs and other plants, and in moist habitats on the borders of deserts (Chap. 2.1.2).

As in terrestrial arthropods, two strategies of low-temperature survival (Chap. 9.2) are found among nematodes (Sayre 1964; Pickup 1990a; Wharton and Block 1993). Some species are killed by freezing and avoid ice formation by supercooling at temperatures below the melting point of their body fluids. Other species are freeze tolerant and survive the formation of ice within their tissues. As discussed below, in some species part of the population survives freezing while other individuals are killed and depend on supercooling for survival.

The cold tolerance of nematodes has recently been reviewed by Wharton (1995), pointing out the biological differences between nematodes and terrestrial arthropods. Special techniques have to be employed to separate the nematodes from the substrate, and some of the treatments may affect the physiological state of the animals. Methods for studying cold tolerance of nematodes have to be modified compared with those used for arthropods.

In principle, nematodes are aquatic organisms and depend on a high moisture content of their environment for growth and development. During freezing, ice formed from moisture in contact with the body surface may inoculate the body fluids through the cuticle or via external openings. Freezing in nematodes is often initiated by exogenous ice nucleation when external water freezes (Wharton and Brown 1991; Wharton and Block 1993; Wharton 1994). Nematodes have much less potential for avoiding or preventing inoculative freezing than do terrestrial arthropods, and inoculation is likely to be the main mechanism of ice formation. In some species, means of preventing inoculation have evolved. As pointed out by Wharton (1995), an eggshell or a sheath around other stages may prevent exogenous ice nucleation. In this case the enclosed embryo or larva will supercool in the presence of external ice, surviving by freeze avoidance.

11.1.1 Strategies of Cold Hardiness

Supercooling has been demonstrated in a number of nematode species and has been related to a freeze avoiding strategy (Fig. 11.1). According to Wharton et al. (1984), the plant parasite *Ditylenchus dipsaci* and the free-living nematode *Panagrellus silusae* are freeze sensitive and avoid freezing by supercooling. *D. dipsaci* possesses some ability to prevent inoculative freezing when frozen in contact with water, while *P. silusae* cannot prevent ice seeding and is killed when the external water freezes. In the free-living nematode *Panagrellus redivivus*, the supercooling capacity is enhanced by acclimation to low temperatures (Mabbet and Wharton 1986). These changes may be related to the accumulation of trehalose or other low molecular weight carbohydrates. According to Ash and Atkinson (1982), the trehalose concentration of the eggs of *Nematodirus battus* increases during chilling, parallel to a decrease in mean supercooling point (SCP).

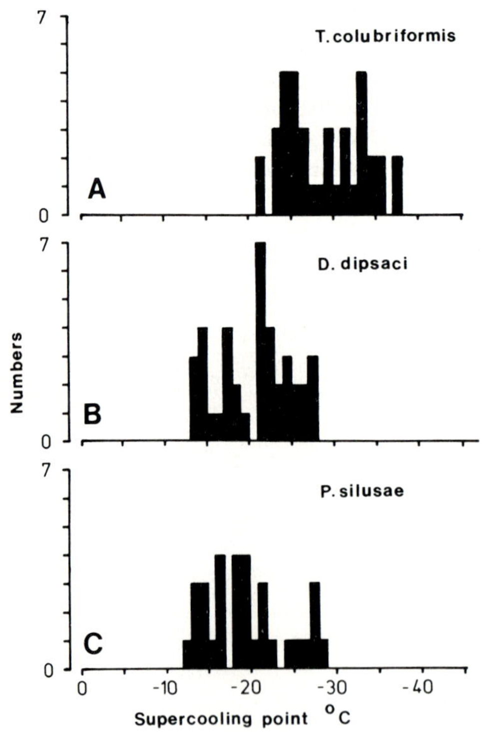

Fig. 11.1A-C. The distribution of supercooling points in three species of nematodes. **A** Free-living third stage juveniles of the sheep parasite *Trichostrongylus colubriformis*. **B** Fourth stage juveniles of the plant parasite *Ditylenchus dipsaci* **C** The free-living nematode *Panagrellus silusae*. (Wharton et al. 1984)

Although a great capacity for supercooling has been recorded, individuals of the same species or specific stages within a species may also survive freezing. Thus, the free-living stages of the sheep parasite *Trichostrongylus colubriformis* can survive extended supercooling when submerged in liquid paraffin (Wharton et al. 1984; Wharton and Allen 1989). When frozen in contact with water, part of the nematode population remains supercooled, while a proportion of the population survives freezing caused by ice nucleation. In the third instar larvae, the probability of inoculation is reduced by an external sheath formed from the cuticle of the previous larval stage. A larger proportion of the nematodes is inoculated by ice if the sheath is removed or injured, but the survival rate of frozen larvae is increased.

In a similar way, the second stage larvae of the potato-cyst nematode *Globodera rostochiensis* are protected from ice inoculation within the egg shell (Perry and Wharton 1985). The percentage recovery declines with decreasing temperatures below zero, but more than 50 % survive even in contact with ice at -20 °C. After hatching, free juveniles cannot survive freezing of their body fluids resulting from ice nucleation across the cuticle.

The risk of inoculative freezing may be greatly reduced in dry soils, as shown by Forge and MacGuidwin (1992a) in the overwintering stages of the plant parasite Meloidogyne hapla. When second stage juveniles were frozen at -1 to -4 °C, survival was greater in soil at water potentials of -1910 to -520 kPa than in soil at higher water potentials (Fig. 11.2). The increased survivorship probably results from a reduction of the pore space being filled with ice, thereby reducing the chances of inoculative freezing. Low water potentials may indirectly induce physiological changes that make nematodes more able to survive frozen conditions. Acclimation to low temperatures also increases the ability of second-instar juveniles of *M. hapla* to survive external freezing (Forge and MacGuidwin 1992b).

11.1.2 Antarctic Nematodes

In his studies on nematodes from Signy Island in the Maritime Antarctic, Pickup (1990a,b,c) found large supercooling capacities in a number of species. Before cooling, excess water around the nematodes was removed and the worms were immersed in liquid paraffin to prevent water loss by evaporation. By this technique the mean SCP of the free-living predatory nematode *Coomansus gerlachi* was found to vary from -20 °C in mid-winter to -9 °C in the summer (Pickup 1990b). The changes in SCPs from low to high were partly explained by the presence of food in the gut. Specimens fed on moss showed a shift to high SCPs while starvation resulted in a shift to low values. In *C. gerlachi* part if the population was killed by freezing while others survived. The probability to survive freezing by endogenous ice nucleation increased at high subzero temperatures; more than 50 % of adult and late juvenile specimens survived at SCPs above -8.7 °C.

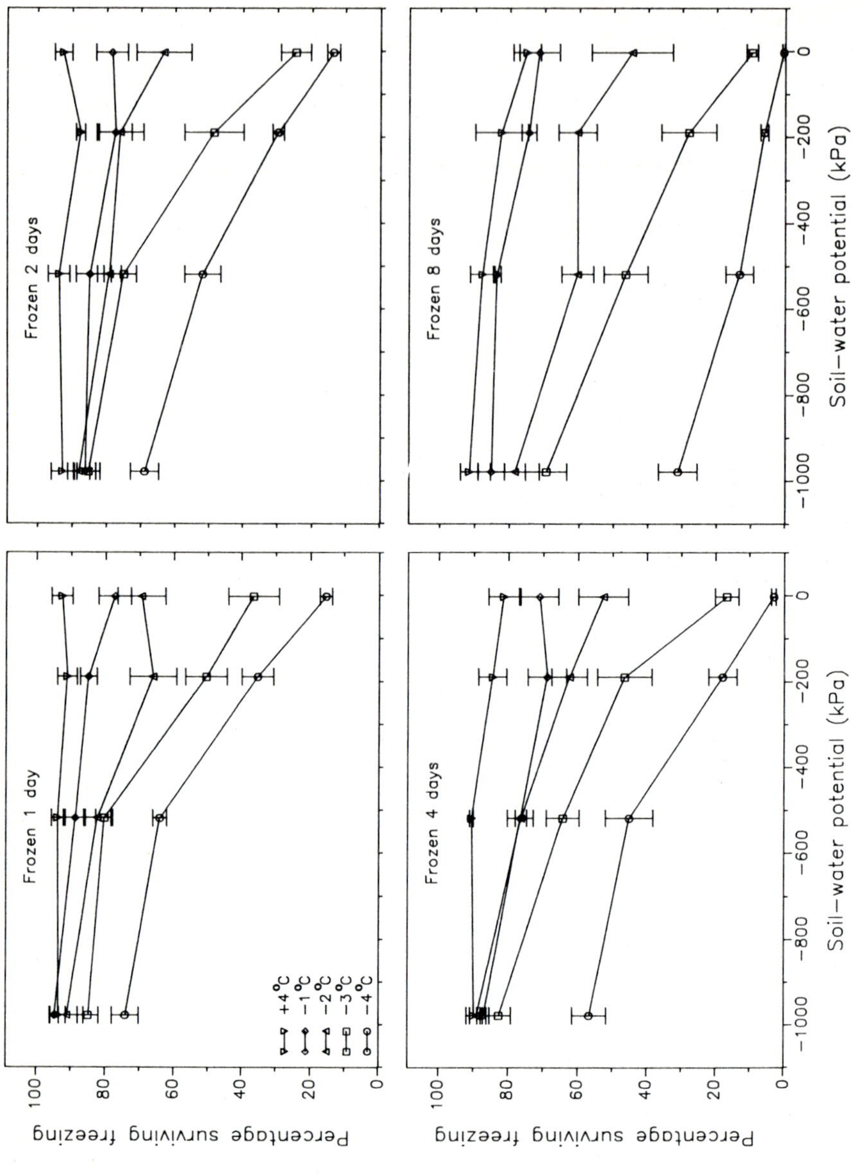

Fig. 11.2. Survivorship of second stage juveniles of the nematode *Meloidogyne hapla* in soil with a pre-freezing water potential of -1010, -530, -167 or -2.5 kPa, frozen at -1 to -4 °C for 1-8 days. Each *point* is the mean (± SE) of five samples. (Forge and MacGuidwin 1992a)

Similarly, a combination of freeze tolerance and freeze avoidance has been demonstrated in three *Eudorylaimus* spp. inhabiting fellfield *Andreaea* mosses at Signy Island (Pickup 1990a). All these species are relatively large free-living nematodes; adults of both *Eudorylaimus conipes* and *E. pseudocarteri* are more than 2 mm in length. Within all three species, individuals with low SCPs were killed by freezing (Fig. 11.3), but their supercooling capacity is sufficient to withstand the extremes of their natural habitat. Individuals with high SCPs, e.g. above -15 °C, showed increasing survival rates with increasing temperatures.

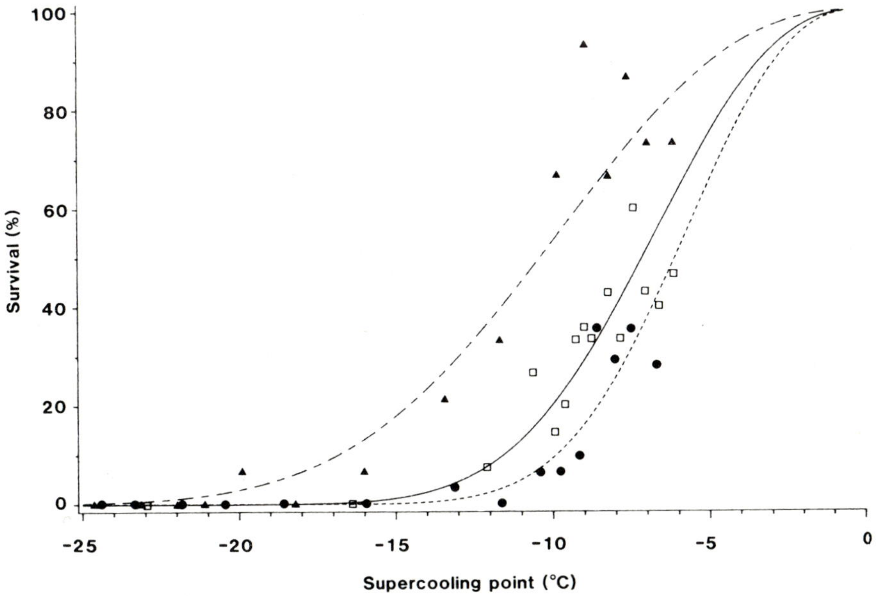

Fig. 11.3. The effect of supercooling on the survival of freezing in the antarctic nematodes *Eudorylanus spaulli (filled circles* and *dotted line*), *E. coniceps (filled triangles* and *brolen line)*, and *E. pseudocartei (open squares* and *solid line*). For each species, 15 *points* are plotted representing means of equal numbers of individuals in rank order of supercooling point. The *lines* represent the fit of the Comperts model to the data. (Pickup 1990a)

Seasonal changes in supercooling capacity were also apparent in *Plectus antarcticus* from fellfields at Signy Island (Pickup 1990c), while *Teratocephalus tilbrooki* and *Ditylenchus* sp. showed less variations. *Ditylenchus* sp., inhabiting the aerial thalli of *Usnea lichens*, had mean SCPs as low as -30 °C, and is able to survive winter temperatures in this range. The species was invariably killed by freezing. *P. antarcticus* also depends on supercooling to survive, and some winter mortality is likely to occur. A fraction of the specimens of *T. tilbrooki* tolerated freezing when frozen at ca. -11 °C.

Another antarctic nematode, *Panagrolaimus davidi,* from the McMurdo Sound region, also avoids freezing by supercooling when free of surface moisture (Wharton and Brown 1991; Wharton 1994). In air, no freezing was observed at temperatures down to -60 °C. Approximately 50-60 % of the nematodes survived in the supercooled state, and mortality may be due to desiccation rather than to cold stress. Enclosed in liquid paraffin, some specimens had SCPs below -25 °C. Although all specimens froze at -30 and -40 °C, between 10 and 25 % of them survived. In water, the nematodes were seeded by exogenous ice nucleation. A large proportion tolerated freezing at -5 °C and a small proportion tolerated temperatures down to -40 °C. The degree of survival depended on both previous acclimation temperature and the age of the culture. In long-term exposures at -20 and -80 °C, 10 % and 25 %, respectively, of the nematodes survived 28 days suspended over a sludge of potassium sulphate.

In nature, contact with ice will result in exogenous ice nucleation preventing supercooling to low temperatures (Wharton and Brown 1991). Recently, exogenous ice nucleation has been demonstrated in nine species of nematodes inhabiting vegetation at Signy Island (Wharton and Block 1993). Observations of the freezing processes under a cryomicroscope revealed that exogenous ice nucleation was predominant when the animals were frozen in contact with a 250 mosmol physiological saline solution. Most of the species froze shortly after the surrounding water had frozen. In the predatory nematode *Coomansus gerlachi*, nucleation has been observed to start at the anterior end, with the ice front travelling backwards through the body (Fig. 11.4). *Dytilenchus* sp. was the only species showing some ability to restrict exogenous ice nucleation. More than 50 % survival was recorded in most of the species after freezing to -15 °C. The only exception was *Panagrolaimus* sp. where approximately 80 % of the specimens were killed.

11.1.3 Freeze Tolerance During Anhydrobiosis

The nematode *Aphelencus avenae* enters anhydrobiosis following slow dehydration (Chap. 5.2.2), and the effects of variable water content on its freeze tolerance have been studied by O'Dell and Crowe (1979). Worms with water contents below 21 % of fresh weight were relatively unaffected by exposure to -196 °C for 15 min. With water contents above 38 %, however, freezing at this temperature was fatal to most of the nematodes. As suggested by O'Dell and Crowe (1979), ice formation does not occur when the water content is below 21 %, since most of the water is then bound.

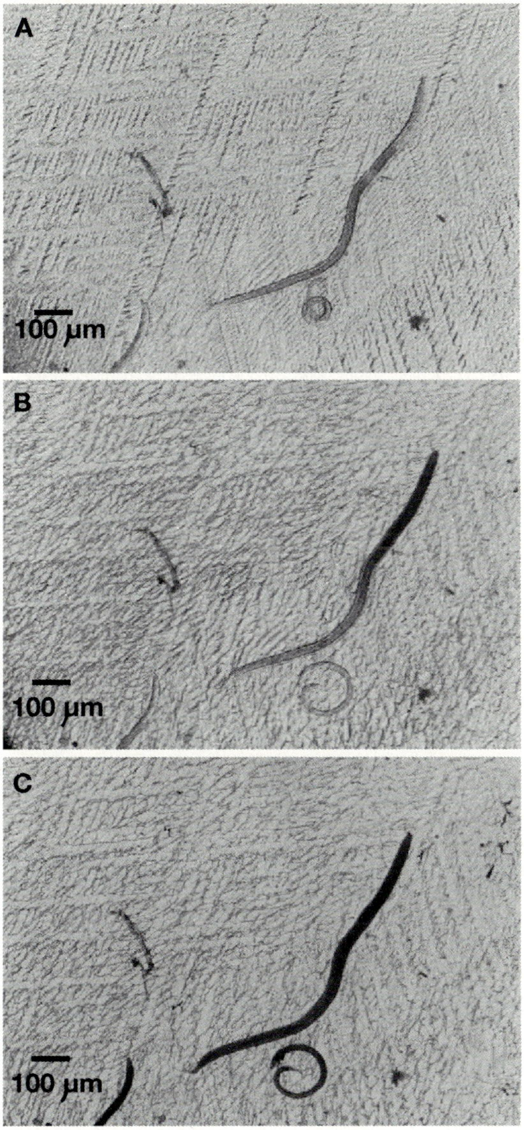

Fig. 11.4A-C. Photomicrographs of *Coomansus gerlachei* during freezing in salt buffer whilst cooling at 0.5 °C min^{-1}. Freezing occurred at -8 °C. **A** The medium has frozen but the nematode remains unfrozen. **B** Freezing of the nematode has been initiated by exogenous ice nucleation, the ice front starting at the anterior end of the nematode and travelling back through the body. **C** The nematode is completely frozen; *bar* 100 µm. (Wharton and Block 1993)

Few data are available, but extreme freeze tolerance may also be found in other species of nematodes during anhydrobiosis. For instance, anhydrobiotic *Dytilenchus* sp. from the maritime Antarctic survived exposures to -40 and -80 °C for 48 h (Pickup and Rothery 1991).

Desiccation will reduce the risk of inoculative freezing because the melting point of the body fluid is lowered. According to Wharton (1994), various nematodes are more freeze tolerant when partly desiccated. Juveniles of *Elaphostrongylus rangifer*, which is a parasite of reindeer, survive exposure to -20 °C at 0 % RH (Lorentzen and Halvorsen 1986).

11.1.4 Freeze Tolerance or Freeze Avoidance?

Several of the experiments on supercooling in nematodes were conducted in liquid paraffin following the removal of surface water (e.g. Pickup 1990a, b, c; Wharton and Brown 1991). At first glance, these conditions appear to be of limited ecological relevance. From the examples discussed in Chapter 5.1.2, however, it appears that freeze avoidance is the strategy of survival in several species. It is not fully known how nematodes can remain unfrozen in a frozen environment, but a few species are able to restrict exogenous ice nucleation. Mortality due to inoculative freezing may also be reduced at low water potentials in frozen porous soils (Forge and MacGuidwin 1992a); in which case increased supercooling capacity will enhance survival.

In many species of nematodes, part of the population survives freezing, while a proportion depends on supercooling. The reason for this difference within a particular species or population is not clearly understood. Most species of antarctic nematodes are able to survive freezing at relatively high sub-zero temperatures, but high mortalities have been recorded when freezing occurs following endogenous ice nucleation below -15 to -20 °C. Nevertheless, a proportion of some species survives freezing even at temperatures as low as -40 °C. In nature, the nematodes are in contact with ice, and in most species exogenous inoculation takes place at temperatures at which freezing is tolerated. It remains to be seen if the species that survive to be frozen at high subzero temperatures, e.g. -10 or -15 °C, also tolerate further cooling, corresponding with natural temperature fluctuations in their normal environment. In the Antarctic, nematodes must tolerate freezing temperatures for months, but more information is needed about the survival of different species during long-term exposures to different temperature regimes.

By combining freeze avoidance and freeze tolerance, the cold hardiness strategies of nematodes appear to differ from those of terrestrial arthropods. A more distinct division into freeze avoiding and freeze tolerant species is frequently found in the latter group (Chap. 9). Several freeze avoiding insects, however, survive freezing at temperatures close to the melting point of their haemolymph, and the degree of freeze tolerance varies widely within the same species. Many nematodes are apparently better adapted to survive relatively low freezing temperatures, and further studies on the physiology of this mechanism would be of great interest.

11.2 Tardigrades

11.2.1 Freezing in Liquid Gases

Tardigrades in anhydrobiosis are extremely resistant to physiological stress (Ramløv 1989; Wright et al. 1992). In their dehydrated state they may tolerate exposures to high temperatures, intensive UV radiation, and the freezing temperatures of liquid gases. As an example, Rahm (1924) maintained dehydrated tardigrades in liquid air for up to 20 months.

On account of their high tolerance during anhydrobiosis, freezing tolerance in tardigrades is often misinterpreted as being restricted to this state (Ramløv and Westh 1992). Rahm (1923, 1924), however, also reported that *Ramazzottins oberhaeuseri, Milnesium tardigradum* and *Microbiotus* sp. survived exposures to -253 °C in their hydrated state.In their experiments, Ramløv and Westh (1992) found that hydrated specimens of *Adorybiotus coronifer* survived exposures to -196 °C for 15 min, but longer periods were not tested. Survival rate, however, was greatly affected by cooling rate. No tardigrades survived when dropped directly into liquid nitrogen, corresponding to a cooling rate of approximately 1500 °C min^{-1}. In contrast, about 50 % survived exposure at -196 °C when cooled at a rate of 30 °C min^{-1}. Highest rates of survival were found in specimens with high trehalose content. Ramløv and Westh (1992) also showed that tardigrades precooled at a cooling rate of 30 °C min^{-1} to temperatures in the range of -10 to -90 °C were able to survive subsequent shock cooling to -196 °C. The tardigrades all froze below -5 °C, and the results suggest that following initial ice formation, the freezing temperature is of little importance for survival. Although it is not known if some bound water remains unfrozen at -196 °C, Ramløv and Westh (1992) suggested that the freeze tolerance of *A. coronifer* is closely related to its ability to survive total desiccation during anhydrobiosis.

11.2.2 Freezing at Environmental Temperatures

Although tolerance to extremely low temperatures has been a matter of great attention, little is known about the relation of tardigrades to low temperatures more representative of those of their natural environments. Westh and Kristensen (1992), however, pointed out that tardigrades from Greenland are able to tolerate very low winter temperatures. Of the two species studied by these authors, *Amphibolus nebulosus* inhabits dense populations of limnic mosses and algae, and overwinters as non-anhydrobiotic frozen cysts or, in a few cases, as eggs and adults. *Adorybiotus coronifer* typically inhabits drought-resistant mosses in the Arctic, and overwinters in a frozen or anhydrobiotic state. This species has been recorded at 5600 m in the Himalayas, and is common among dry mosses at Öland in the Baltic Sea.

Both species have a moderate capacity for supercooling. Using differential scanning calometry, Westh and Kristensen (1992) measured their crystallization temperatures (=SCPs) to be in the range of -6 to -7 °C. No difference appeared between summer- and winter-acclimated *Adorybiotus coronifer* or between active and encysted *Amphibolus nebulosus*. In both species freezing was initiated by ice nucleating agents (INAs), but differences in their heat stability indicated that different substances are present in the two species. During freezing, ice growth proceeded very rapidly in both and, as pointed out by Westh and Kristensen (1992), this represents a fundamental difference from insects where ice growth often proceeds more slowly.

11.2.3 Antarctic Tardigrades

A number of species of tardigrades have been reported from the Antarctic (Chap. 2.2.3) but very little is known about their ability to survive low temperatures. On continental antarctic nunataks, overwintering tardigrades are exposed to some of the most severe temperature conditions on earth. In some locations air temperatures may reach a minimum of -50 to -60 °C, but tardigrades and other invertebrates may be protected from such extremes in their microhabitats, e.g. by a snow cover.

In a recent study, Sømme and Meier (1995) showed that three species of tardigrades from nunataks in Dronning Maud Land have a considerable capacity to survive prolonged exposure at freezing temperatures. The tardigrades *Echiniscus jenningsi*, *Macrobiotus furciger* and *Diphascon chilenense* were tested in both their hydrated and dehydrated states at -20, -80 and -180 °C.

Hydrated tardigrades showed no increase in mortality after 590 days in a freezer at -22 °C. At -80 °C a clear decrease was found in the survival rates of *Echiniscus jenningsi* with exposures from 7 to 150 days, but mortality did not increase in *Macrobiotus furciger* and *Diphascon chilenense* under similar conditions. All hydrated specimens of *E. jenningsi* were killed after 1 day at -180 °C, but a few specimens of the two other species survived. Following 2 days at this temperature the mortality was 100 % in both species.

Dehydrated tardigrades were kept at -22 °C for almost 8 years, and large proportions of all three species survived this exposure. In *Echiniscus jenningsi*, there was a significant decrease in survival from 635 to 3040 days of exposure, and only 15 % survived the longest period. The survival of *Macrobiotus furciger* and *Diphascon chilenense* did not change, and approximately half of the tardigrades were alive following both periods at -22 °C.

In contrast to hydrated specimens, dehydrated tardigrades also survived exposure to -180 °C. Although there were large variations, mortality rates did not change significantly during 2 to 14 days at this temperature. Before exposure, samples of soil containing the tardigrades were dried at 5 % RH at 3 °C for 5 days. Mean survival rates for the different exposure times varied from 44 to 53 % in *Echiniscus jenningsi*, from 26 to 51 % in *Macrobiotus furciger*, and from 32 to 43 % in *Diphascon chilenense*.

In conclusion, it appears that all species are able to survive in their hydrated state for extended periods at temperatures down to -80 °C. Judging from the differences in survival rates at -180 °C, dehydrated specimens are even better adapted to extremely low environmental temperatures. Higher mortalities were found in *Echiniscus jenningsi* than in *Macrobiotus furciger* and *Diphascon chilenense*, but it appears that the cold hardiness capacity of all three species is sufficient for survival during overwintering on antarctic nunataks.

11.3 Rotifers

Bdelloid rotifers are known to enter anhydrobiosis (Chap. 5.4) and in this state they are extremely tolerant to low temperatures (Gilbert 1974). Rahm (1923) found that several species in dried mosses survived 8 h at -270 °C, and Becquerel (1950) showed that *Habrotrocha constricta* and *Philodina roseola* survived for 2 h at temperatures close to the absolute zero. Survival in anhydrobiotic rotifers is increased by slow cooling rates.

At Cape Royds in the Antarctic, Murray (1910) noted that the bdelloid rotifer *Philodina gregaria* withstood repeated freezing and thawing at weekly intervals for several months. This species is abundant in antarctic lakes and ponds which are frozen most of the year. Rotifers could be obtained alive during the winter by thawing ice from the ponds, and Murray (1910) found that *Adineta grandis* survived at temperatures down to -40 °C.

The ability of hydrated rotifers to survive freezing appears to be variable (Gilbert 1974), but relatively few experiments have been carried out. Soil-living bdelloids have been found to survive exposures at -4 °C for 1 week, but survival rates of isolated specimens at -8 ° was poorer. On the other hand, Rahm (1923) reported that wet, moss-dwelling rotifers could endure slow cooling down to -253 °C. In *Philodina acuticornis odiosa*, cooled to temperatures of -60 to -150 °C, Koehler (1967) found highest survival rates at cooling rates of 3 to 5 °C min^{-1}. Glycerol solutions at a concentration of 2-3 % and a 7.5 % dimethyl sulphoxide solution greatly increased survival rates, as did the availability of ample supplies of bacterial food upon thawing (Koehler 1967; Koehler and Johnson 1969). Almost complete recovery was obtained under proper experimental conditions. In addition to normal swimming and feeding activities, survivors of the freezing experiments had a normal reproductive capacity and a normal life span.

11.4 Lumbricid Earthworms

11.4.1 Effect of Dehydration

There are few investigations of the cold tolerance of earthworms, allthough species of the family Lumbricidae are widely distributed in the northern hemisphere. According to Holmstrup (1992, 1994), post-embryonic individuals of lumbricid earthworms die in frozen soil. Those that survive the winter have presumably migrated to deeper frost-free layers. In a Finnish forest, Huhta (1980) observed how a sudden drop in temperature killed all post-embryonic earthworms, while egg capsules or cocoons were more cold hardy. Holmstrup et al. (1990) also provided evidence that cocoons are the most cold-hardy stage in nine species of earthworms from Denmark.

Holmstrup (1992, 1994) found that fully hydrated cocoons of five species of lumbricid earthworm had mean SCPs in the range -5 to -12 °C, depending on the developmental stage of the embryo. Following dehydration with 50 -60 % weight loss, the mean SCPs were lowered to the range -12 to -16 °C, and in some specimens of *Dendrobaena octaedra* to -25 °C. The capacity of supercooling may not be of ecological relevance, since the cocoons are in contact with ice under natural conditions, and may be subjected to inoculative freezing. The experiments demonstrated, however, that both embryos and newly hatched larvae were killed by freezing at temperatures corresponding to their SCPs.

The cold hardiness of earthworm cocoons is closely associated with their tolerance to dehydration (Chap. 8.3.2). In an atmosphere above ice at -3 °C, Holmstrup (1992) found that cocoons of *Dendrobaena octaedra* lost about 60 % of their fresh weight, at which point the vapour pressure of their body fluid was in equilibrium with the ice (Fig. 8.1). A weight loss of this order actually corresponds to ca. 80 % of the water of fully hydrated cocoons. Under natural conditions the cocoons will lose water when the surrounding soil begins to freeze, and Holmstrup (1994) showed that the rate of water loss was higher in contact with frozen soil than above ice at -3 °C. When the vapour pressure of the cocoon's body fluid is in equilibrium with the surrounding ice, the risk of inoculative freezing is presumably greatly reduced.

In conclusion, Holmstrup (1992, 1994) demonstrated that the cocoons are dehydrated when exposed to subzero temperatures in a frozen environment because of the difference in water vapour pressure between supercooled water and ice. Following a gradual decrease in melting points, this equilibrium was established in *Dendrobaena octaedra* within 15 days in an atmosphere above ice at temperatures ranging from -2.1 to -6.6 °C (Holmstrup and Westh 1994). In fully hydrated cocoons about 15 % of the total water is osmotically inactive or bound, but a proportion is turned into osmotically active water during dehydration. At -8 °C, which the cocoons survive, 97 % of osmotically active water was withdrawn from the cocoons.

11.4.2 Long Term Survival

Holmstrup (1994) also showed that the cocoons of earthworms are able to survive long exposures at freezing temperatures. Lumbricid earthworms of the species *Dendrobaena octaedra*, *Dendrodrilus rubidus tenuis*, *Aporrectodea caliginosa* and *Allolobophora chlorotica* were collected in Jutland, Denmark and *Dendrodrilus rubidus norvegicus* from the alpine region of Hardangervidda, Norway. Among these species, *A. caliginosa* and *A. chlorotica* are deep soil dwellers, while the remaining species live closer to the surface. The survival rates of the earthworm cocoons during prolonged storage at -3 and -8 °C are shown in Fig. 11.5. All species survived several days at both temperatures. The survival rates of the deep-dwelling species *Aporrectodea caliginosa* and *Allolobophora chlorotoca* were among the lowest, but at -8 °C there were also high mortalities in *Dendrodrilus rubidus tenuis* and *Dendrodrilus rubidus norvegica*. *Dendrobaena octaedra* was by

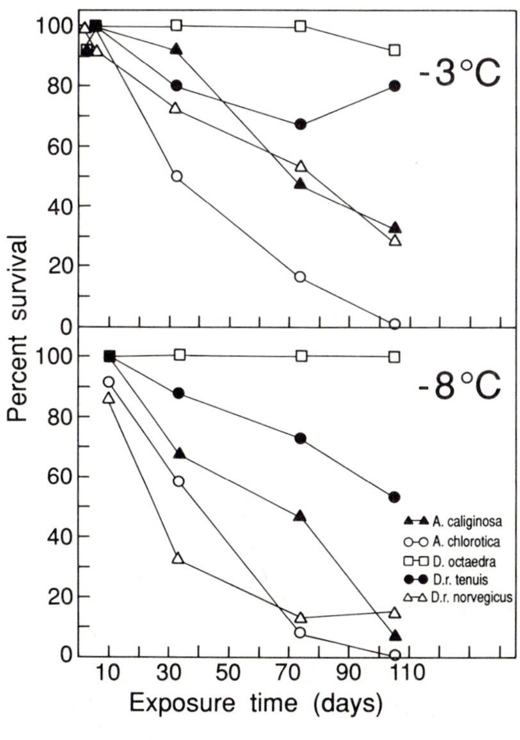

Fig. 11.5. Survival rates of lumbricid earthworm cocoons exposed to subzero temperatures in frozen soil, expressed as percent hatchability of 15 cocoons. (Data from Holmstrup 1992)

far the most cold-tolerant species. It is remarkable that although *D.r. norvegicus* was collected from an alpine location, it is not more cold-tolerant than the closely related *D.r.tenuis* from a lowland habitat. The reason for this is probably that the microhabitats of the alpine species are frequently protected from low temperatures by dense layers of snow.

As pointed out by Holmstrup (1994), the surface-dwelling species are potentially most exposed to low temperatures. They are also known to have the northernmost distribution, being recorded from Greenland, Siberia and northern Fennoscandia. *Aporrectodea caliginosa* and *Allolobophora chlorotica* are rarely found in very cold regions, and their survival rates may be increased by retreating to deeper layers of the soil.

11.4.3 A Protective Dehydration Mechanism

From the results presented, Holmstrup and Westh (1994) pointed out that the embryos of *Dendrobaena octaedra* within their cocoons do not conform with the traditional classification of cold hardiness as freeze tolerance or freeze avoidance. The cocoons do not tolerate ice formation (Holmstrup 1992, 1994) and are therefore not freeze tolerant. On the other hand, their tolerance of subzero temperatures is not based on prolonged supercooling, as is the case in many freeze avoiding terrestrial arthropods. The avoidance of freezing in the earthworm cocoons is based on their ability to vapour equilibrate with the frozen soil water. During this process, the melting point of the cocoon fluid is adjusted according to the temperature of its surroundings. Holmstrup and Westh (1994) called this a 'protective dehydration mechanism', and suggested that it is a previously undescribed type of cold hardiness. The protective dehydration mechanism resembles the protective dehydration of cells in freeze tolerant insects by the formation of extracellular ice, but in the case of the earthworm cocoons the dehydration effect is produced by extra-organismal ice in the surroundings.

11.5 Conclusions

In nematodes, rotifers, tardigrades and lumbricid earthworm, there is a close association between resistance to drought and cold. Anhydrobiosis is common in the former three groups, and extremely low temperatures may be tolerated in this state. Fully hydrated tardigrades also survive freezing, but in nematodes the situation is more complex. A proportion of a nematode population may survive freezing, while another proportion is freeze susceptible.

Embryos in their cocoons and post-embryonic stages of lumbricid earthworms are killed by freezing. Due to rapid dehydration of the cocoons in a frozen environment, an equilibrium is established between the melting point of the cocoons and the surrounding temperature. Holmstrup and Westh (1994) termed this a protective dehydration mechanism, and pointed out that it is different from freeze avoidance in terrestrial arthropods based on supercooling. The cold tolerance strategy of lumbricid earthworm cocoons clearly illustrates how overwintering in frozen soil may have a desiccating effect on many organisms.

12 Summary and Concluding Remarks

12.1 Deserts

Temperate and subtropical deserts are situated in belts at certain latitudes around the world (Fig. 1.1), and comprise areas of great diversity with regard to topography, climate and vegetation. Highest temperatures on an annual basis are found in subtropical deserts, which are characterized by the lack of snowfall in winter although frost may occur during the night. Temperate inland deserts, such as those of central Asia, are often situated in mountainous areas with hot summers and cold winters. From a biological point of view, these deserts offer some of the most harsh environments, where adaptations to both extreme drought and cold are required. Special conditions are found in the cool coastal deserts of South America and Africa, where moisture is provided by condensation of vapour through the cooling of warmer air masses.

Polar deserts provide environments of extreme cold. In both the Arctic and the Antarctic the ground is frozen for most of the year. In spite of enormous quantities of snow and ice, liquid water is not available to terrestrial plants and animals. For this reason polar regions are both cold and arid during the winter. In their frozen world, invertebrates live at the risk of dehydration if the water vapour pressure of their body fluids is higher than that of the surrounding ice.

Precipitation is low in most arctic and antarctic areas, and often of a magnitude comparable with that of hot arid deserts. In summer, moisture is provided by the melting of snow and ice, creating microclimatic conditions in which living organisms grow and develop. Due to the permafrost, the upper layer of the ground often remains waterlogged in spring and early summer. Since air temperatures are low, evaporation rates are lower in polar areas than in hot temperate and subtropical deserts. Following a flush of water during spring, the aridity of polar deserts increases during the summer. For their survival, invertebrates have to tolerate the driest conditions that occur in their microhabitats. As in other environments, adaptations in living organisms must include tolerance of the most extreme conditions that may be experienced.

The microclimate is of particular importance for invertebrates in both hot and cold arid environments. In hot deserts, more favourable conditions of temperature and humidity are found in the vegetation, under rocks and at various depths in the soil. During the polar summers, the main problem is to absorb as much heat as possible, rather than to avoid extreme temperatures. Typically, temperatures show great fluctuations in most arctic and antarctic habitats, but are increased by several degrees in vegetation and the upper soil layers.

12.2 The Fauna

The great variations in the climates of the different geographical areas are reflected in the fauna. The fauna of hot deserts may be of ancient origin where arid climates have prevailed for long periods. It may also have evolved in relatively recent times following climatic changes, such as in the Sahara, which had a wetter climate during the Pleistocene. Hot deserts are bordered by areas with more favourable climates and higher biodiversity. A gradual invasion of the desert by invertebrates may take place by species that are able to adapt to arid environments.

The distribution of many invertebrate taxa of hot deserts is still poorly known. For this reason it is difficult to compile lists of species for comparison with faunas from different regions. Where more systematic investigations have been carried out, as in southern California (Table 2.1) it appears that thousands of insect species may be present, in addition to a high diversity of arachnids.

The fauna of the Arctic is of relatively recent origin, in the sense that most, if not all, species have invaded the region since the last glaciation less than 10 000 years ago. The relatively low diversity of some taxa may be explained by their late arrival, and also by the difficulties encountered by many species in adapting to the cold and arid environment.

In the Antarctic, the situation is different, since the continent has been covered by ice for millions of years. Some invertebrate species are apparently relics from periods with warmer climate. Due to the present harsh climate and geographical isolation, invasion by new species has been limited. As a result of these conditions, the fauna of the entire antarctic region (Table 2.2) is poorer than those of any other regions of the world.

A striking characteristic of hot deserts is the high diversity of large predaceous arthropods, which comprise a high proportion of the total arthropod fauna. Typical representatives are spiders, scorpions and solifugids. Among insects, termites and ants are extremely abundant in hot arid environments. Beetles dominate as well, and among them tenebrionids are particularly well adapted to arid conditions.

In contrast, polar deserts have low diversity of spiders and lack several of the insect taxa that are numerous in hot deserts. Another striking characteristic of polar areas is the low diversity of beetles. A few arctic and subarctic species belong to the Curculionidae, which is a widespread family in hot deserts. The world's most southern beetles inhabiting some of the subantarctic islands, belong to the family Perimylopidae, which is closely related to the Tenebrionidae.

In addition to parasitic insects living on their hot-blooded hosts, Diptera and Hymenoptera are the dominant insect orders of the Arctic (Table 2.3). Juvenile stages of several species of higher flies and nematocerous Diptera are found in wet soil, as well as in lakes and rivers. They are apparently well adapted to survive even when their surroundings are frozen.

In some respects there are similarities between the fauna of hot and cold deserts. Microarthropods are widely distributed in temperate and subtropical deserts, and represent a dominant component of the invertebrate fauna of polar regions. Both mites and collembolans have a long evolutionary history and considerable ability to disperse over time. In the Antarctic, mites and collembolans are the only terrestrial arthropods of the continental zone. They are also numerous in arctic regions, with regard to both number of species and population densities, and are among the invertebrates found nearest to the peaks of the world's highest mountains. In hot deserts, microarthropods have a patchy distribution, often associated with vegetation and plant litter. Among both mites and collembolans, closely related taxa may be found in hot and cold deserts. As an example, species of the genus *Folsomia* (Fig. 2.3) are common in the Australian deserts and numerous at Svalbard in the north.

Similarly, nematodes appear to be widely distributed in both polar regions and hot deserts. Nematodes occur in high population densities in arctic soils and are also found in extreme continental antarctic locations. In spite of their fragile appearance, they are among the most numerous soil animals of hot deserts (Fig. 2.4). The survival of nematodes in different types of hostile environment is partly based on their ability to enter anhydrobiosis.

Some of the inland deserts of North America and central Asia are exceptional in having hot summers and cold winters. Although numerous faunistic studies are available, e.g. on the insect fauna at high altitudes, there are few investigations of adaptations of invertebrates to this cold and arid environment. Species from these areas may offer interesting examples of the relation between tolerance of cold and aridity.

12.3 Behaviour

Through their behaviour, insects and other terrestrial arthropods are able to avoid extreme climatic conditions. Behavioural adaptations have been closely studied in a number of hot desert species. Utilization of microclimatic niches is important, and many species are active only during seasonal or daily periods with more favourable conditions. Burrowing, thermoregulation and rhythms of activity are dominant patterns in the behaviour of desert arthropods. In some respects, behavioural adaptation is of primary importance for survival, but at the same time a combination of behavioural, physiological and phenological adaptations is required. Life in deserts is only possible through an array of adaptive strategies.

A striking feature of arctic areas is the apparent lack of insect activity when weather conditions are unfavourable. On calm and sunny days, however, the surroundings teem with insect life. Through their opportunistic behaviour, terrestrial arctic arthropods are able to utilize warm and calm periods. Many species actively select the warmest parts of the vegetation to absorb heat. The soil fauna is mainly composed of small collembolan and mite species, but information on their behaviour is limited.

In larger arctic species, such as butterflies and larvae of the moth *Gynaephora groenlandica* (Fig. 6.3), thermoregulation is of great importance. Behavioural orientation resulting in absorbtion of insolation is also common among alpine butterflies and grasshoppers from different geographical regions. In the beetles of tropical alpine habitats, extreme diel temperature fluctuations are avoided by seeking shelter under rocks and in vegetation.

12.4 Life Cycles

In hot deserts, many terrestrial arthropods take advantage of more favourable conditions, such as rainy seasons, for breeding. The climate is usually highly unpredictable, and special adaptations in the life cycles of desert arthropods are required. Flexibility is a key word, and growth and development are often delayed by the lack of moisture and fresh food.

A number of the species in hot deserts complete their life cycles in 1 year, while many small species may be multivoltine. Some of the larger arthropods, such as scorpions and millipedes, require more time to develop and reproduce. In many species, development is halted during periods of drought and high temperatures. Diapause is known in hot desert insects, but is probably not as common as it is in temperate species since the cues for entering diapause are weaker, in particular in subtropical deserts. Other forms of dormancy are common, but further studies on the physiological aspects of these adaptations are required.

In the polar areas, growth and development are limited to a few summer months. Prolonged life cycles are typical of most arctic and antarctic species, and overwintering takes place in several developmental stages. In some species from less severe arctic habitats, univoltine life cycles are maintained through special adaptations. Diapause is apparently rare among polar species. This may be because photoperiodic clues for its induction are weak due to the continuous daylight during the summer.

12.5 Terrestrial Arthropods and Aridity

In collembolans, a close relation between the aridity of the habitat and rate of water loss has been demonstrated. Most species are highly sensitive to desiccation, but others, that are adapted to arid environments, have lower rates of water loss at low atmospheric humidities. In polar and alpine areas, several species of springtails inhabit extremely dry habitats, such as crustose lichens on rocks. The lack of collembolans in the dry valleys of southern Victoria Land suggests that moisture plays a major role in the limitation of terrestrial life. Collembolans are present in coastal bryophyte systems and in inland chalikosystems where melt water is available sporadically during the short summer.

Springtails from hot arid environments of southern France and Australia (Fig. 5.5) are known to enter anhydrobiosis. When water is lost, they shrink to a tight, tun-like structure. Many species probably survive the dry season as drought-resistant eggs.

Mites are often found in association with collembolans, but anhydrobiosis is not known in this taxon. As in collembolans, there are great differences in rate of water loss, and rates are lowest in species from arid microhabitats. Very little is known about the physiology of tolerance to desiccation in microarthropods, and further studies in this field should be encouraged.

Among insects, tolerance to arid conditions has been studied in a number of species. Beetles are often chosen as experimental animals. Physiological mechanisms as well as behavioural adaptations are principle factors for survival in hot arid environments. Water may be lost in several ways (Fig. 4.1) - through faeces, by the release of defensive substances and by transpiration. There are great differences among species with regard to cuticular water flux, but in some desert tenebrionids cuticular transpiration accounts for only a small fraction of the total water loss. Most of the evaporative water loss in such species takes place through respiration.

Waterproofing barriers are provided by the epicuticular lipids and by other components of the cuticle. The epidermal cells probably contribute actively to the control of cuticular water loss. Substantial wax blooms can be observed on the surface of many desert insects and arachnids. The wax deposits make the cuticle less permeable to evaporation. Although the wax may be worn off or even disappear in cold weather, some tenebrionid beetles have the capacity to regenerate the wax at high temperatures and low humidities.

Carabid beetles from hot deserts and alpine areas usually lose water more rapidly than do tenebrionids and curculionids. For this reason, carabids are more dependent on behavioural avoidance of stressful situations. Among other arthropods, desert cockroaches are well equipped to resist water loss, and scorpions often have very low rates of transpiration.

Subantarctic beetles exhibit different degrees of resistance to water loss and avoid desiccation through behavioural adaptations. Although most species inhabit moist environments, they may be temporarily exposed to arid conditions. Adaptations are found in both larvae and adults of subantarctic weevils and perimylopid beetles.

Compensation for water loss is possible in several ways. Most insects are able to absorb water from the rectum, and a high degree of efficiency is attained by species in which a cryptonephridial system is present. Due to the formation of a strong osmotic gradient, water flows from the contents of the rectum and into the Malpighian tubules. Water is transfered from the tubules to the haemolymph by a process that is not completely understood. Cryptonephridial systems are found in species that are highly resistant to arid conditions, and are particularly well developed in tenebrionids and chrysomelids from hot deserts.

The cryptonephridial system apparently evolved for absorption of water from the faeces, but water vapour can also be absorbed from the air. A threshold as low as 81 % RH has been observed in some desert tenebrionids. Desert cockroaches absorb water vapour through a special organ on the hypopharynx, and water vapour absorption is also important in isopods.

Further compensation for water loss may take place through the removal or transfer of osmotically active substances within the body. In this way, higher water content is maintained in the most vital organs. Upon rehydration, ions, amino acids and other substances are returned to the haemolymph. Osmoregulation is important for maintaining physiological and biochemical processes, and is also achieved through the active absorption of substances into the perirectal fluid of the cryptonephridial system.

In conclusion, the physiology of tolerance to arid environments is best known in beetles from hot deserts. Comparison with polar species is difficult to make, since less information is available. Microarthropods, which comprise a dominant element of polar faunas, differ greatly in rates of water loss, but little is known about their physiological mechanisms of water conservation.

12.6 Cold Hardiness

12.6.1 Terrestrial Arthropods

Different strategies of cold hardiness have evolved in insects and other terrestrial arthropods and are of major importance for survival during overwintering. In polar deserts the winters are exceptionally long, and extremely low temperatures are recorded. In temperate deserts, freezing temperatures may dominate for weeks or months, increasing in severity with increasing latitude and elevation above sea level. Even on the equator, freezing temperatures occur regularly on clear nights at high altitudes.

For this reason, cold hardiness is an integrated factor in the adaptations of terrestrial arthropods from most kinds of arid environments, perhaps with the exception of some subtropical deserts where freezing temperatures rarely occur. As pointed out in the previous section, species from polar areas have to tolerate exposure to arid conditions. Similarly, species from many hot deserts have to possess a certain degree of resistance to cold.

Cold hardiness in terrestrial arthropods is in principle based on two main strategies (Table 9.3). Most species are freeze avoiding, while some species tolerate ice formation and are termed freeze tolerant. Freeze avoiding species depend on supercooling to survive temperatures below the melting point of their body fluids. The ability to supercool is greatly increased by the accumulation of low molecular weight cryoprotectant substances, and in some species by reduced water content. The solutes may appear in concentrations of several mol. Mean supercooling points in the range of -30 to -40 °C are not unusual in species from exposed habitats, and even lower values have been recorded in some.

The presence of thermal hysteresis proteins (antifreeze proteins) has been demonstrated in a number of insects. They apparently function in stabilizing the supercooled state, and may reduce the chances of inoculative freezing in species that overwinter in contact with ice. In freeze avoiding insects there is a risk of desiccation unless an equilibrium is established between the water vapour pressure of the body fluids and that of the surroundings.

Many insect species are able to survive extracellular ice formation, while freeze tolerance is not known from springtails, mites and spiders. The degree of freeze tolerance within different species is related to the environmental conditions. As in freeze avoiding species, cold hardiness is enhanced by low molecular weight solutes. Equivalent concentrations of cryprotectant substances have greater effect on freeze tolerant than on freeze avoiding species. It is generally assumed that nucleation of ice following supercooling at low temperatures involves a risk of intracellular freezing, which is fatal. In a number of species, this is avoided due to the presence of ice-nucleating agents, e.g. proteins or peptides that initiate extracellular ice formation in the haemolymph at higher temperatures, e.g. in the range of -5 to -8 °C

In protected microhabitats, the vapour pressure of frozen insects is in equilibrium with that of their frozen surroundings. As long as this equilibrium is maintained, the risk of desiccation is avoided. For this reason freeze tolerance is considered to be a better adaptation to arid environments than freeze avoidance. It has been suggested that this explains the relatively large number of freeze tolerant insect species found in the Arctic.

As outlined above, cold hardiness comprises several kinds of adaptational strategies. The most remarkable adaptation, and the one that has been least studied, is perhaps the ability of polar species to survive low temperatures for extended periods. In many arctic and antarctic habitats, terrestrial arthropods are exposed to low subzero temperatures for most of the year. Their survival is in striking contrast to that of summer-adapted temperate species or of species from tropical and subtropical regions that may succumb after short exposure to low temperatures above zero. Relatively little is known about the reasons for these differences. In long-term survivors fundamental biochemical processes are apparently protected even at low temperatures. The nature of these protective mechanisms, as well as the reasons for cold death without freezing, and as a function of time, are poorly understood.

Terrestrial arthropods from temperate deserts are mostly freeze avoiding. The supercooling capacity of these species is relatively small, but corresponds well with the temperature conditions in their hibernacula. Moderate degrees of freeze tolerance have been demonstrated in one species of scorpions and in some tenebrionid beetles. Close to the equator, certain high altitude grasshoppers survive nocturnal freezing cycles.

12.6.2 Tardigrades and Nematodes

Tardigrades are known to survive extremely low temperatures during anhydrobiosis, but relatively few studies have been carried out on specimens in the hydrated state. Hydrated tardigrades are apparently freeze tolerant and survive long exposures at temperatures in the range of those found in their natural habitats. Exposures to temperatures of liquid gases may also be tolerated in the hydrated state, at least for short periods.

The general division of cold hardiness in freeze avoidance and freeze tolerance is not easily applied to nematodes. In many species, a proportion of the population survives freezing, while others depend on supercooling. Under natural conditions, inoculative freezing probably takes place in most hydrated specimens. In antarctic species, those that survive freezing may tolerate very low temperatures. More information is needed with regard to long-term survival and the effect of low temperatures on population densities in hydrated nematodes, and to determine if specimens in anhydrobiosis are better adapted for overwintering in polar regions.

12.6.3 Lumbricid Earthworms

Lumbricid earthworms are known to overwinter in cocoons although the hydrated embryos are killed by freezing. In frozen soils, the cocoons lose water rapidly, but the embryos within them survive high degrees of desiccation. During dehydration, an equilibrium is established between the melting point of the body fluid and the ambient temperature. Since the cocoons are not freeze tolerant and do not depend on supercooling, their strategy of cold hardiness is, in principle, different from those of terrestrial arthropods and has been termed a protective dehydration mechanism.

12.7 Combined Tolerance of Cold and Drought

The relation between the survival strategies of terrestrial arthropods to cold and aridity has been emphasized by several authors. The question has been raised as to whether these strategies are in fact overlapping adaptations. As described in the previous chapters, the survival of terrestrial arthropods in hot and cold deserts requires adaptations to both types of environmental stress. Some of the strategies that are recognized as giving protection against cold clearly also protect the organism against desiccation.

In a frozen environment, the difference in water vapour pressure between ice and a supercooled organism may result in serious desiccation. Due to the presence of solutes, however, the osmolality of the body fluids increases during transpiration. For this reason, the vapour pressure is reduced and the saturation deficit eventually eliminated. Glycerol and other low molecular weight substances, which are known as cryoprotectants, appear in high concentrations and contribute to the reduction of water vapour pressure. In some insects and the antarctic mite *Alaskozetes antarcticus*, the synthesis of cryoprotectants is triggered by reduced water content. As already mentioned (Chap. 12.6.1), the water vapour pressure of frozen insects is in equilibrium with the frozen surroundings. In this respect, freezing protects against desiccation, and freeze tolerance is enhanced through the accumulation of glycerol and other solutes. The lumbricid earthworm cocoons, mentioned in the previous chapter, offer a special case of combined resistance to desiccation and cold. Due to high water losses the melting point of the body fluids is equilibrated with the ambient freezing temperature, and fatal freezing of the embryo is avoided.

Reduced water content associated with increased haemolymph osmolality is also important for both cold hardiness and tolerance of drought. Appropriate hibernacula protect against extreme temperature fluctuations and promote more favourable moisture conditions.

Further studies on the overlapping adaptations to cold and drought should be encouraged. It should be kept in mind that some methods of maintaining water balance probably do not affect cold hardiness. Since glycerol and other solutes reduce rates of water loss in overwintering species, it might be argued that these substances should also be found in species from hot deserts and other arid environments. Solutes such as glycerol and sugars are indeed present in some diapausing temperate insects during the summer, but appear not to be accumulated in hot desert arthropods. The reason for this may be that the accumulation of polyols and sugars depends on the blocking of certain metabolic pathways at low temperatures. Accumulation also takes place at higher temperatures in diapausing overwintering insect eggs, in which metabolic rates are reduced, but may not be possible at normal metabolic rates in desert arthropods under high temperature regimes.

12.8 Anhydrobiosis

Tardigrades, rotifers and nematodes are typical inhabitants of polar areas, and nematodes in particular are also numerous in hot deserts. The aridity of their habitats makes it impossible for such small animals to survive without high tolerance of desiccation. Their strategies for survival are highly specialized and entirely different from those of most terrestrial arthropods. Under periods of drought, water is gradually lost from tardigrades, rotifers and nematodes. Only a small proportion remains as bound water, and the animals enter a state of anhydrobiosis.

Among nematodes, anhydrobiosis has been demonstrated in a number of species from different geographical regions and different trophic levels. The inactive state is associated with morphological changes, such as coiling, and the formation of cysts and aggregations. In hot deserts, many species probably spend most of their time in anhydrobiosis. In the Antarctic, anhydrobiosis has been demonstrated among nematodes inhabiting fellfields and mosses that are subjected to regular periods of drought.

Tardigrades require a relatively slow drying regime to enter anhydrobiosis. They are associated with mosses in hot deserts, but are much more numerous in polar areas. During dehydration, a characteristic tun is formed (Fig. 5.3), and the tardigrades can remain in anhydrobiosis for several years.

Both nematodes and tardigrades accumulate trehalose in preparation for anhydrobiosis. Due to its stabilizing effect on phospholipids, trehalose is known to protect cell membranes during desiccation, and also to inhibit damage to proteins. The role of trehalose in tardigrades and nematodes is not yet fully understood, but recent studies on nematodes have shown that other factors are involved as well. Low rates of water loss are important in some species.

During anhydrobiosis, tardigrades, rotifers and nematodes may survive freezing temperatures as low as those of liquid gases. The temperature at which freezing takes place is apparently of little importance since no freezable water is available. In this respect, there is a close relationship between resistance to drought and cold. At the molecular level, however, there are fundamental differences between freezing and dehydration. During freezing of cells and protein solutions, a fraction of the body water remains unfrozen, and proteins may be stabilized by a variety of solutes. During extreme dehydration the non-freezable water is removed as well, and only highly specific substances can stabilize the molecules.

References

Abushama FT (1984) Epigeal insects. In: Cloudsley-Thompson JL (ed) Key environments, Sahara Desert. Pergamon Press, Oxford, pp 129-144

Addison JA (1977) Population dynamics and biology of Collembola on Truelove Lowland. In: Bliss LC (ed) Truelove Lowland, Devon Island, Canada: a high arctic ecosystem. University of Alberta Press, Edmonton, pp 363-382

Ahearn GA (1970a) The control of water loss in desert tenebrionid beetles. J Exp Biol 53:573-595

Ahearn GA (1970b) Water balance in the whipscoripion *Mastigoproctus giganteus* (Lucas) (Arachinida, Uropygi). Comp Biochem Physiol 35A:339-353

Ahearn GA, Hadley NF (1969) The effects of temperature and humidity on water loss in two desert tenebrionid beetles, *Eleodes armata* and *Cryptoglossa verrucosa.* Comp Biochem Physiol 30:739-749

Aldryhim YN, Mills III CW, Aldawood AS (1992) Ecological distribution and seasonality of darkling beetles (Coleoptera: Tenebrionidae) in the central region of Saudi Arabia. J Arid Environ 23:415-422

Alexander G, Hilliard JR Jr (1964) Life history of *Aeropedellus clavata* in the alpine tundra of Colorado. Ann Entomol Soc Am 57:310-317

Alexandrova VD (1988) Vegetation of the Soviet polar deserts. Cambridge University Press, Cambridge

Allan T, Warren A (1993) Deserts. The encroaching wilderness. Mitchell Beasley, London

Andersen J, Zachariassen KE, Maloiy GMO, Kamau JMZ (1986) Adaptation of carabid beetles to dry habitats of East Africa. J Trop Ecol 2:127-138

Andrewartha HG, Birch LC (1954) The distribution and abundance of animals. University of Chicago Press, Chicago

Aroian RV, Carta L, Kaloshian I, Sternberg PW (1993) A free-living *Panagrolainus* sp. from Armenia can survive in anhydrobiosis for 8.7 years. J Nematol 25:500-502

Asahina E (1969) Frost resistance in insects. Adv Insect Physiol 6:1-49

Ash CPJ, Atkinson HF (1986) *Nematodirus battus*: development of cold hardiness in dormant eggs. Exp Parasitol 62:24-28

Aunaas T, Baust JG, Zachariassen KE (1983) Ecophysiological studies on arthropods from Spitsbergen. Polar Res 1:235-240

Azocar A, Monasterio M (1980) Estudio de la varabilidad mesu y microclimatica en el paramo de Mucubaji. In: Monasterio M (ed) Estudios ecologicos en los paramos Andinos. Universidad de Los Andes, Merida, pp 225-262

Bachmetjew P (1901) Experimentelle entomologische Studien vom physikalish-chemischen Standpunkt aus. Band 1. Wilhelm Engelmann, Leipzig

Bakken H (1985) Cold-hardiness in the alpine beetles *Patrobus septentrionis* and *Calathus melanocephalus.* J Insect Physiol 31:447-453

Bale JS (1987) Insect cold hardiness: freezing and supercooling - an ecophysiological perspective. J Insect Physiol 33:899-908

Bale JS, Harrington R, Clough MS (1988) Low temperature mortality of the peach-potato aphid *Myzus persica.* Ecol Entomol 13:121-129

Barra JA, Poinsot-Balaguer N (1977) Modifications ultrastructurales accompagnant l'anhydrobiose chez un Collembole: *Folsomides variabilis.* Rev Écol Biol Sol 14:189-197

Barra JA, Poinsot-Balaguer N (1983) Mise en évidence d'un complexe hygrophile dans les canaux transcuticulaires chez *Folsomides angularis* en anhydrobiose. Pedobiologia 25:279-286

Barry RG, Hare KF (1974) Arctic climate. In: Ives JD, Barry RG (eds) Arctic and alpine environments. Methuen, London, pp 17-54

Bauer R (1993) Coenological and phenological study of granite boulder dwelling Collembola. Zool Anz 230:237-248

Bauer R, Christian E (1993) Adaptation of three springtail species to granite boulder habitats (Collembola). Pedobiologia 37:280-290

Baumann H (1922) Die Anabiose der Tardigraden. Zool Jahresber 45:501-556

Baust JG, Edwards J (1979) Mechanisms of freeze tolerance in an Antarctic midge (*Belgica antarctica*). Physiol Entomol 4:1-5

Baust JG, Miller LK (1970) Seasonal variations in glycerol content and its influence on cold-hardiness in the Alaskan carabid beetle, *Pterostichus brevicorne*. J Insect Physiol 16:979-990

Baust JG, Nishino M (1991) Freezing tolerance in the goldenrod gall fly (*Eurosta solidaginis*). In: Lee RE Jr, Denlinger DL (eds) Insects at low temperature. Chapman and Hall, New York, pp 260-275

Baust JG, Rojas RR (1985) Insect cold hardiness: facts and fancy. J Insect Physiol 31:755-759

Baust JG, Grandee R, Condon G, Morrissey RE (1979) The diversity of overwintering strategies utilized by separate populations of gall insects. Physiol Zool 52:572-580

Beck SD (1980) Insect photoperiodism. 2nd edn. Academic Press, New York

Becquerel P (1950) La suspension de la vie au-dessous de 1/20 °K absolu par démagnétisation adiabatique de l'alun de fer dans le vide le plus élevé. C R Acad Sci 231:261-263

Behrens W (1984) Environmental aspect of insect dormancy. In: Hoffman KH (ed) Environmental physiology and biochemistry of insects. Springer, Berlin Heidelberg New York, pp 67-94

Bengtsson SA, Fjellberg A, Solhøy T (1974) Abundance of tundra arthropods in Spitsbergen. Entomol Scand 5:137-142

Birkemoe T, Dózsa-Farkas K (1994) New records of Enchytraidae (Oligochaela) from Spitsbergen, Svalbard. Fauna Norv Ser A 15:35-44

Bjerke R, Lundheim R, Rein JO, Strømme JA, Zachariassen KE (1989) Water balance and osmotic regulation of the East African scorpion *Lychas burdoi* (Simon). Fauna Norv Ser B 36:33-36

Bliss LC (1975) Devon Island, Canada. Ecol Bull (Stockholm) 20:17-60

Block W (1980) Survival strategies in polar terrestrial arthropods. Biol J Linn Soc 14:29-38

Block W (1984a) Terrestrial microbiology, invertebrates and ecosystems. In: Laws RW (ed) Antarctic ecology, vol 1. Academic Press, London, pp 162-236

Block W (1984b) A comparative study of invertebrate supercooling at Signy Island, Maritime Antarctic. Br Antarct Surv Bull 64:67-76

Block W (1985a) Survival on land. Biologist 32:133-138

Block W (1985b) Ecological and physiological studies of terrestrial arthropods in the Ross Dependency 1984. Br Antarct Surv Bull 68:115-122

Block W (1991) To freeze or not to freeze? Invertebrate survival of sub-zero temperatures. Funct Ecol 5:284-290

Block W, Christensen B (1985) Terrestrial Enchytraeidae from South Georgia and the maritime Antarctic. Br Antarct Surv Bull 69:65-70

Block W, Duman JG (1989) The presence of thermal hysteresis producing antifreeze proteins in the Antarctic mite *Alaskozetes antarcticus*. J Exp Zool 250:229-231

Block W, Sømme L (1983) Low temperature adaptations in beetles from the sub-Antarctic island of South Georgia. Polar Biol 2:109-114

Block W, Burn AJ, Richard KJ (1984) An insect introduced to the maritime Antarctic. Biol J Linn Soc 23:33-39

Block W, Harrisson PM, Vannier G (1990) A comparative study of patterns of water loss from two Antarctic springtails (Insecta, Collembola). J Insect Physiol 36:181-187

Block W, Webb NR, Coulson S, Hodkinson ID, Worland MR (1994) Thermal adaptation in the arctic collembolan *Onychiurus arcticus* (Tullberg). J Insect Physiol 40:715-722

Böcher J (1975) Notes on the reproductive biology and egg-diapause in *Nysius groenlandicus* (Zett.) (Heteroptera: Lygaeidae). Vidensk Medd Dan Naturhist Foren 138:21-38

Böcher J (1976) Population studies on *Nysius groenlandicus* (Zett.) (Heteroptera: Lygaeidae) in Greenland with particular reference to climatic factors, especially the snow cover. Vidensk Medd Dan Naturhist Foren 139:61-89

Böcher J (1988) The Coleoptera of Greenland. Medd Grønl Biosci 26:1-100

Böcher J (1992) Insekter og andre små dyr. Nat Verden 8:296-300

Booth RG, Usher MB (1985) Relationship between Collembola and their environment in a maritime Antarctic moss-turf habitat. In: Fitter AH (ed) Ecological interactions in soil: plants microbes and animals. Blackwell, Oxford, pp 279-284

Booth RG, Usher MB (1986) Arthropod communities in a maritime Antarctic moss-turf habitat: life history strategies of the prostigmatid mites. Pedobiologia 29:209-218

Brodo F (1990) Crane flies (Diptera: Tipulidae) of the arctic islands. In: Harington CR (ed) Canada's missing dimension: Science and history in the Canadian arctic islands, vol II. Canadian Museum of Nature, Ottawa, pp 471-484

Broza M, Borut A, Pener MP (1976) Osmoregulation in the desert tenebrionid beetle *Trachydema philistina* Reiche during dehydration and subsequent rehydration. Isr J Med Sci 12:868-871

Bunnell FL, MacLean SF Jr, Brown J (1975) Barrow, Alaska, USA. Ecol Bull (Stockholm) 20:73-124

Burn AJ (1981) Feeding and growth in the Antarctic collembolan Cryptopygus antarcticus. Oikos 36:59-64

Burn AJ (1984) Life cycle strategies in two Antarctic Collembola. Oecologia 64:223-229

Butler MG (1982) A 7-year life cycle for two *Chironomus* species in arctic Alaskan tundra ponds (Diptera: Chironomidae). Can J Zool 60:58-70

Callaghan T, Sømme L, Sonesson M (1993) Impacts of climate change at high latitudes on terrestrial plants and invertebrates. Directorate for Nature Management, Trondheim, Norway

Cannon RJC (1986) Effects of contrasting relative humidities on the cold tolerance of an antarctic mite. J Insect Physiol 32:523-534

Cannon RJC (1987) Effects of low-temperature acclimation on the survival and cold tolerance of an antarctic mite. J Insect Physiol 33:509-521

Cannon RJC, Block W (1988) Cold tolerance of microarthropods. Biol Rev 63:23-77

Cannon RJC, Schenker R (1985) Cold tolerance of a cryptostigmatid mite at Signy Island, Maritime Antarctic. Br Antarct Surv Bull 67:1-5

Cannon RJC, Block W, Collett GD (1985) Loss of supercooling ability in *Cryptopygus antarcticus* (Collembola: Isotomidae) associated with water uptake. Cryo-Lett 6:73-80

Carlisle DM, Ellis PE, Betts E (1965) The influence of aromatic shrubs on sexual maturation in the desert locust *Schistocerca gregaria*. J Insect Physiol 11:1541-1558

Carlsson G (1962) Studies on Scandinavian blackflies. Opusc Entomol Suppl 21:1-280

Carpenter JF, Crowe JH (1988) Modes of stabilization of a protein by organic solutes during desiccation. Cryobiology 25:459-470

Cassier P (1968) Détermination de la période photosensible chez le criquet migrateur africain, *Locusta migratoria migratorioides* (R. et F.), phase gregaire. Insect Soc 15:3-30

Cepeda-Pizarro JG, Maldonado MB, Vilches JG, Rojas MA, Pereira NP (1992) A litter-bag study of mite densities (Actinedida and Oribatida) in Atriplex litter and soil of a desert site in northern Chile. J Arid Environ 23:177-188

Chappell MA (1983) Metabolism and thermoregulation in desert and montane grasshoppers. Oecologia 56:126-131

Cheke RA (1990) A migrant pest of the Sahel: the Senegalese grasshopper *Oedaleus senegalensis*. Philos Trans R Soc Lond B328:539-553

Chernov JI (1985) The living tundra. Cambridge University Press, Cambridge

Chernov JI, Striganova BR, Ananjeva SI (1977) Soil fauna of the polar desert at Cape Cheluskin, Taimyr Peninsula, USSR. Oikos 29:175-179

Chown SL (1993) Desiccation resistance in six sub-Antarctic weevils (Coleoptera: Curculionidae): humidity as an abiotic factor influencing assemblage structure. Funct Ecol 7:318-325

Chown SL, van Drimmelen M (1992) Water balance and osmoregulation in weevil larvae (Coleoptera: Curculionidae; Brachycerinae) from three different habitats in sub-Antarctic Marion Island. Polar Biol 12:527-532

Christiansen K (1964) Bionomics of Collembola. Annu Rev Entomol 9:147-178

Clegg JS (1979) Metabolism and the intracellular environment: the vicinal water-network model. In: Drost-Hansen W, Clegg JS (eds) Dry biological systems. Academic Press, New York, pp 117-153

Clench HK (1966) Behavioural thermoregulation in butterflies. Ecology 47:1021-1034

Cloudsley-Thompson JL (1962) Some aspects of the physiology and behaviour of *Dinothrombium* (Acari). Entomol Exp Appl 5:69-73

Cloudsley-Thompson JL (1964) On the function of the sub-elytral cavity in desert Tenebrionidae (Col.). Entomol Mon Mag 100:148-151

Cloudsley-Thompson JL (1967) The water relations of scorpions and tarantulas from the Sonoran Desert. Entomol Mon Mag 103:217-220

Cloudsley-Thompson JL (1973) Factors affecting the supercooling of tropical Arthropoda, especially locusts. J Nat Hist 7:471-480

Cloudsley-Thompson JL (1977) Adaptational biology of the Solifugae (Solpugida). Bull Br Arachnol Soc 4:61-71

Cloudsley-Thompson JL (1978) Diurnal rhythms of supercooling in locusts. Int J Biometeorol 22:112-115

Cloudsley-Thompson JL (1984) Arachnids. In: Cloudsley-Thompson JL (ed) Key environments. Sahara Desert. Pergamon Press, Oxford, pp 175-204

Cloudsley-Thompson JL (1989) Temperature and the activity of ants and other insects in Central Australia. J Arid Environ 16:185-192

Cloudsley-Thompson JL (1991) Ecophysiology of desert arthropods and reptiles. Springer, Berlin Heidelberg New York

Cloudsley-Thompson JL, Chadwick MJ (1964) Life in deserts. Dufour, Philadelphia

Cloudsley-Thompson JL, Crawford CS (1970a) Lethal temperatures of some arthropods of the southwestern United States. Entomol Mon Mag 106:26-29

Cloudsley-Thompson JL, Crawford CS (1970b) Water and temperature relations, and diurnal rhythms of scolopendromorph centipedes. Entomol Exp Appl 13:187-193

Coe MJ (1967) The ecology of the alpine zone of Mount Kenya. Dr W Junk, The Hague

Coe MJ (1978) The decomposition of elephant carcasses in the Tsavo (East) National Park, Kenya. J Arid Environ 1:71-86

Coenen-Stass D (1981) Some aspects of the water balance of two desert woodlice, *Hemilepistus aphganicus* and *Hemilepistus reaumuri* (Crustacea, Ispoda Oniscoidea) Comp Biochem Physiol 70A:405-419

Coenen-Stass D (1989) Investigations on microclimatic and climatic preference of the desert wood-louse *Hemilepistus reaumuri* (Audouin & Savigny, 1826) (Oniscidea Isopoda Crustacea). Monit Zool Ital (N.S.) Monogr 4:425-434

Cohen AC, Cohen JL (1981) Microclimate, temperature and water relation of two species of desert cockroaches. Comp Biochem Physiol 69A:165-167

Conradi-Larsen EM, Sømme L (1973) Anaerobiosis in the overwintering beetle *Pelophila borealis.* Nature (Lond) 245:388-390

Constantinou C, Cloudsley-Thompson JL (1986) Supercooling in the various instars of the mealworm beetle *Tenebrio molitor* L. J Nat Hist 20:649-651

Convey P (1993) Growth and survival strategy of an Antarctic mite. Ecography 17:97-107

Cooper PD (1983) Components of evaporative water loss in the desert tenebrionid beetles *Eleodes armata* and *Cryptoglossa verrucosa*. Physiol Zool 56:47-55

Corbet PS (1972) The microclimate of arctic plants and animals, on land and in fresh water. Acta Arct 18:1-43

Corbet PS, Danks HV (1975) Egg-laying habit of mosquitoes in the high arctic. Mosq News 35:8-14

Coutchié PA, Crowe JH (1979) Transport of water vapor by tenebrionid beetles. I. Kinetics. Physiol Zool 52:67-87

Coutchié PA, Machin J (1984) Allometry of water vapor absorption in two species of tenebrionid beetle larvae. Am J Physiol 247:R230-R236

Crawford CS (1972) Water relations in a desert millipede *Orthoporus ornatus* (Girard) (Spirostreptidae) Comp Biochem Physiol 42A:521-535

Crawford CS (1981) Biology of desert invertebrates. Springer, Berlin Heidelberg New York

Crawford CS (1991a) Animal adaptations and ecological processes in desert dune fields. J Arid Environ 21:245-250

Crawford CS (1991b) The community ecology of macroarthropod detritivores. In: Polis GA (ed) The ecology of desert communities. University of Arizona Press, Tucson, pp 89-112

Crawford CS, Cloudsley-Thompson JL (1971) Water relation and desiccation-avoiding behaviour in the vinegaroon *Mastigoproctus giganteus* (Arachnida: Uropygi). Entomol Exp Appl 14:99-106

Crawford CS, Riddle WA (1974) Cold hardiness in centipedes and scorpions in New Mexico. Oikos 25:86-92

Crawford C, Riddle WA (1975) Overwintering physiology of the scorpion *Diplocentrus spitzera*. Physiol Zool 48:84-92

Crawford CS, Riddle WA, Pugach S (1975) Overwintering physiology of the Centipede *Scolopendra polymorpha*. Physiol Zool 48:290-294

Crawford CS, Goldenberg S, Warburg MR (1986) Seasonal water balance in *Archispirostreptus syriacus* (Diplopoda: Spirostreptidae) from mesic and xeric Mediterranean environments. J Arid Environ 10:127-136

Crowe JH (1972) Evaporative water loss by tardigrades under controlled relative humidities. Biol Bull 142:407-416

Crowe JH (1975) The physiology of cryptobiosis in tardigrades. Mem Ist Ital Idrobiol Suppl 32:37-59

Crowe JH, Clegg JS (1978) Dry biological systems. Academic Press, New York

Crowe JH, Madin KAC (1975) Anhydrobiosis in nematodes: evaporative water loss and survival. J Exp Zool 193:323-334

Crowe JH, Carpenter JF, Crowe LM, Anchordoguy TJ (1990) Are freezing and dehydration similar stress vectors? A comparison of modes of interaction of stabilizing solutes with biomolecules. Crybiology 27:219-231

Crowe JH, Hoekstra FA, Crowe LM (1992) Anhydrobiosis. Annu Rev Physiol 54:579-599

Czermak B (1981) Autökologie und Populationsdynamik hochalpiner Araneen unter besonderer Berücksichtigung von Verteilung, Individuendichte und Biomasse in Grassheidebiotopen. Veröff Österr MaB Hochgebirgsprogr Hohe Tauern 4:101-152

Danilevski AS (1961) Photoperiodism and seasonal development of insects. State University, Leningrad (in Russian)

Danks HV (1978) Models of seasonal adaptation in the insects. I. Winter survival. Can Entomol 110:1167-1205

Danks HV (1981) Arctic arthropods. Entomological Society of Canada, Ottawa

Danks HV (1987) Insect dormancy: an ecological perspective. Biological Survey of Canada, Ottawa

Danks HV (1990) Arctic insects: instructive diversity. In: Harington CR (ed) Canada's missing dimension: science and history in the Canadian arctic islands, vol II. Canadian Museum of Nature, Ottawa, pp 444-470

Danks HV(1991) Winter habitats and ecological adaptations for winter survival. In: Lee RE Jr, Denlinger DL (eds) Insects at low temperature. Chapman and Hall, New York, pp 231-259

Danks HV, Oliver DR (1972) Seasonal emergence of some high arctic Chironomidae (Diptera). Can Entomol 104: 661-686

Dartnall HJG, Hollowday ED (1985) Antarctic rotifers. Br Antarct Surv Sci Rep 100:1-46

Dastych H (1984) The Tardigrada from Antarctica with a description of several new species. Acta Zool Cracov 27:377-436

Davey MC, Clarke KJ (1991) Spatial distribution of microalgae on Antarctic fellfield soil. Antarct Sci 3:257-263

Dearn JM (1977) Variable life history characteristics along an altidudinal gradient in three species of Australian grasshoppers. Oecologia 28:67-85

Denlinger DL (1986) Dormancy in tropical insects. Annu Rev Entomol 31:239-264 Australian grasshopper. Oecologia 28:67-85

Denlinger DL (1991) Relation between cold hardiness and diapause. In: Lee RE Jr, Denlinger DL (eds) Insects at low temperature. Chapman and Hall, New York, pp 174-198 de Wilde J (1969) Diapause and seasonal synchronization in the adult Colorado beetle (*Leptinotarsa decemlineata* Say). Symp Soc Exp Biol 23:263-284

DeZordo (1979a) Ökologische Untersuchungen an Wirbellosen des zentralalpinen Hochgebirges (Obergurgl, Tirol) III. Lebenszyklen und Zönotik von Coleopteren. Veröff Univ Innsbruck 118, Alpine Biol Stud 11:1-131

DeZordo (1979b) Phänologie von Carabiden im Hochgebirge Tirols (Obergurgl, Österreich) (Insecta: Coleoptera). Ber Nat Med Ver Innsbruck 66:73-83

Dingle H (1978) Migration and diapause in tropical, temperate, and island milkweed bugs. In: Dingle H (ed) Evolution of insect migration and diapause. Springer, Berlin Heidelberg New York, pp 254-276

Downes JA (1965) Adaptation of insects in the Arctic. Annu Rev Entomol 10:257-274

Doyère PLN (1842) Memoirs sur les Tardigrades. Ann Sci Nat 18:5

Dudeney JR (1987) The antarctic atmosphere. In: Walton DWH (ed) Antarctic Science. Cambridge University Press, Cambridge, pp 191-248

Duman JG, Montgomery SL (1991) Subzero temperature adaptations in arthropods from the summit of Mauna Kea, Hawaii. J Exp Zool 259:409-412

Duman JG, Wu DW, Xu L, Tursman D, Olsen TM (1991a) Adaptations of insects to subzero temperatures. Q Rev Biol 66:387-410

Duman JG, Xu L, Neven LG, Tursman D, Wu DW (1991b) Hemolymph proteins involved in insect sub zero-temperature tolerance: ice nucleators and antifreeze proteins. In: Lee RE Jr, Denlinger DA (eds) Insects at low temperature. Chapman and Hall, New York, pp 94-127

Edney EB (1977) Water balance in land arthropods. Springer, Berlin Heidelberg New York

Edwards JS (1986) How small ectotherms thrive in the cold without really trying. Cryo-Lett 6:388-390

Edwards JS, Baust J (1981) Sex ratio and adult behaviour of the Antarctic midge *Belgica antarctica* (Diptera, Chironomidae). Ecol Entomol 6:239-243

Eger JE Jr, Witz JA, Hartstack AW Jr, Sterling WL (1982) Survival of pupae of *Heliothis virescens* and *Heliothis zea* (Lepidoptera; Noctuidae) at low temperatures. Can Entomol 14:289-301

Eisenbeis G (1982) Physiological absorption of liquid water by Collembola: absorption by the ventral tube at different salinities. J Insect Physiol 28:11-20

Eisenbeis G (1989) Allometric relationships of the body surface area to body mass in different life forms of soil arthropods. 2 Surface area and climate. In: Dallai R (ed) 3rd Int seminar on Apterygota., University of Siena, Siena, pp 387-400

El-Kifl AH, Ghabbour SI (1984) Soil fauna. In: Cloudsley-Thompson JL (ed) Key environments. Sahara Desert. Pergamon Press, Oxford, pp 91-104

Elkins NZ, Withford WG (1982) the role of microarthropods and nematodes in decomposition in semi-arid ecosystems. Oecologia (Berl) 55:303-310

Ellenby C (1968) Desiccation survival in the plant parasitic nematodes *Heterodera rostochiensis* Wollenweber and Ditylenchus dipsaci (Kuhn) Filipjev. Proc R Soc Lond B169:203-213

Ellenby C (1969) Dormancy and survival in nematodes. Soc Exp Biol Symp 23:83-97

Erpenbeck A, Kirchner W (1983) Zur Kälteresistenz der kleinen roten Waldameise *Formica polyctena* Foerst. (Hymenoptera, Formicidae). Z Angew Entomol 96:217-281

Evans AAF, Perry RN (1976) Survival strategies in nematodes. In: Croll NA (ed) The organization of nematodes. Academic Press, New York, pp 383-424

Evenari M (1985) The desert environment. In: Evenari M, Noy-Meir I, Goodall DW (eds) Hot deserts and arid shrublands, A. Ecosystems of the world 12A. Elsevier, Amsterdam, pp 1-22

Evenari M, Noy-Meir I, Goodall DW (eds) (1985) Hot deserts and arid shrublands, A. Ecosystems of the world 12A. Elsevier, Amsterdam

Evenari M, Noy-Meir I, Goodall DW (eds) (1986) Hot deserts and arid shrublands, B. Ecosystems of the world 12B. Elsevier, Amsterdam

Farrow RA (1990) Flight and migration in acriodids. In: Chapman RF, Joern A (eds) Biology of grasshoppers. John Wiley, New York, pp 227-314

Fielding M (1951) Observations on the length of dormancy in certain plant infesting nematodes. Proc Helminthol Soc Wash 18:110-112

Fields PG, McNeil JN (1986) Possible dual cold-hardiness strategies in *Cisseps fulvicollis* (Lepidoptera: Arctiidae). Can Entomol 118:1309-1311

Fishpool LDC, Popov GB (1984) The grasshopper faunas of the savannas of Mali, Niger, Benin and Togo. Bull Inst Fondam Afr Noire Ser A 43:275-410

Fitzpatric EA (1979) Radiation. In: Goodall DW, Perry RA (eds) Arid-land ecosystems, vol 1. Cambridge University Press, Cambridge, pp 347-371

Fitzsimons JM (1971) Temperature and three species of Antarctic arthropods. Pac Insect Monogr 25:127-135

Forge TA, MacGuidwin AE (1992a) Effects of water potential and temperature on survival of the nematode *Meloidogyne hapla* in frozen soil. Can J Zool 70:1553-1560

Forge TA, MacGuidwin AE (1992b) Impact of thermal history on tolerance of Meloidogyne hapla second-stage juveniles to external freezing. J Nematol 24:262-268

Franco PJ, Edney EB, McBrayer JF (1979) The distribution and abundance of soil arthropods in the northern Mojave Desert. J Arid Environ 2:137-149

Freckman DW (1986) The ecology of dehydration in soil organisms In: Leopold AC (ed) Membranes, metabolism and dry organisms. Comstock, Ithaca, pp 157-168

Freckman DW, Mankau R (1977) Distribution and trophic structure of nematodes in desert soils. Ecol Bull 25:511-514

Freckman DW, Virginia RA (1989) Plant feeding nematodes to depths of 12 meters in mesquite dominated desert ecosystems. Ecology 70:1665-1678

Freckman DW, Virginia RA (1991) Nematodes in the McMurdo Dry Valleys of southern Victoria Land. Antarct J US 26:233-234

Freckman DW, Womersley C (1982) Physiological adaptations of nematodes in Chihuahuan desert soils. In: Lebrun P, André HM, DeMedts A, Grégoire-Wibo C, Wanthy G (eds) New trends in soil biology. Dieu-Brichart, Ottignies-Louvain-la-Neuve, Belgium, pp 395-403

Garay I (1981) Le peuplement de microarthropodes dans la litière sur pied de *Espeletia lutescens* et *Espeletia timotensis*. Rev Écol Biol Sol 18:209-219

Gehrken U (1984) Winter survival of an adult bark beetle *Ips acuminatus* Gyll. J Insect Physiol 30:421-429

Gehrken U (1992) Inoculative freezing in the adult beetles *Ips acuminatus* and *Rhagium inquisitor*. J Insect Physiol 38:519-524

Gehrken U, Sømme L (1994) Tolerance to desiccation in beetles from the High Atlas Mountains. Comp Biochem Physiol 109A:913-922

Gehrken U, Strømme A, Lundheim R, Zachariassen KE (1991) Inoculative freezing in overwintering tenebrionid beetle, *Bolitophagus reticulata* Panz. J Insect Physiol 37:683-687

Gilbert JJ (1974) Dormancy in rotifers. Trans Am Microsc Soc 93:490-513

Gillis JE, Possai KW (1983) Thermal niche partitioning in the grasshoppers *Arphia conspersa* and *Trimerotropis suffusa* from a montane habitat in central Colorado. Ecol Entomol 8:155-161

Goodall DW, Perry RA, Howes KMW (eds) (1979) Arid-land ecosystems: structure, functioning and management, vol 1. Cambridge University Press, Cambridge

Greenaway P (1984) Survival strategies in desert crabs. In: Cogger HG, Cameron EE (eds) Arid Australia. Australian Museum, Sidney, pp 145-152

Greenslade P (1981) Survival of Collembola in arid environments: observations in South Australia and the Sudan. J Arid Environ 4:219-228

Greenslade P (1982) Origins of the collembolan fauna of arid Australia. In: Barker R, Greenslade PJM (eds) Evolution of the flora and fauna of arid Australia. Peacock, Adelaide, pp 267-272

Greenslade PJM, Greenslade P (1984) Soil surface insects of the Australian arid zone. In: Cogger HG, Cameron EE (eds) Arid Australia. Australian Museum, Sidney, pp 153-176

Greven (1972) Vergleichlende Untersuchungen am Integument von Hetero- und Eutardigraden. Z Zellforsch Micros Anat 135:517-538

Grigg GC (1973) Some consequences of the shape and orientation of magnetic termite mounds. Aust J Zool 25:87-94

Grimstone AV, Mullinger AM, Ramsay JA (1968) Further studies on the rectal complex of the mealworm, *Tenebrio molitor* L. (Coleoptera, Tenebrionidae). Philos Trans R Soc B253:343-382

Hadley NF (1970) Water relations of the desert scorpion *Hadrurus arizonensis*. J Exp Biol 53:547-558

Hadley NF (1984) Cuticle: ecological significance. In: Bereiter-Hahn J, Matoltsy AG, Richards KS (eds) Biology of the integument, vol I. Invertebrates. Springer, Berlin, Heidelberg, New York, pp 658-693

Hadley NF (1985) The adaptive role of lipids in biological systems. John Wiley, New York

Hadley NF (1990) Environmental physiology. In: Polis GA (ed) The biology of scorpions. Stanford University Press, Stanford, CA, pp 321-340

Hadley NF (1994) Water relation of terrestrial arthropods. Academic Press, San Diego

Hadley NF, Massion DD (1985) Oxygen consumption, water loss and cuticular lipids of high and low elevation populations of the grasshopper *Aeropedellus clavatus* (Orthoptera: Acrididae). Comp Biochem Physiol 80A:307-311

Hadley NF, Toolson EC, Quinlan MC (1989) Regional differences in cuticular permeability in the desert cicada *Diceroprocta apache*: implications for evaporative cooling. J Exp Biol 141:219-230

Hadley NF, Knisley CB, Schultz TD, Pearson DL (1990) Water relation of tiger beetle larvae (*Cicindela marutha*): correlations with habitat microclimate and burrowing activity. J Arid Environ 19:189-197

Hammer M (1944) Studies on the oribatids and collemboles of Greenland. Medd Grønl 141:1-210

Hansen T (1973) Variation in glycerol content in relation to cold-hardiness in the larvae of *Petrova resinella* L. (Lepidoptera, Tortricidae). Eesti NSV Tead Akad Toim Biol 22:105-112

Hanssen-Bauer I, Solås MK, Steffensen EL (1990) The climate of Spitsbergen. The Norwegian Meteorological Institute, Rep No 39/90

Harrisson PM, Block W, Worland MR (1990) Moisture and temperature dependent changes in cuticular permeability of the antarctic springtail Parisotoma octooculata (Willem). Rev Écol Biol Sol 27:435-448

Harrisson PM, Rothery P, Block W (1991) Drying processes in the Antarctic collembolan *Cryptopygus antarcticus* (Willem). J Insect Physiol 37:883-890

Heathcote RL (1983) The arid lands: their use and abuse. Longman, London

Heatwole H, Harrington S (1989) Heat tolerances of some ants and beetles from the pre-Saharan steppe of Tunisia. J Arid Environ 16:69-77

Heinrich B (1993) The hot-blooded insects. Strategies and mechanisms of thermoregulation. Harvard University Press, Cambridge, MA

Higa LM, Wormersley CZ (1993) New insights into the anhydrobiotic phenomenon - the effects of trehalose content and differential rates of evaporative water loss on the survival of *Aphelenchus avenae* J Exp Zool 267:120-129

Hinton HE (1968) Reversible suspension of metabolism and the origin of life. Proc R Soc Lond B171:43-57

Holdgate MW (1964) Terrestrial ecology in the maritime Antarctic. In: Carrick R, Holdgate MV, Prévost J (eds) Biologie Antarctique. Hermann, Paris, pp 181-194

Holm E, Edney EB (1973) Daily activity in Namib Desert arthropods in relation to climate. Ecology 54:45-56

Holmstrup M (1992) Cold hardiness strategy in cocoons of the lumbricid earthworm *Dendrobaena octaedra* (Savigny). Comp Biochem Physiol 102A:49-54

Holmstrup M (1994) Physiology of cold hardiness in cocoons of five earthworm taxa (Lumbricidae: Oligochaeta). J Comp Physiol B164:222-228

Holmstrup M, Westh P (1994) Dehydration of earthworm cocoons exposed to cold: a novel cold hardiness mechanism. J Comp Physiol B164:312-315

Holmstrup M, Hansen BT, Nielsen A, Østergaard IK (1990) Frost tolerance of lumbricid earthworm cocoons. Pedobiologia 34:361-366

Holt JA (1987) Carbon mineralization in semi-arid northeastern Australia: the role of termites. J Trop Ecol 3:255-263

Hoshikawa K, Tsutsui H, Honma K, Sakagami SF (1988) Cold resistance in four species of beetles overwintering in the soil, with notes on the overwintering strategies of some soil insects. Appl Entomol Zool 23:273-281

Hovanitz W (1950) The biology of *Colias* butterflies. I. The distribution of the North American species. Wasmann J Biol 8:49-75

Hughes RD, Greenham PM, Tyndale-Briscoe M, Walker JM (1972) A synopsis of observations on the biology of the Australian bushfly (*Musca vetusstisma* Walker). J Aust Entomol Soc 11:311-331

Huhta V (1980) Effects of the cold winter on populations of Enchytraeidae and Lumbricidae. In: Sokolov VE (ed) Animal adaptations to winter conditions. USSR Academy of Sciences, Moscow, pp 141-145

Humble LM, Ring RA (1985) Inoculative freezing of a larval parasitoid within its host. Cryo-Lett 6:59-66

Hunter DM, Gregg PC (1984) Variation in diapause potential and strength in eggs of the Australian plague locust *Chortoicetes terminifera* (Walker) (Orthoptera: Acrididae) J Insect Physiol 30:867-870

Janetschek H (1970) Environments and ecology of terrestrial arthropods in the High Antarctic. In: Holdgate MW (ed) Antarctic ecology, vol II. Academic Press, London, pp 871-885

Jennings PG (1976a) Tardigrada from the Antarctic Peninsula and Scotia Ridge region. Br Antarct Surv Bull 44:77-95

Jennings PG (1976b) Tardigrada from Signy Island, South Orkney Island, with a note on the Rotifera. Br Antarct Surv Bull 44:1-25

Joern A, Gaines SB (1990) Population dynamics and regulation in grasshoppers. In: Chapman RF, Joern A (eds) Biology of grasshoppers. John Wiley, New York, pp 415-482

Johnston JS, Heed WB (1976) Dispersal of desert-adapted *Drosophila*: the saguaro-breeding *D. nigrospiracula*. Am Nat 110:629-651

Kaufmann T (1969) Life history of *Upis ceramboides* at Fairbanks, Alaska. Ann Entomol Soc Am 62:922-923

Kaufmann T (1971) Hibernation in the arctic beetle *Pterostichus brevicornis* in Alaska. J Kans Entomol Soc 44:81-92

Keilin D (1959) The problem of anabiosis or latent life: history and current concepts. Proc R Soc Lond B150:149-191

Kennedy AD (1993) Water as a limiting factor in the Antarctic environment: a biogeographical synthesis. Alp Arctic Res 25:308-315

Kevan PG (1975) Sun-tracking solar furnaces in high arctic flowers: significance for pollination and insects. Science 189:723-726

Kevan PG, Danks HV (1986) Adaptations of Arctic insects. In: Sage B (ed) The Arctic and its wildlife. Croom Helm, London, pp 55-57

Kevan PG, Shorthouse JD (1970) Behavioural thermoregulation by high arctic butterflies. Arctic 23:268-279

Kingsolver JG (1983) Thermoregulatory strategies in Colias butterflies. I. The determinants of body temperature and flight activity: elevational patterns and mechanistic limitations. Ecology 64:534-546

Kingsolver JG (1985) Butterfly thermoregulation: organismic mechanisms and population consequences. J Res Lepid 24:1-20

Kirchner W (1973) Ecological aspect of cold resistance in spiders (A comparative study). In: Weiser W (ed) Effects of temperature on ectothermic organisms. Springer, Berlin, Heidelberg, New York, pp 271-279

Koch C (1962) The Tenebrionidae of southern Africa XXXI. Comprehensive notes on the tenebrionid fauna of the Namib Desert. Ann Transvaal Mus 24:63-104

Koehler JK (1967) Studies on the survival of the rotifer Philodina after freezing and thawing. Cryobiology 3:392-399

Koehler JK, Johnson LK (1969) Food supply as a factor in the survival of frozen and thawed rotifers. Cryobiology 5:375-378

Kukal O (1990) Energy budget for activity and growth of a high-arctic insect, Gynaephora groenlandica (Wocke) (Lepidoptera: Lymantriidae). In: Harington CR (ed) Canada's missing dimension: science and history in the Canadian arctic islands, vol II. Canadian Museum of Nature, Ottawa, pp 458-510

Kukal O (1991) Behavioral and physiological adaptations to cold in a freeze-tolerant arctic insect. In: Lee RE Jr, Denlinger DE (eds) Insects at low temperature. Chapman and Hall, New York, pp 276-300

Kukal O, Kevan PG (1987) The influence of parasitism on the life history of a high arctic insect, *Gynaephora groenlandica* (Wocke) (Lepidoptera: Lymantriidae). Can J Zool 65:156-163

Kukal O, Heinrich B, Duman JG (1988) Behavioural thermoregulation in the freeze tolerant arctic caterpillar Gynaephora groenlandica. J Exp Biol 138:181-193

Layne JR Jr, Lee RE Jr, Huang JL (1989) Inoculation triggers freezing at high subzero temperatures in a freeze-tolerant frog (*Rana sylvatica*) and insect (*Eurosta solidaginis*). Can J Zool 68:506-510

Leather SR, Walters KFA, Bale JS (1993) The ecology of insect overwintering. Cambridge University Press, Cambridge

Lee RE Jr (1991) Principles of insect low temperature tolerance. In: Lee RE Jr, Denlinger DL (eds) Insects at low temperature. Chapman and Hall, New York, pp 17-46

Lee RE Jr, Baust JG (1985) Low temperature acclimation in the desert spider, *Agelenopsis aperta* (Aranea, Agelenidae). J Arachnol 13:129-136

Lee RE Jr, Denlinger DL (eds) 1991) Insects at low temperatures. Chapman and Hall, New York

Lee RE Jr, McGrath JJ, Morason RT, Taddeo RM (1993) Survival of intracellular freezing, lipid coalescence and osmotic fragility in fat body cells of the freeze tolerant gall fly *Eurosta solidaginis*. J Insect Physiol 39:445-450

Leech RE (1966) The spiders (Araneida) of Hazen Camp 81°49' N, 71°18' W. Quaest Entomol 2:153-212

Leinaas HP, Fjellberg A (1985) Habitat structure and life history strategies of two partly sympatric and closely related lichen feeding collembolan species. Oikos 44:448-458

Leinaas HP, Sømme L (1984) Adaptations in *Xenylla maritima* and *Anurophorus laricis* (Collembola) to lichen habitats on alpine rocks. Oikos 43:197-206

Lewis JGE (1972) The life histories and distribution of the centipedes *Rhysida nudatogoensis* and *Estmostigmus trigonopodus* (Scolopendromorpha: Scolopendridae) in Nigeria. J Zool (Lond) 167:399-414

Lewis JGE (1984) Woodlice and myriapods. In: Cloudsley-Thompson JL (ed) Key environments. Sahara Desert. Pergamon Press, Oxford, pp 115-127

Logan RF (1968) Causes, climates and distribution of deserts. In: Brown GW Jr (ed) Desert biology. Academic Press, New York, pp 21-50

Lomnicki A (1963) The distribution and abundance of ground-surface inhabiting arthropods above the timberline in the region of Zolta Turnia in the Tatra Mountains. Acta Zool Cracov 8:183-249

Lorentzen G, Halvorsen O (1986) Survival of the first stage larva of the metastrongylid nematode *Elaphostrongylus rangiferi* under various conditions of temperature and humidity. Holarct Ecol 6:301-304

Louw S (1983) The diversity and daily and seasonal activity of ground-living Tenebrionidae (Coleoptera) in the southern Namib and Kalahari ecosystems. Cimbebasia A7(3):35-56

Lovelock JE (1953) The mechanism of the cryoprotective effect of glycerol against freezing and thawing. Biochem Biophys Acta 11:28-36

Lozina-Lozinskii LK (1974) Studies in cryobiology. John Wiley, New York

Lundberg H (1980) Effects of weather on foraging-flights of bumblebees (Hymenoptera, Apidae) in a subalpine/alpine area. Holarct Ecol 3:104-110

Lundheim R, Zachariassen KE (1993) Water balance of overwintering beetles in relation to strategies for cold tolerance. J Comp Physiol B163:1-4

Luyet BJ (1966) Anatomy of the freezing process in physical systems. In: Meryman HT (ed) Cryobiology. Academic Press, London, pp 115-138

Mabbett K, Wharton DA (1986) Cold-tolerance and acclimation in the free-living nematode, *Panagrellus redivivus*. Rev Nematol 9:167-170

Machin J (1967) Structural adaptation for reducing water-loss in three species of terrestrial snail. J Zool (Lond) 152:55-65

Machin J (1979) Atmospheric water absorption in arthropods. Adv Insect Physiol 14:1-48

Machin J (1983) Water vapour absorption in insects. Am J Physiol 244:R187-R192

Machin J, O'Donnell MJ (1991) Rectal complex ion activites and electrochemical gradients in larvae of the desert beetle, *Onymacris*: comparison with *Tenebrio*. J Insect Physiol 37:829-838

MacKay WP (1991) The role of ants and termites in desert communities. In: Polis GA (ed) The ecology of desert communities. University of Arizona Press, Tucson, pp 113-150

MacKenzie AP (1977) Non-equilibrium freezing behaviour of aqueous systems. Philos Trans R Soc London B278:167-189

MacLean SF Jr (1975) Ecological adaptations of tundra invertebrates. In: Vernberg FJ (ed) Physiological adaptations to the environment. Intext Educational Publishing, New York, pp 269-300

MacLean SF Jr (1983) Life cycle and distribution of psyllids (Homoptera) in arctic and subarctic Alaska. Oikos 40:445-451

Maddrell SHP (1971) The mechanisms of insect excretory systems. Adv Insect Physiol 8:199-331

Madge DS (1964) The water-relation of *Belba geniculosa* Oudms and other species of oribatid mites. Acarologia 6:199-233

Madin KC, Crowe JH (1975) Anhydrobiosis in nematodes: carbohydrate and lipid metabolism during dehydration. J Exp Zool 193:335-342

Mani MS (1968) Ecology and biogeography of high altitude insects. Junk, The Hague

Marsh AC (1986) Ant species richness along a climatic gradient in the Namib Desert. J Arid Environ 11:235-241

Marsh AC (1988) Activity patterns of some Namib Deserts ants. J Arid Environ 14:61-73

Masaki S (1980) Summer diapause. Annu Rev Entomol 25:1-25

Masaki S (1983) Introduction. In Brown VK, Hodek I (eds) Diapause and life cycle strategies in insects. Dr W Junk, The Hague, pp 1-5

Masaki S, Nagase A (1992) Genetics, variability and origin of tropical, subtropical and temperate life cycles in the *Dianemobius fascipes* species-complex (Orthoptera; Gryllidae). Stud Neotrop Fauna Environ 27:69-82

Maslen NR (1979) Six new nematode species from the maritime Antarctic. Nematologica 25:288-308

Maslen NR (1981) The Signy Island terrestrial reference site: XII. Population ecology of nematodes with additions to the fauna. Br Antarct Surv Bull 53:57-75

Matthée JJ (1951) The structure and the physiology of the eggs of *Locustana pardalina* (Walker) (Acrididae) by high temperature. J Entomol Soc South Afr 41:25-30

May RM (1949) La Vie des Tardigrades. Gallimard, Paris

Maxwell B (1992) Arctic climate: potential for change under global warming. In: Chapin III FS, Jeffries RL, Reynolds JF, Shaver GR, Svoboda J (eds) Arctic ecosystems in a changing climate. Academic Press, San Diego, pp 11-34

Mazur P (1984) Freezing in living cells: mechanisms and implications. Am J Physiol 247:C125-C147

McClain E, Praetorius RL, Haurahan SA, Seely MK (1984) Dynamics of wax bloom of a seasonal Namib Desert tenebrionid, *Cauracara phalangium* (Coleoptera: Adesmiini). Oecologia 63:314-319

McGinnies WG (1979) Arid-land ecosystems - common features throughout the world. In: Goodall DW, Perry RA (eds) Arid-land ecosystems: structure, functioning and management, vol 1. Cambridge University Press, Cambridge, pp 299-316

McKay GA, Findlay BF, Thompson HA (1970) A climatic perspective of tundra areas. In: Fuller WA, Kevan PG (eds) Productivity and conservation in northern circumpolar lands. IUCN 16:10-33

Medel RG, Vásquez RA (1994) Comparative analysis of harvester ant assemblages of Argentinean and Chilean desert zones. J Arid Environ 26:363-371

Meidell EM (1983) Diapause, aerobic and anaerobic metabolism in alpine adult *Melasoma collaris* (Coleoptera). Comp Biochem Physiol 74A:101-105

Meigs P (1953) World distribution of arid and semi-arid homoclimates. UNESCO Arid Zone Res Ser 1:203-209

Merivee E (1978) Cold-hardiness in insects. Acad Sci Estonian SSR, Tallin

Meryman HT (1974) Freezing injury and its prevention in living cells. Annu Rev Biophys 3:341-363

Meyer E (1983) Struktur und Dynamik einer Population von *Epichnopterix ardua* Mann (Lep.: Psychidae) in einem Krummseggenrasen der Hohen Tauern (Kärnten, Österreich). Zool Jahrb Syst 110:165-177

Meyer E, Eisenbeis G (1985) Water relations in millipeds from some alpine habitat types (Central Alps, Tyrol) (Diplopoda). Bijdr Dierkd 55:131-142

Mikhail WZA (1993) Effects of soil structure on soil fauna in a desert wadi in Southern Egypt. J Arid Environ 24:321-331

Miller LK (1978) Freezing tolerance in relation to cooling rate in an adult insect. Cryobiology 15:345-349

Miller K (1982) Cold-hardiness strategies of some adult and immature insect overwintering in interior Alaska. Comp Biochem Physiol 73A:595-604

Miller PL (1969) On the occurrence and some characteristics of *Cyrtopus fastuosus* Bigot (Dipt. Stratiomyidae) and *Polypedilum* sp. (Dipt., Chironomidae) from temporary habitats in western Nigeria. Entomol Mon Mag 106:233-238

Molgaard P (1982) Temperature observations in high arctic plants in relation to microclimate in the vegetation of Peary Land, North Greenland. Arctic Alp Res 14:105-115

Monasterio M (ed) (1980) Estudios ecologicos en los paramos Andinos. Universidad de Los Andes, Merida

Morason RT, Allenspach AL, Lee RE Jr (1974) Comparative ultrastructure of fat body cells of freeze-susceptible and freeze-tolerant *Eurosta solidaginis* larvae after chemical fixation and high pressure freezing. J Insect Physiol 40:155-164

Morrissey RE, Baust JG (1976) The ontogeny of cold tolerance in the gall fly, Eurosta solidaginis. J Insect Physiol 22:431-437

Morton SR (1985) Granivory in arid regions: comparisons of Australia with North and South America. Ecology 66:1859-1866

Murray J (1910) Antarctic Rotifera. In: Murray J (ed) British Antarctic Expeditions 1907-1909, vol. 1. Heinemann, London, pp 41-65

Murray MD, Orton MN, Cameron AS (1967) The Antarctic flea Glaciopsyllus antarcticus Smith and Dunnet. Antarct Res Ser 10:393-395

Naidu SG, Hattingh J (1986) Water balance and osmoregulation in *Stips stali*, a nocturnal tenebrionid beetle from the Namib Desert. J Insect Physiol 32:891-896

Nakamura I, Ae SA (1977) Prolonged diapause of *Papilio alexanor.* Arid zone adaptation directed by larval host plants. Ann Entomol Soc Am 70:481-484

Needham GR, Teel PD (1986) Water balance by ticks between bloodmeals. In: Sauer Jr, Hair JA (eds) Morphology, physiology and behavioural biology of ticks. Ellis Horwood, Chichester, UK, pp 100-151

Neven LG, Duman JG, Beals JM, Castellino FJ (1986) Overwintering adaptations of the stag beetle, *Ceruchus piceus*: removal of ice nucleators in winter to promote supercooling. J Comp Physiol B156:707-716

Nicholas WL, Stewart AC (1985) Cryptobiotic nematodes in dry soil from Kinchega National Park, NSW. Austr Nematol Newsl 4:18-21

Nicolson SW (1980) Water balance and osmoregulation in *Onymacris plana*, a tenebrionid beetle from the Namib Desert. J Insect Physiol 26:315-320

Nicolson SW, Louw GN, Edney EB (1984) Use of a ventilated capsule and tritiated water to measure evaporative water losses in a tenebronid beetle. J Exp Biol 108:477-481

Noble-Nesbitt J (1970) Water balance in the firebrat, *Thermobia domestica* (Packard). The site of water uptake from the atmosphere. J Exp Biol 52:193-200

Noble-Nesbitt J (1973) Rectal uptake of water in insects. In: Bolis L, Schmidt-Nielsen K, Maddrell SHP (eds) Comparative physiology. Elsevier-North Holland, Amsterdam, pp 333-351

Norris MJ (1959) The influence of day-length on imaginal diapause in the red locust, *Nomadacris septemfasciata* (Serv.). Entomol Exp Appl 2:154-168

Nurminen M (1973) Distribution of northern Enchytraeids (Oligochaeta). Ann Zool Fenn 10:483-486

O'Dell SJ, Crowe JH (1979) Freezing in nematodes: the effects of variable water contents. Cryobiology 16:534-541

O'Donnell MJ, Machin J (1988) Water vapor absorption by terrestrial organisms. Adv Comp Environ Physiol 2:47-90

O'Donnell MJ, Machin J (1991) Ion activites and electrochemical gradients in the mealworm rectal complex. J Exp Biol 155:375-402

Ohyama Y (1977) Population density of free-living mites in the ice-free areas around Syowa Station, East Antarctica. Antarct Res 60:47-56

Oliver DR (1968) Adaptations of Arctic Chironomidae. Ann Zool Fenn 5:111-118

Østbye E, Sømme L (1972) The overwintering of *Pelophila borealis* Payk. I. Survival rates and cold-hardiness. Nor Entomol Tidsskr 19:165-168

Ottesen P, Sømme L (1987) Adaptations to high altitudes in beetles form Tenerife. Vieraea 17:217-226

Paarmann W (1979) A reduced number of larval instars, as an adaptation of the desert carabid beetle *Thermophilum* (*Anthia*) *sexmaculatum* F. (Coleoptera, Carabidae) to its arid environment. In: den Boer PJ, Thiele HV, Weber F (eds) On the evolution of behaviour in carabid beetles. Misc Papers 18, Agricultural University of Wageningen, Wageningen, The Netherlands

Page WW (1980) Occurrence of the grasshopper pest *Zonocerus variegatus* (L.) (Orthoptera: Pyrgomorphidae) in south-western Nigeria in relation to wet and dry seasons. Bull Entomol Res 70:621-633

Pan C, Shimada K (1991) Cold hardiness of four antarctic terrestrial mites in the active season at King George Island. J Insect Physiol 37:325-331

Parker LW, Santos PF, Phillips J, Whitford WG (1984) Carbon and nitrogen dynamics during the decomposition of litter and roots of a Chihuahuan Desert annual, *Lepidium lasiocarpum*. Ecol Monogr 54:339-360

Payne MM (1926) Freezing and survival of insects at low temperatures. Q Rev Biol 1:270-282

Perry RN, Wharton DA (1985) Cold tolerance of hatched and unhatched second stage juveniles of the potato cyst-nematode *Globodera rostochiensis*. Int J Parasitol 15:441-445

Pickup J (1990a) Strategies of cold-hardiness in three species of Antarctic dorylaimid nematodes. J Comp Physiol B 160:167-173

Pickup J (1990b) Seasonal variation in the cold-hardiness of a free-living predatory Antarctic nematode, *Coomansus gerlachi* (Mononchidae). Polar Biol 10:307-315

Pickup J (1990c) Seasonal variation in the cold hardiness of three species of free-living Antarctic nematodes. Funct Ecol 4:257-264

Pickup J, Rothery P (1991) Water-loss and anhydrobiotic survival in nematodes from Antarctic fellfields. Oikos 61:379-388

Pienkowski RL (1976) Behaviour of the adult alfalfa weevil in diapause. Ann Entomol Soc Am 69:155-157

Pierre F (1958) Écologie et peuplement Entomologique des Sables Vifs du Sahara Nord-Occidental. Centre National de la Recherche Scientifique, Paris

Pilato G (1979) Correlations between cryptobiosis and other biological characteristics in some soil animals. Boll Zool 46:319-332

Poinsot N (1968) Cas d'anhydrobiose chez le Collembole *Subisotoma variabilis* Gisin. Rev Écol Biol Sol 4:585-586

Poinsot-Balaguer N (1976) Dynamique des communantes de Collemboles en milieu xerique mediterraneen. Pedobiologia 16:1-17

Polis GA (1980) Seasonal patterns and age-specific variation in the surface activity of a population of desert scorpions in relation to environmental factors. J Anim Ecol 49:1-18

Polis GA (ed) (1990) The biology of scorpions. Stanford University Press, Stanford, CA

Polis GA (1991a) Desert communities: an overview of patterns and processes. In: Polis GA (ed) The ecology of desert communities. University of Arizona Press, Tucson, pp 1-26

Polis GA (1991b) Food webs in desert communities: complexity via diversity and omnivory. In: Polis GA (ed) The ecology of desert communities. University of Arizona Press, Tucson, pp 383-437

Polis GA, Yamashita T (1991) The ecology and importance of predaceous arthropods in desert communities. University of Arizona Press, Tucson, pp 180-222

Popov GB, Wood TG, Haggis MJ (1984) Insect pests of the Sahara. In: Cloudsley-Thompson JL (ed) Key environments. Sahara Desert. Pergamon Press, Oxford, pp 145-174

Powell JA (1974) Occurrence of prolonged diapause in ethmiid moths. Pan Pac Entomol 50:220-225

Pryor ME (1962) Some environmental features of the Hallett Station, Antarctica, with special reference to soil arthropods. Pac Insects 4:681-728

Pugh PJA (1993) A synonymic catalogue of the Acari from Antarctica, the sub-Antarctic Islands and the Southern Ocean. J Nat Hist 27:323-421

Pullin AS, Bale JS (1988) Cause and effects of pre-freeze mortality in aphids. Cryo-Lett 9:101-103

Rahm PG (1923) Biologische und physiologische Beitrage zur Kenntnis der Moosfauna. Z Allg Physiol 20:1-34

Rahm PG (1924) Weitere physiologische Versuche mit niederen Temperaturen. Verh D Zool Ges 29:106-111

Ramløv H (1989) Fænomenet cryptobiose specielt med henblik på tardigraden *Adorybiotus coronifer* (The phenomenon of cryptobiosis, specially with regard to the tardigrade *Adorybiotus coronifer*) MSc Thesis, University of Copenhagen, Copenhagen (in Danish)

Ramløv H, Westh P (1992) Survival of the cryptobiotic eutardigrade *Adorybiotus coronifer* during cooling to -196 °C: effect of cooling rate, trehalose level and short-term acclimation. Cryobiology 19:125-130

Ramløv H, Bedford J, Leader J (1992) Freezing tolerance of the New Zealand alpine weta, *Hemideina maori* Hutton (Orthoptera; Stenopelmatidae). J Therm Biol 17:51-54

Ramsey JA (1964) The rectal complex of the mealworm, *Tenebrio molitor* L. (Coleptera, Tenebrionidae). Philos Trans R Soc Lond B248:279-314

Réaumur RAF (1736) Mémoirs pour servir à l'Histoire des Insectes 2. d'Imprimerie Royal, Paris

Remmert H (1980) Arctic animal ecology. Springer, Berlin Heidelberg New York

Remmert H (1981) Body size of terrestrial arthropods and biomass of their populations in relation to the abiotic parameters of their milieu. Oecologia 50:12-13

Ricci CN (1987) Ecology of bdelloids: how to be successful. Hydrobiologia 147:117-127

Richards KW (1973) Biology of *Bombus polaris* Curtis and *B. hyperboreus* Schönherr at Lake Hazen, Northwest Territories (Hymenoptera: Bombini). Quaest Entomol 9:115-157

Riddle WA, Pugach S (1976) Cold hardiness in the scorpion, *Paruroctonus aquilonalis*. Cryobiology 12:248-253

Riddle WA, Crawford CS, Zetone AM (1976) Patterns of hemolymph osmoregulation in three desert arthropods. J Comp Physiol B112:295-305

Riechert SE (1979) Development and reproduction in desert animals. In: Goodall DW, Perry RA, Howes KMW (eds) Arid-land ecosystems: structure functioning and management, vol 1. Cambridge University Press, Cambridge, pp 797-822

Ring RA (1977) Cold hardiness of the bark beetle *Scolytus ratzeburgi* Jans. (Coleoptera, Scolytidae). Norw J Entomol 24:125-136

Ring RA (1980) Insects and their cells. In: Smith AMJ, Farrant J (eds) Low temperature preservation in medicine and biology. Pitman Medical, London, pp 187-217

Ring RA (1981) The physiology and biochemistry of cold tolerance in arctic insects. J Therm Biol 6:219-229

Ring RA (1982) Freezing tolerant insects with low supercooling points. Comp Biochem Physiol 73A:605-612

Ring RA, Danks HV (1994) Desiccation and cryoprotection: overlapping adaptations. Cryo-Lett 15:181-190

Ring RA, Tesar D (1980) Cold hardiness of the arctic beetle *Pytho americanus* Kirby Coleoptera, Pythidae (Salpingidae). J Insect Physiol 26:763-774

Ring RA, Tesar D (1981) Adaptation to cold in Canadian arctic insects. Cryobiology 18:199-211

Ring RA, Block W, Sømme L, Worland MR (1990) Body water content and desiccation resistance in some arthropods from subantarctic South Georgia. Polar Biol 10:581-588

Rissing SW, Pollock GB (1989) Behavioural ecology and community organization of desert seed-harvester ants. J Arid Environ 17:167-173

Robinson W (1928) Determination of the natural undercooling and freezing points of insects. J Agric Res 37:749-755

Rudolph D (1982) Occurrence, properties and biological implications of the active uptake of water vapor from the atmosphere in Psocoptera. J Insect Physiol 28:111-121

Rudolph D (1983) The water-vapor uptake system of the Phtiraptera. J Insect Physiol 29:15-25

Ryan JK (1981) Invertebrate faunas at IBP tundra sites. In: Bliss LC, Heal OW, Moore JJ (eds) Tundra ecosystems: a comparative analysis. Cambridge University Press, Cambridge, pp 517-539

Saini RS (1964) Histology and physiology of the cryptonephridial system in insects. Trans R Entomol Soc Lond 116:347-392

Salt RW (1936) Studies on the freezing process in insects. Tech Bull Minn Agric Exp Stn 116:1-41

Salt RW (1961) Principles of insect cold-hardiness. Annu Rev Entomol 6:55-74

Salt RW (1962) Intracellular freezing in insects. Nature (Lond) 193:1207-1208

Salt RW (1966) Effect on cooling rate on the freezing temperatures of supercooled insects. Can J Zool 44:655-659

Sarmiento G (1986) Ecological features of climate in high tropical mountains. In: Vuilleumier F, Monasterio M (eds) High altitude tropical biogeography. Oxford University Press, Oxford, pp 11-45

Sayre RM (1964) Cold-hardiness of nematodes. I. Effects of rapid freezing on the eggs and larvae of *Meloidogyne incognita* and *M. hapla*. Nematologica 10:168-179

Schatz H (1985) The life cycle of an alpine oribatid mite, *Oromurcia sudetica* Willmann. Acarologia 26:95-100

Schenker R (1983) Effects of temperature acclimation on cold-hardiness of alpine micro-arthropds. Rev Écol Biol Sol 20:37-47

Schmoller R (1970) Life history of alpine tundra Arachnida in Colorado. Am Midl Nat 83:119-133

Scholander PF, Flagg W, Hock RJ, Irving L (1953) Studies on the physiology of frozen plants and animals in the arctic. J Cell Comp Physiol 42:1-56

Schwarz AMJ, Green TGA, Seppelt RD (1992) Terrestrial vegetation at Canada Glacier, Southern Victoria Land, Antarctica. Polar Biol 12:397-404

Schwarz AMJ, Green JD, Green TGA, Seppelt RD (1993) Invertebrates associated with moss communites at Canada Glacier, southern Victoria Land, Antarctica. Polar Biol 13:157-162

Shachak M (1980) Energy allocation and life history strategy of the desert isopod H. reaumuri. Oecologia 456: 404-413

Shachak M, Chapman EA, Orr Y (1976a) Some aspects of the ecology of the desert snail *Sphincterochila boissieri* in relation to water and energy flow. Isr J Med Sci 12:887-891

Shachak M, Chapman EA, Steinberger Y (1976b) Feeding, energy flow and soil turnover in desert isopod, *Hemilepistus reaumuri*. Oecologia 24:57-69

Shapiro AM (1979) The phenology of *Pieris napi microstriata* (Lepidoptera: Pieridae) during and after the 1975-77 Californian drought, and its evolutionary significance. Psyche 86:1-10

Shapiro AM (1985) Behavioral and ecological observations of Peruvian high-Andean pierid butterflies (Lepidoptera). Stud Neotrop Fauna Environ 20:1-13

Shimada K (1989) Ice-nucleating activity of the alimentary canal of the freeze-tolerant prepupae of *Trichiocampus populi*. (Hymenoptera: Tenthredinidae). J Insect Physiol 35:113-120

Shimada K, Riihimaa H (1988) Cold acclimation, inoculative freezing and slow cooling; essential factors contributing to the freezing-tolerance in diapausing larvae of *Chymomyza costata* (Diptera: Drosophilidae). Cryo-Lett 9:5-10

Shimada K, Pan C, Ohyama Y (1992) Variation in summer cold-hardiness of the oribatid mite *Alaskozetes antarcticus* from contrasting habitats on King George Island. Polar Biol 12:701-706

Shook RS (1978) Ecology of the wolf spider *Lycosa carolinensis* Walkenauer (Aranea: Lycosidae) in a desert community. J Arachnol 6:53-64

Slifer EH, King RL (1961) The inheritance of diapause in grasshopper eggs. J Hered 52:39-44

Smith AP (1979) The function of dead leaves in *Espeletia schultzii* (Compositae), an Andean giant rosette plant. Biotropica 1:43-47

Smith RIL, Walton DWH (1975) South Georgia, subantarctic. Ecol Bull 20:399-423

Solhøy T (1975) Dynamics of Oribatei populations on Hardangervidda. In: Wielgolaski FE (ed) Fennoscandian tundra ecosystems. Part 2. Animals and system analysis. Springer, Berlin Heidelberg New York, pp 60-65

Sømme L (1964) Effects of glycerol on cold hardiness in insects. Can J Zool 42:89-101
Sømme L (1974a) Anaerobiosis in some alpine Coleoptera. Nor Entomol Tidsskr 21:155-158
Sømme L (1974b) The overwintering of *Pelophila borealis* Payk.-III. Freezing tolerance. Nor Entomol Tidsskr 21:131-134
Sømme L (1978) Notes on the cold hardiness of prostigmatid mites from Vestfjella, Dronning Maud Land. Norw J Entomol 25:51-55
Sømme L (1981) Cold tolerance of alpine, Arctic and Antarctic Collembola and mites. Cryobiology 18:212-220
Sømme L (1982) Supercooling and winter survival in terrestrial arthropods Comp Biochem Physiol 73A:519-543
Sømme L (1986a) Ecology of *Cryptopygus sverdrupi* (Insecta: Collembola) from Dronning Maud Land, Antarctica. Polar Biol 6:179-184
Sømme L (1986b) Terrestrial arthropods of Bouvetøya. Polar Res 4:231-233
Sømme L (1986c) Tolerance to low temperatures and desiccation in insects from Andean paramos. Arct Alp Res 18:253-257
Sømme L (1989) Adaptation of terrestrial arthropods to the alpine environment. Biol Rev 64:367-407
Sømme L (1993) The terrestrial arthropod fauna of Svalbard. Arct Insect News 4:2-4
Sømme L (1994) The adaptation of alpine terrestrial arthropods to desiccation. Acta Oecol 15:55-62
Sømme L, Block W (1982) Cold-hardiness of Collembola at Signy Island, maritime Antarctic. Oikos 38:168-176
Sømme L, Block W (1991) Adaptations to alpine and polar environments in insects and other terrestrial arthropods. In: Lee RE Jr, Denlinger DL (eds) Insects at low temperature. Chapman and Hall, New York, pp 318-359
Sømme L, Conradi-Larsen EM (1977) Cold-hardiness of collembolans and oribatid mites from windswept mountain ridges. Oikos 29:118-126
Sømme L, Conradi-Larsen EM (1979) Frost resistance in alpine, adult *Melasoma collaris* (Coleoptera). Oikos 33:80-84
Sømme L, Meier T (1995) Cold-hardiness of Tardigrada from Dronning Maud Land, Antarctica. Polar Biol 15:221-224
Sømme L, Zachariassen KE (1981) Adaptations to low temperature in high altitude insects from Mount Kenya. Ecol Entomol 6:199-204
Soriano A (1983) Deserts and semi-desert of Patagonia. In: West NE (ed) Temperate deserts and semi-deserts. Ecosystems of the world 5. Elsevier, Amsterdam, pp 423-460
Spallanzani L (1769) Nouvelles recherches sur les decouvertes microscopique et sur la generation des corps organises, vol. 1
Spaull VW (1973) Quantitative and qualitative distribution of soil nematodes of Signy Island, South Orkney Islands. Br Antarct Surv Bull 34:177-184
Steigen AL (1975) Energetics in a population of *Pardosa palustris* (L.) (Aranea, Lycosidae) on Hardangervidda. In: Wielgolaski FE (ed) Fennoscandian tundra ecosystems, Part 2. Animals and system analysis. Springer, Berlin Heidelberg New York, pp 129-144
Steiner G, Albin FE (1946) Resuscitation of the nematode *Tylenchus polyhypnus*, n.sp., after almost 39 years' dormancy. J Wash Acad Sci 36:97-99
Stonehouse B (1989) Polar ecology. Blackie, Glasgow
Storey KB, Storey JM (1988) Freeze tolerance in animals. Physiol Rev 68:27-84
Strathdee AT, Bale JS, Block WC, Webb NR, Hodkinson ID, Coulson SJ (1993) Extreme adaptive life-cycle in a high arctic aphid, *Acyrthosiphon svalbardicum*. Ecol Entomol 18:254-258
Sturm H (1978) Zur Ökologie der andinen Paramoregion. Biogeographica 14. Junk, The Hague
Sudzuki M (1979) On the microfauna of the Antarctic region, III. Microbiota of terrestrial interstices. Mem Natl Inst Polar Res Spec Issue 11:104-126
Tanno K (1967) Freezing injury in fat body cells of the poplar sawfly. In: Asahina E (ed) Cellular injury and resistance in freezing organisms. Institute of Low Temperature Science, Sapporo, pp 245-257

Tanno K (1968) Frost resistance in the poplar sawfly, *Trichiocampus populi* Okamoto. IV. Intracellular freezing in fat-cell and injury occurring upon metamorphoses. Low Temp Sci Ser B 26:71-78

Tanno K (1975) Supercooling in arctic Collembola in early summer. Low Temp Sci Ser B33:39-43

Tanno K (1977) Ecological observations and frost-resistance in overwintering prepupae, *Sciara* sp. (Sciaridae) Low Temp Sci Ser B35:63-74

Tauber MJ, Tauber CA (1973) Nutritional and photoperiodic control of the seasonal reproductive cycle in *Chrysopa mohave* (Neuroptera). J Insect Physiol 19:729-736

Tauber MJ, Tauber CA, Masaki S (1986) Seasonal adaptations of insects. Oxford University Press, New York

Thomas DB (1983) Tenebrionid beetle diversity and habitat complexity in the eastern Mojave desert. Coleopt Bull 37:135-145

Tilbrook PJ (1967) Arthropod ecology in the maritime Antarctic. In: Gressitt JL (ed) Entomology of Antarctica. Antarctic Research Series 10. American Geophysical Union, Washington, DC, pp 331-356

Tjeder B (1963) Three subapterous craneflies from Alaska. Opusc Entomol 28:229-241

Trewartha GT, Horn LH (1980) An introduction to climate (5th edn). McGraw-Hill, New York

Ushatinskaya RS (1957) Principles of cold resistance in insects. Academy of Sciences of the USSR, Moscow (in Russian)

Ushatinskaya RS (1990) Latent life and anabiosis. Hayka, Moscow (in Russian)

Usher MB, Dastych H (1987) Tardigrada from the maritime Antarctic. Br Antarct Surv Bull 77:163-166

Uvarov BP (1966) Grasshoppers and locusts: a handbook of acridology. vol 1. Cambridge University Press,

Van der Laak S (1982) Physiological adaptations to low temperature in freezing-tolerant *Phyllodecta laticollis* beetles. Comp Biochem Physiol 73A:613-620 Cambridge

Vannier G (1978) La résistance à la desiccation chez les premiers arthropodes terrèstres. Bull Soc Ecophysiol 3:13-42

Vannier G (1982) The importance of ecophysiology for both biotic and abiotic studies of soil. In: Lebrun P, André HM, De Medts A, Grégoire-Wibo C, Wauthy G (eds) New trends in soil biology. Dieu-Brichart, Ottignies-Louvain-La-Neuve, Belgium

Vannier G, Thibaud JM (1978) Réduction ou perte totale de la capacité de régulation hydrique chez des espèces de collemboles cavernicoles appartenant à la famille des Tomoceridae. Bull Soc Ecophysiol 3:124-126

Verhoef HA, Witteveen J (1980) Water balance in Collembola and its relation to habitat selection; cuticular water loss and water uptake. J Insect Physiol 26:201-208

von Almen H, Zettel J (1984) Contribution to the cold hardiness of *Entomobrya nivalis* (Collembola) from the Swiss Prealps. Zool Jahrb Syst 111:231-244

Wall BJ (1970) Effects of dehydration and rehydration on *Periplaneta americana*. J Insect Physiol 16:1027-1042

Wallace MMH (1968) The ecology of *Sminthurus viridis* (Collembola) II. Diapause in the aestivating egg. Aust J Zool 16:871-883

Wallwork JA (1972) Distribution patterns and population dynamics of micoarthropods of a desert soil in southern California. J Anim Ecol 41:291-310

Wallwork JA (1973) Zoogeography of some terrestrial micro-Arthropoda in Antarctica. Biol Rev 48:223-259

Wallwork JA (1982) Desert soil fauna. Praeger, New York

Wallwork JA, Kamill BW, Whitford WG (1984) Lifestyles of desert litter-dwelling microarthropods: a reappraisal based on the reproductive behaviour of cryptostigmatid mites. S Afr J Sci 80: 163-169

Walter H, Box EO (1983) The Pamir - an ecologically well-studied high-mountain desert biome. In: West NE (ed) Temperate deserts and semi-deserts. Ecosystems of the world 5. Elsevier, Amsterdam, pp 237-269

Walton DWH (1984) The terrestrial environment. In: Laws RM (ed) Antarctic ecology, vol 1. Academic Press, London, pp 1-60

Wardhaug KG (1986) Diapause strategies in the Australian plaque locust (*Chortoicetes terminifera* Walker). In: Taylor F, Karban R (eds) The evolution of insect life cycles. Springer, Berlin Heidelberg New York, pp 89-104

West CC (1982) Life histories of three species of sub-Antarctic oribatid mites. Pedobiologia 23:59-67

West NE (1983a) Comparisons and contrasts between the temperate deserts and semi-deserts of three continents. In: West NE (ed) Temperate deserts and semi-deserts. Ecosystems of the world 5. Elsevier, Amsterdam, pp 461-472

West NE (ed) (1983b) Temperate deserts and semi-deserts. Ecosystems of the world 5. Elsevier, Amsterdam

Westh P, Kristensen RM (1992) Ice formation in the freeze-tolerant eutardigrades *Adorybiotus coronifer* and *Amphibolus nebulosus* studied by differential scanning calorimetry. Polar Biol 12:693-699

Westh P, Ramløv H (1991) Trehalose accumulation in the tardigrade *Adorybiotus coronifer* during anhydrobiosis. J Exp Zool 258:303-311

Wharton DA (1995) Cold tolerance strategies in nematodes. Biol Rev 70 (in press)

Wharton DA, Allen GS (1989) Cold tolerance mechanisms of the free-living stages of *Trichostrongylus colubriformis* (Nematoda: Trichostrongylidae). J Exp Biol 45:353-370

Wharton DA, Barclay S (1993) Anhydrobiosis in the free-living antarctic nematode *Panagrolaimus davidi* (Nematoda: Rhabditidae) Fundam Appl Nematol 16:17-22

Wharton DA, Block W (1993) Freezing tolerance in some Antarctic nematodes. Funct Ecol 7:578-584

Wharton DA, Brown IM (1989) A survey of terrestrial nematodes from the McMurdo Sound region, Antarctica. NZ J Zool 16:467-470

Wharton DA, Brown IM (1991) Cold tolerance mechanisms of the antarctic nematode *Panagrolaimus davidi.* J Exp Biol 155:629-641

Wharton DA, Young SR, Barrett J (1984) Cold tolerance in nematodes. J Comp Physiol B154:73-77

Whitman DW, Orsak L (1985) Biology of *Taeniopoda eques* (Orthoptera: Acrididae) in southeastern Arizona. Ann Entomol Soc Am 78:811-825

Whitmore DH, Gonzalez R, Baust JG (1985) Scorpion cold hardiness. Physiol Zool 58:526-537

Wigglesworth VB (1972) The principles of insect physiology. 7th edn. Chapman and Hall, London

Wisdom CS (1991) Patterns of heterogeneity in desert herbivorous insect communities. In: Polis GA (ed) The ecology of desert communities. University of Arizona Press, Tuscon, pp 151-179

Womersley C (1981) Biochemical and physiological aspects of anhydrobiosis. Comp Biochem Physiol 70B:669-678

Womersley C (1987) A reevaluation of strategies employed by nematode anhydrobiotes in relation to their natural environment. In: Veech JA, Dickson DW (eds) Vistas on nematology. Society of Nematologists, Hyattsville, MD, pp 165-173

Womersley C, Ching C (1989) Natural dehydration regimes as a prerequisite for the successful induction of anhydrobiosis in the nematode *Rotylenchulus reniformis.* J Exp Biol 143:359-372

Wood FE Jr, Nordin JH (1976) Studies on the low temperature induced biogenesis of glycerol by adult *Protophormia terranovae.* J Insect Physiol 22:1665-1673

Worland MR, Block W (1986) Survival and water loss in some Antarctic arthropods. J Insect Physiol 32:579-584

Wright JC (1988) Structural correlates of permeability and tun formation in tardigrade cuticle: an image analysis study. J Ultrastruct Mol Struct Res 101:23-29

Wright JC (1989a) Desiccation tolerance and water-retentive mechanisms in tardigrades. J Exp Biol 142:267-292

Wright JC (1989b) The tardigrade cuticle II. Evidence for a dehydration-dependent permeability barrier in the intracuticle. Tissue Cell 21:263-279

Wright JC (1991) The significance of four xeric parameters in the ecology of terrestrial tardigrades. J Zool (Lond) 224:59-77

Wright JC, Machin J (1990) Water vapour absorption in terrestrial isopods. J Exp Biol 154:13-30

Wright JC, Machin J (1993) Atmospheric water absorption and the water budget of terrestrial isopods (Crustacea, Isopoda, Oniscidae). Biol Bull 184:243-253

Wright JC, Westh P, Ramløv H (1992) Cryptobiosis in Tardigrada. Biol Rev 67:1-29

Yoder JA, Denlinger DL (1992) Water vapour uptake by diapausing eggs of a tropical walking stick. Physiol Entomol 17:97-103

Young AM (1982) Population biology of tropical insects. Plenum Press, New York London

Young SR, Block W (1980) Experimental studies on cold tolerance in an Antarctic terrestrial mite. J Insect Physiol 26:189-200

Zachariassen KE (1980) The role of polyols and nucleating agents in cold-hardy beetles. J Comp Physiol 140:227-234

Zachariassen KE (1985) Physiology of cold tolerance in insects. Physiol Rev 65:799-832

Zachariassen KE (1991a) Routes of transpiratory water loss in a dry-habitat tenebrionid beetle. J Exp Biol 157:425-437

Zachariassen KE (1991b) The water relation of overwintering insects. In: Lee RE Jr, Denlinger DL (eds) Insects at low temperature. Chapman and Hall, New York, pp 47-63

Zachariassen KE, Einarson S (1993) Regulation of body fluid compartments during dehydration of the tenebrionid beetle *Rhytinola praelonga*. J Exp Biol 182:283-289

Zachariassen KE, Hammel HT (1976) Nucleating agents in the haemolymph of insects tolerant to freezing. Nature (Lond) 262:285-287

Zachariassen KE, Husby JA (1982) Antifreeze effect of thermal hysteresis agents protects highly supercooled insects. Nature (Lond) 298:865-867

Zachariassen KE, Maloiy GMO (1989) Water balance of beetles as an indicator of environmental humidity. Fauna Norv Ser B36:27-31

Zachariassen KE, Hammel HT, Schmidek W (1979) Osmotically inactive water in relation to tolerance to freezing in *Eleodes blanchardi* beetles. Comp Biochem Physiol 63A:203-206

Zachariassen KE, Baust JG, Lee RE Jr (1982) A method for the quantitative determination of ice nucleating agents in insect hemolymph. Cryobiology 19:180-184

Zachariassen KE, Andersen J, Maloiy NO, Kamau JMZ (1987) Transpiratory water loss and metabolism of beetles from arid areas in East Africa. Comp Biochem Physiol 86A:403-408

Zachariassen KE, Andersen J, Kamau JMZ, Maloiy GMO (1988) Water loss in insects from arid and humid habitats in East Africa. Acta Entomol Bohemoslov 85:81-93

Zak JC, Freckman DW (1991) Soil communities in desert: micoarthropods and nematodes. In: Polis GA (ed) The ecology of desert communities. University of Arizona Press, Tucson, pp 55-88

Zaslavski VA (1988) Insect development. Photoperiodic and temperature control. Springer, Berlin Heidelberg New York

Zettel J (1982) Age dependence of cold hardiness in *Isotoma hiemalis* and some other springtails. Cryo-Lett 3:311

Zettel J (1984) Cold-hardiness strategies and thermal hysteresis in Collembola. Rev Écol Biol Sol 21:189-203

Subject Index